Carbohydrate-active e

Structure, function and applications

Edited by

Kwan-Hwa Park

CRC Press

Boca Raton Boston New York Washington, DC

WOODHEAD PUBLISHING LIMITED

Oxford Cambridge New Delhi

Published by Woodhead Publishing Limited, Abington Hall, Granta Park,
Great Abington, Cambridge CB21 6AH, UK
www.woodheadpublishing.com

Woodhead Publishing India Private Limited, G-2, Vardaan House, 7/28 Ansari Road,
Daryaganj, New Delhi – 110002, India

Published in North America by CRC Press LLC, 6000 Broken Sound Parkway NW,
Suite 300, Boca Raton, FL 33487, USA

First published 2008, Woodhead Publishing Limited and CRC Press LLC
Reprinted 2009, 2010

British Library Cataloguing in Publication Data
A catalogue record for this book is available from the British Library.

Library of Congress Cataloging in Publication Data
A catalog record for this book is available from the Library of Congress.

Woodhead Publishing ISBN 978-1-84569-519-4 (book)
Woodhead Publishing ISBN 1-84569-575-0 (e-book)
CRC Press ISBN 978-1-4398-0158-1
CRC Press order number: N10036

CONTENTS

Part II Functions and applications of carbohydrate-active enzymes

This book is based on papers presented at the 2008 Agricultural Biotechnology Symposium on "Carbohydrate-Active Enzymes: structure, function, and applications" held on September 26–27, 2008 in Seoul National University, Seoul, Korea. This symposium was organized by the Center for Agricultural Biomaterials (CAB), Seoul National University, Seoul, Korea.

CONTRIBUTORS

J. M. Andersen Enzyme and Protein Chemistry, Department of Systems Biology, Technical University of Denmark, Denmark

A. Blennow Plant Biology Laboratory, Faculty of Life Sciences, University of Copenhagen, Denmark

Hyunju Cha Center for Agricultural Biomaterials, and Department of Food Science and Biotechnology, School of Agricultural Biotechnology, Seoul National University, Korea

Jaeho Cha Department of Microbiology, Pusan National University, Busan, Korea

Seiya Chiba Research Faculty of Agriculture, Hokkaido University, Sapporo, Japan

C. Christiansen Enzyme and Protein Chemistry, Department of Systems Biology, Technical University of Denmark, Denmark Plant Biology Laboratory, Faculty of Life Sciences, University of Copenhagen, Denmark

A. Dilokpimol Enzyme and Protein Chemistry, Department of Systems Biology, Technical University of Denmark, Denmark

C. Finnie Enzyme and Protein Chemistry, Department of Systems Biology, Technical University of Denmark, Denmark

Kenji Fukuda Research Faculty of Agriculture, Hokkaido University, Sapporo, Japan

Shinya Fushinobu Department of Biotechnology, University of Tokyo, Tokyo, Japan

Suk-Jin Ha Graduate School of Biotechnology, and Institute of Life Science and Resources, Kyung Hee University, Yongin, Korea

M. Abou Hachem Enzyme and Protein Chemistry, Department of Systems Biology, Technical University of Denmark, Denmark

P. Hägglund Enzyme and Protein Chemistry, Department of Systems Biology, Technical University of Denmark, Denmark

Masafumi Hidaka National Food Research Institute, Ibaraki and Department of Biotechnology, University of Tokyo, Tokyo, Japan

H Yuji Honda National Food Research Institute, Ibaraki and Ishikawa Prefectural University, Ishikawa, Japan

Hironori Hondoh Research Faculty of Agriculture, Hokkaido University, Sapporo, Japan

Štefan Janeček Institute of Molecular Biology, Slovak Academy of Sciences, Bratislava Department of Biotechnologies, Faculty of Natural Sciences, University of SS. Cyril and Methodius, Trnava, Slovakia

J. M. Jensen Enzyme and Protein Chemistry, Department of Systems Biology, Technical University of Denmark, Denmark

Jong-Hyun Jung Graduate School of Biotechnology, and Institute of Life Science and Resources, Kyung Hee University, Yongin, Korea

Min-Sung Kang Research Faculty of Agriculture, Hokkaido University, Sapporo, Japan

Bong Gwan Kim Department of Food and Nutrition, Yonsei University, Seoul, Korea

Doman Kim Laboratory of Functional Carbohydrate Enzymes and Microbial Genomics, School of Biological Sciences and Technology and The Research Institute for Catalysis, Institute of Bioindustrial Technology, Chonnam National University, Gwangju, Korea

Hye Min Kim Department of Food and Nutrition, Yonsei University, Seoul, Korea

Jung Wan Kim Department of Biology, University of Incheon, Incheon, Korea

Su Kyung Kim Department of Food Science and Biotechnology, School of Agricultural Biotechnology, Seoul National University, Korea

Tae-Jip Kim Department of Food Science and Technology, Chungbuk National University, Cheongju, Korea

Yong-Ro Kim Center for Agricultural Biomaterials and Department of Biosystems & Biomaterial Science Engineering, Seoul National University, Korea

Young-Min Kim Laboratory of Functional Carbohydrate Enzymes and Microbial Genomics, School of Biological Sciences and Technology and The Research Institute for Catalysis, Institute of Bioindustrial Technology, Chonnam National University, Gwangju, Korea

Young-Wan Kim Department of Food and Biotechnology, Korea University, Jochiwon, Korea

Atsuo Kimura Research Faculty of Agriculture, Hokkaido University, Sapporo, Japan

Motomitsu Kitaoka National Food Research Institute, Ibaraki, Japan

Takashi Kuriki Biochemical Research Laboratory, Ezaki Glico Co., Ltd., Osaka, Japan

Byong Hoon Lee Department of Microbiology/Immunology, McGill University, Montreal, Canada

Heeseob Lee Department of Food Science and Nutrition, Pusan National University, Busan, Korea

Jin-Ha Lee Research Faculty of Agriculture, Hokkaido University, Sapporo, Japan

Seungjae Lee, Center for Agricultural Biomaterials, Seoul National University, Seoul, Korea

Soo-Bok Lee Department of Food and Nutrition, Yonsei University, Seoul, Korea

Sung Ho Lee Global Grain Applications and Technical Services, Genencor®, a Danisco Division, Beloit, WI, USA

Dan Li Center for Agricultural Biomaterials, Seoul National University, Korea and Jilin Key Lab of Agricultural Products Processing, Changchun University, Changchun, China

Xiaolei Li Jilin Key Lab of Agricultural Products Processing, Changchun University, Changchun, China

K. Maeda Enzyme and Protein Chemistry, Department of Systems Biology, Technical University of Denmark, Denmark

San San Mar Research Faculty of Agriculture, Hokkaido University, Sapporo, Japan

Young-Hwan Moon Laboratory of Functional Carbohydrate Enzymes and Microbial Genomics, School of Biological Sciences and Technology and The Research Institute for Catalysis, Institute of Bioindustrial Technology, Chonnam National University, Gwangju, Korea

Haruhide Mori Research Faculty of Agriculture, Hokkaido University, Sapporo, Japan

Hiroyuki Nakai Research Faculty of Agriculture, Hokkaido University, Sapporo, Japan

M. M. Nielsen Enzyme and Protein Chemistry, Department of Systems Biology, Technical University of Denmark, Denmark

Mamoru Nishimoto Research Faculty of Agriculture, Hokkaido University, Sapporo, Japan

Hiromi Nishiura Biochemical Research Laboratory, Ezaki Glico Co., Ltd., Osaka, Japan

Koji Nomura Biochemical Research Laboratory, Ezaki Glico Co., Ltd., Osaka, Japan

Masayuki Okuyama Research Faculty of Agriculture, Hokkaido University, Sapporo, Japan

Cheon-Seok Park Graduate School of Biotechnology, and Institute of Life Science and Resources, Kyung Hee University, Yongin, Korea

Jong-Tae Park Center for Agricultural Biomaterials, and Department of Food Science and Biotechnology, School of Agricultural Biotechnology, Seoul National University, Seoul, Korea

Kwan-Hwa Park Center for Agricultural Biomaterials, and Department of Food Science and Biotechnology, School of Agricultural Biotechnology, Seoul National University, Seoul, Korea

John F. Robyt Department of Biochemistry, Biophysics, and Molecular Biology, Iowa State University, Ames, USA

Soo In Ryu Department of Food and Nutrition, Yonsei University, Seoul, Korea

Wataru Saburi Research Faculty of Agriculture, Hokkaido University, Sapporo, Japan

Yoshiyuki Sakano Tokyo University of Agriculture and Technology, Tokyo, Japan

Dong-Ho Seo Graduate School of Biotechnology, and Institute of Life Science and Resources, Kyung Hee University, Yongin, Korea

Eun Seung Seo Enzyme and Protein Chemistry, Department of Systems Biology, Technical University of Denmark, Denmark

Jay K. Shetty Global Grain Applications and Technical Services, Genencor®, Danisco Division, Palo Alto, CA, USA

Mee Son Research Faculty of Agriculture, Hokkaido University, Sapporo, Japan

H. N. Song Korea Research Institute of Bioscience and Biotechnology, Daejeon, Korea

Kazuhisa Sugimoto Biochemical Research Laboratory, Ezaki Glico Co., Ltd., Osaka, Japan

Birte Svensson Enzyme and Protein Chemistry, Department of Systems Biology, Technical University of Denmark, Denmark

M. B. Vester-Christensen Enzyme and Protein Chemistry, Department of Systems Biology, Technical University of Denmark, Denmark

Jintanart Wongchawalit Research Faculty of Agriculture, Hokkaido University, Sapporo, Japan

Eui Jeon Woo Korea Research Institute of Bioscience and Biotechnology, Daejeon, Korea

Takeshi Yamamoto Research Faculty of Agriculture, Hokkaido University, Sapporo, Japan

Sang-Ho Yoo Department of Food Science and Technology, Sejong University, Seoul, Korea

S. M. Yoon Korea Research Institute of Bioscience and Biotechnology, Daejeon, Korea

PREFACE

Recent advances in biochemistry and biotechnology have provided significant progress on basic research and applications of carbohydrate active enzymes. However, the mechanism of the catalytic reaction has not been fully understood, as the enzymes often showed unusual substrate specificity and mode of action. Therefore, this symposium has emphasized the enzymatic reaction mechanism, structure–function relationship and role in the living organism. The Center for Agricultural Biotechnology (CAB) at Seoul National University has organized the "Annual Symposia on Agricultural Biotechnology" including carbohydrate enzymes since 1990. Out of this symposium, a number of excellent results on the new types of carbohydrate enzymes and their applications have been reported. This volume, "Carbohydrate-active enzymes: structure, function, and application", based on the 2008 Agricultural Biotechnology Symposium in Seoul, September 26–27, 2008 organized by CAB, summarizes the current information on carbohydrate enzymes by international experts. This book has primarily focused on the classification, structure, specific mechanisms of amylolytic enzymes, metabolism and applications in the hope that it will stimulate the readers and drive work for future research. We hope that readers will find this book useful for the current status of some carbohydrate enzymes that have not been well investigated.

We would like to thank all the authors for their magnificent work, time and devotion. I am grateful to Professor In-Won Lee, former director of CAB who initiated this symposium specifically on carbohydrate enzymes and led the organizing committee and editorial members to publish this book. I would also like to thank Professor Tae Wha Moon, director of CAB and other members of CAB for their continuous support. My special thanks are extended to Professors Sang-Ho Yoo, Suyong Lee, Myo Jeong Kim, Young Wan Kim, Hee Seop Lee, Young Jin Choi for editorial efforts. I am grateful to my co-workers, who have contributed to this book in various ways in particular Professor Pan Sik Chang and Professor Yong-Ro Kim.

Kwan-Hwa Park

PART I

STRUCTURE–FUNCTION RELATIONSHIP OF CARBOHYDRATE-ACTIVE ENZYMES

BIOSYNTHESIS OF POLYSACCHARIDES

John F. Robyt

ABSTRACT

The mechanisms involved in the biosynthesis of six polysaccharides is described in the following order: (1) Introduction to the first purported biosynthesis of polysaccharides, glycogen and starch by phosphorylases; (2) biosynthesis of *Salmonella* O-antigen polysaccharide; (3) biosynthesis of bacterial cell wall polysaccharide, peptido-murein; (4) biosynthesis of dextran by B-512FMC dextransucrase; (5) biosynthesis of bacterial cellulose and xanthan; (6) biosynthesis of starch in starch granules. The structures of the six polysaccharides are quite diverse. There are four β-linked hetero-polysaccharides (2), (3), and (5), and two α-linked homo-polysaccharides (4) and (6). The first five are biosynthesized by prokaryote bacteria and the sixth polysaccharide (starch) was shown to be biosynthesized by eight different eukaryotic plant sources. All six of the polysaccharides have been shown to be biosynthesized by a common mechanism in which the monomer or repeating unit is added to the reducing-end of a growing polysaccharide chain in a two catalytic-site insertion mechanism. The β-linked polysaccharides are covalently α-linked to a lipid pyrophosphate, bactoprenol pyrophosphate, at the active-site of the synthesizing enzymes; the α-linked polysaccharides are β-linked directly to the synthesizing enzymes. When the monomer or repeating unit is inserted between the growing polysaccharide and the lipid pyrophosphate or the enzyme, the configuration of the linkage of the polysaccharide is inverted, giving the correct stereochemistry for the specific polysaccharide. Eventually, the polysaccharides are released from the active-sites by an acceptor reaction with water or with another carbohydrate.

Key words: cellulose synthase; dextransucrase; insertion mechanism; primer mechanism; starch synthase

INTRODUCTION

Polysaccharides were the first biopolymers purported to be biosynthesized *in vitro* (Cori and Cori 1939) observed that the reaction of liver phosphorylase with α-D-glucose-1-phosphate (α-Glc-1-P) and glycogen added glucose residues to the nonreducing-ends of glycogen chains. Shortly thereafter, Hanes (1940) reported a similar reaction for potato phosphorylase in which α-Glc-1-P and starch also added glucose residues to the nonreducing-ends of the starch chains. Up to this time, the reaction catalyzed by phosphorylases was with inorganic phosphate (P_i) and glycogen or starch chains to give α-Glc-1-P and a partially degraded polysaccharide. It was found that phosphorylases catalyzed these two reactions with equilibrium constants close to one (Swanson and Cori, 1948). The equilibrium, however, seemed to favor the degradation reaction than the synthetic reaction. The reactions were formulated for glycogen and starch chains, as the following:

$$P_i + \underset{\textit{starch chain}}{\text{G-G-G-G-G-}\cdots} \underset{\text{PHOSPHORYLASE}}{\overset{\text{degradative}}{\underset{\text{synthetic}}{\rightleftharpoons}}} \underset{\alpha\text{-}Glc\text{-}1\text{-}P}{\text{G-P}} + \underset{\substack{\textit{degraded starch chain}\\ \textit{(putative primer)}}}{\text{G-G-G-G-}\cdots}$$

The reactions show that the degradation involves inorganic phosphate that removes glucose residues from the nonreducing-end of the polysaccharide chains to remove glucose residues and form α-Glc-1-P and a partially degraded polysaccharide chain. The reverse, synthetic reaction, involves the transfer of glucose from α-Glc-1-P to α-1→4 glucan chains or to the nonreducing-ends of an α-1→4 linked glucose oligosaccharide. The addition of just α-Glc-1-P to the phosphorylases, however, gave no reaction. It was, thus, recognized that a preformed polysaccharide or oligosaccharide chain was absolutely required to have synthesis by these reactions and the concept of a *required primer* was established.

As the reaction was studied more carefully, it was found that starting with α-Glc-1-P and a starch or glycogen chain, the reaction rapidly slowed down and stopped, as the concentration of P_i increased. It was further found that the synthetic reaction did not occur *in vivo* at all, as the concentration of P_i in animal and plant tissue was 20- to 40-times the concentration of α-Glc-1-P (Trevelyan et al., 1952; Ewart et al,. 1954; Liu and Shannon, 1981) and the *in vivo* conditions greatly favored degradation, rather than synthesis. Further, the addition of phosphorylases to just α-Glc-1-P gave no reaction. It, thus, appeared that phosphorylases only catalyzed the degradation of glycogen and starch and not the synthesis.

The studies of (Cori and Cori, 1939; Hanes, 1940; and Swanson and Cori, 1948), however, led to the development of the hypothesis for a *required primer* chain for the biosynthesis of polysaccharides. With essentially no evidence this concept has stuck in the minds of many people since then and relatively recently, it has been postulated for the mechanism of biosynthesis of polysaccharides, even with a paucity of experimental evidence (Bocca et al., 1997; Ball et al. 1998; Ball and Morell, 2003; and Tomlinson and Denyer, 2003).

Some 20 years after the phosphorylase experiments, (De Fekete et al., 1960; Recondo and Leloir, 1961; Leloir et al., 1961) found that the high-energy donor of glucose for starch biosynthesis was uridine diphospho glucose (UDPGlc) and adenosine diphospho glucose (ADPGlc) and that when ADPGlc was incubated with starch granules, starch chains were biosynthesized. ADPGlc was the better of the two donors. The biosynthetic enzymes, *starch synthase* and *starch branching enzyme* were apparently entrapped in the granules during their synthesis. Many years later, (Robyt and Mukerjea, 2000) found that starch granules that had been in bottles on the laboratory shelves for over 40 years, still retained the ability to incorporate glucose from ADPGlc into starch.

When De Fekete et al. (1960), Recondo and Leloir (1961), and Leloir et al. (1961) incubated starch granules with ADP-[^{14}C] Glc, ^{14}C-glucose was incorporated into the starch. When they solubilized the starch and reacted it with the exo-acting enzyme, β-amylase, they obtained ^{14}C-labeled maltose from which they assumed that the synthesis of starch involved the addition of glucose from ADPGlc to the nonreducing-ends of the starch chains. This experiment has been widely considered as proof that starch chains are biosynthesized by the addition of glucose from ADPGlc to the nonreducing-ends of starch primer chains. This assumption, however, is not necessarily correct in that if the starch chains had been synthesized *de novo* from the reducing-end, rather than from the nonreducing-end of a primer, the synthesized chains would have every glucose residue in the chains labeled, and the subsequent reaction with β-amylase would also give ^{14}C-labeled maltose. See Section 6 for recent studies on how starch is biosynthesized.

MECHANISM FOR THE BIOSYNTHESIS OF *SALMONELLA* O-ANTIGEN POLYSACCHARIDE

The O-antigen surface polysaccharide of *Salmonella anatum* is a hetero-polysaccharide that was the first polysaccharide to have its mechanism of synthesis definitively determined (Dankert, et al. 1966; Wright et al., 1967; Bray and Robbins, 1967; Robbins et al., 1967). The polysaccharide is composed of a linear structure of β-mannosyl-β-rhamnosyl-β-galactosyl repeating sequence. The trisaccharide is biosynthesized from the sugar diphospho nucleotides, GDPMan, TDPRha, and UDPGal. The first reaction is the reaction of UDPGal with a lipid phosphate, bactoprenol phosphate to give bactoprenol pyrophosphoryl-α-D-galactopyranoside (Dankert et al., 1966; Wright et al., 1967)

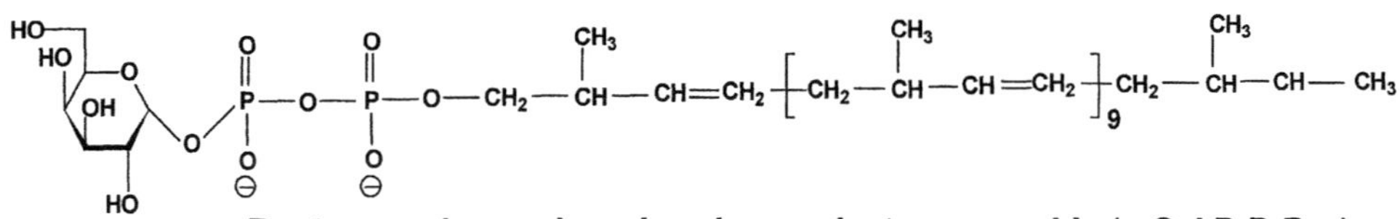

Bactoprenol pyrophosphoryl α-D-galactopyranoside (α-Gal-P-P-Bpr)

Assembly of the trisaccharide then occurs by the enzyme catalyzed addition of L-rhamnose to C4-OH of D-galactose, and the addition of D-mannose from GDPMan to the C4-OH of L-rhamnose to give Man–Rha–Gal–P–P–Bpr. This trisaccharide bactoprenol pyrophosphate is synthesized inside the cell by the addition of the monosaccharides in sequence to the bactoprenol pyrophosphate, which is partially embedded in the lipid bilayer of the cell membrane. The trisaccharide is enveloped by bactoprenol and is then transported through the lipid membrane to the outside of the cell, where polymerization occurs.

Bray and Robbins (1967) showed, by pulse and chase experiments, that the repeating trisaccharide was transferred to the reducing end of a growing chain according to the following reactions:

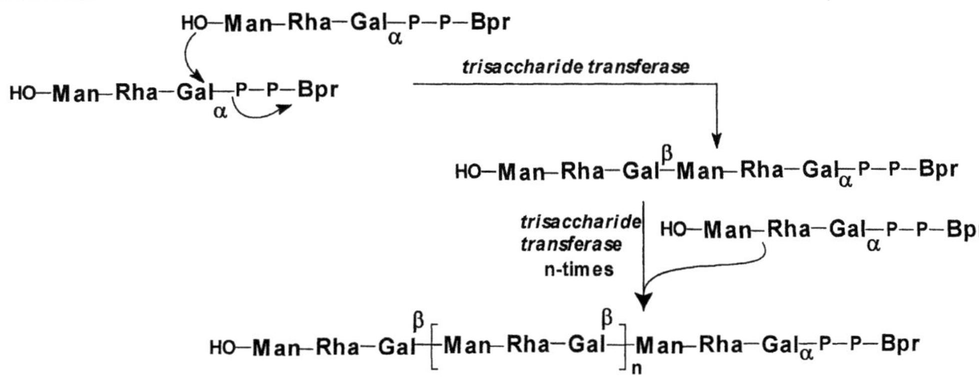

The C4-OH of the D-mannose makes a nucleophilic attack onto the C1 of the D-galactose, giving inversion of the configuration from α to β and the insertion of the trisaccharide between the reducing-end and the bactoprenol pyrophosphate. This reaction occurs repeatedly to give polymerization of the polysaccharide by the addition to the reducing-end.

MECHANISM FOR THE BIOSYNTHESIS OF BACTERIAL CELL WALL POLYSACCHARIDE, MUREIN

Murein is a polysaccharide with a repeating sequence of N-acetyl-D-glucosamine (NAG) linked β-1→4 to N-acetyl-D-muramic acid (NAM) in which a pentapeptide is attached to the carboxyl group of NAM. It also was found that bactoprenol phosphate was involved in the biosynthesis of the bacterial cell wall poly-peptidomurein (Anderson, et al., 1965; Struve and Neuhaus, 1965; Struve et al., 1966):

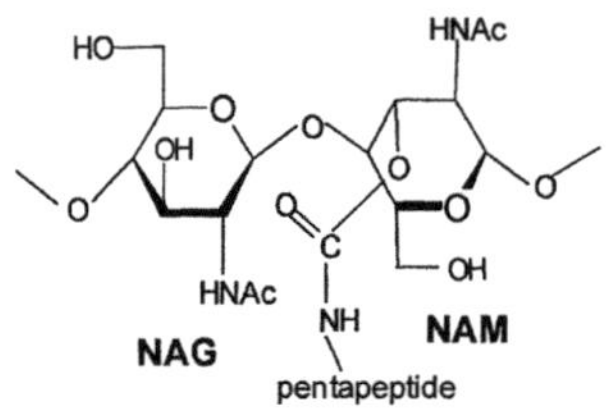

NAG-NAM-pentapeptide repeating unit of bacterial cell wall polysaccharide

The biosynthesis also starts inside the bacterial cell, where UDP-N-acetyl-D-muramic acid reacts with bactoprenol phosphate to give α-N-acetyl-D-muramic acid pentapeptide bactoprenol pyrophosphate plus UMP. N-Acetyl-D-glucosamine is then enzymatically added to C4-OH of the N-acetyl muramic acid in a β-linkage to give NAG-NAM-bactoprenol pyrophosphate, which is then transported through the cell membrane lipid bilayer to the outside of the cell where it is polymerized. Using ^{14}C-N-acetyl-D-glucosamine, it was reported in 1973 that the disaccharide is added to the reducing-end of a growing murein chain by the C4-OH of NAG attacking C1 of NAM at the reducing-end of the growing chain, giving the insertion of the disaccharide between the growing chain and the bactoprenol pyrophosphate (Ward and Perkins, 1973), essentially an identical mechanism, as the biosynthesis of *Salmonella* O-antigen polysaccharide:

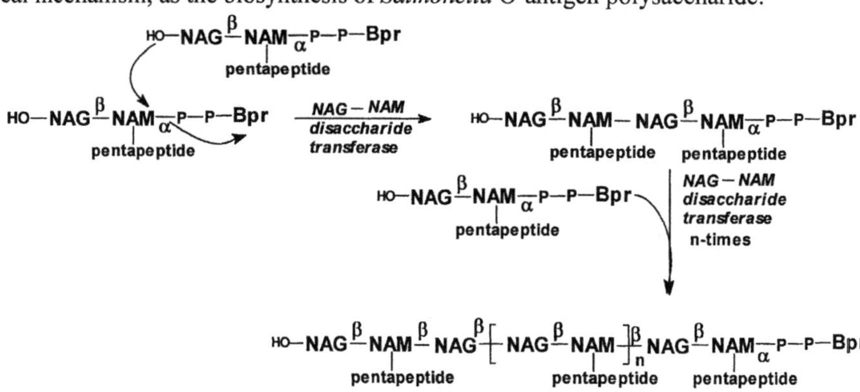

MECHANISM FOR THE BIOSYNTHESIS OF DEXTRAN BY *LEUCONOSTOC MESENTEROIDES* B-512FMC DEXTRANSUCRASE

Shortly after the report of the mechanism for the biosynthesis of the bacterial cell wall polysaccharide, Robyt et al. (1974) reported the mechanism of *L. mesenteroides* B-512F dextransucrase biosynthesis of dextran. In contrast to the O-antigen polysaccharide and bacterial cell wall polysaccharide, Dextran is a homopolysaccharide, with only one monomer unit, glucose, linked by α-1→6 glycosidic bonds in the main chains and two kinds of α-1→3 branch linkages, single glucose units and long α-1→6 linked units. The substrate for dextran synthesis is sucrose, a compound with high-energy glucose, similar to the energy of nucleotide diphospho carbohydrates.

Robyt et al. (1974) studied the mechanism of B-512F dextransucrase, using a pulse with ^{14}C-sucrose and a chase with nonlabeled sucrose and Bio-Gel P2 immobilized-enzyme. The resulting dextran from the pulse and chase reactions were isolated, reduced with $NaBH_4$, and then acid hydrolyzed, giving ^{14}C-glucitol from the reducing-end of the dextran and ^{14}C-glucose from the remainder of the dextran. The chased dextran gave a 100-fold decrease of ^{14}C-glucitol. These experiments definitively showed that the polymerization of dextran was from the addition of glucose to the reducing-end of the dextran chain. It would have been impossible to have obtained ^{14}C-glucitol, if the addition of glucose had been to the nonreducing-end of a primer. Using pulse and chase experiments, Robyt and Martin (1983) showed that the two glucansucrases elaborated by *Streptococcus mutans*, dextransucrase and mutansucrase, also added glucose to the reducing-ends of the dextran and mutan, an α-1→3 linked glucan, chains; Ditson and Mayer (1984) also found that glucose was added to the reducing-end of dextran chains during biosynthesis of dextran by *Step. sangius* dextransucrase.

Robyt and Walseth (1978) also found that when the immobilized dextransucrase was pulsed with ^{14}C-sucrose, and washed several times with buffer and then glucose was added to the immobilized-enzyme, two molecular weight products were formed: (a) a low molecular weight (LMW) product, identified as isomaltose and (b) a high molecular weight (HMW) product, dextran. Similar results were obtained when fructose was added, a LMW product, leucrose, and a HMW product dextran and when maltose was added, a LMW product, panose, and a HMW product, dextran. These experiments definitely show that two covalent complexes were formed during dextran biosynthesis, a glucosyl- and a dextranyl-enzyme intermediates. Parnaik et al. (1983) also found a glucosyl- and a dextranyl-enzyme intermediates for *Streptococcus sangius* dextransucrase.

In a review, Ebert and Schenk (1968) early proposed a two-site insertion mechanism to be the most reasonable and logical for the biosynthesis of dextran, but without supporting experimental evidence. Robyt et al. (1974), Robyt and Walseth (1978), and Robyt and Martin (1983) provided the experimental evidence and further elaborated on the two-site insertion mechanism for dextran biosynthesis, involving both glucosyl- and dextranyl-covalent intermediates. Using equilibrium dialysis experiments, Su and Robyt (1994) provided confirmation of the mechanism for B-512FMC dextransucrase by showing that it has two sucrose binding-sites at the active-site.

Dextransucrase also catalyzes a secondary reaction that takes place when LMW carbohydrates, such as, glucose, fructose, or maltose is present or added to dextransucrase-sucrose digests (Robyt and Eklund, 1983; Fu and Robyt, 1990; Fu and Robyt, 1991). These reactions are called *acceptor reactions.* There are over 30 known LMW carbohydrates and several non-carbohydrates that have primary and/or secondary alcohol groups that give products (Robyt, 1995; Yoon, et al., 2004). Glucose gives isomaltose, fructose gives leucrose, and maltose, gives panose. Isomaltose and panose go on and give a series of isomaltodextrin homologues of exponentially decreasing amounts, as the size of the homologues increase. These acceptor reactions terminate dextran biosynthesis (Robyt and Eklund, 1983 and Su and Robyt, 1993) and inhibit the biosynthesis by competing for the glucose. Water is also an acceptor, although a relatively inefficient one, and it terminates dextran biosynthesis with a certain frequency, by hydrolyzing the dextran-enzyme covalent intermediate, releasing the dextran from the active-site (Robyt and Walseth, 1978).

Carbohydrate enzymologists searched for several years (1954-1976) for a dextran branching enzyme, similar to the known starch branching enzyme (Q-enzyme), but they were never able to find one. Robyt and Taniguchi (1976) showed that dextransucrase itself catalyzes the formation of the branch linkages by an acceptor reaction in which released exogenous dextran chains act as acceptors. The C3-OH group of a glucose residue in the exogenous dextran chain attacks either the covalently linked glucose intermediate to give a single glucose α-1$\rightarrow$3 branch or it attacks the covalently linked dextranyl chain at the active-site to give an α-1$\rightarrow$3 linked dextran chain attached to the exogenous dextran acceptor. Su and Robyt (1994) showed by equilibrium dialysis that there was one acceptor binding-site.

Recently Moulis et al. (2006) claimed that the two-site insertion mechanism was not the mechanism for the biosynthesis of dextran. They used a C- and N-terminal truncated B-512F dextransucrase that was cloned in *E. coli*, and proposed, without any convincing experimental evidence, that dextransucrase first hydrolyzes sucrose by an acceptor reaction with water, giving glucose and fructose and that the glucose and sucrose acted as a initiator primers and that the dextran was thus polymerized by the addition of glucose to the nonreducing-ends of the resulting isomaltodextrin primers. While both the hydrolysis of sucrose and the acceptor reactions of glucose, fructose, and isomaltodextrins are well known, Moulis et al. (2006) did not show any definitive experimental evidence that dextran was polymerized in this way.

Robyt et al. (2008) very recently experimentally found that neither glucose nor sucrose were initiator primers. They added 0.1 μCi of ^{14}C-glucose to a B-512FMC dextransucrase-sucrose digest and only found 40 dpm out of 2.2×10^5 dpm of glucose incorporated into dextran, which is less than 0.02% of the labeled glucose added to the digest, indicating that it was not acting as an initiator primer. It most likely was incorporated in the dextran by the release of a very small amount of dextran from the active-site by an acceptor reaction. Treatment of a HMW dextran (d.p. 521, MW = 84,420 Da) with 0.01 M HCl at 50°C and also with invertase for several hours, did not give any fructose, which would have been expected if sucrose was acting as an initiator primer and therefore located at the reducing-end of the dextran chain.

Robyt et al. (2008) also studied the kinetics of dextran formation in terms of the amount and the number average MW of the dextran. In addition, they also studied the formation of LMW products, formed during the reaction of dextransucrase, as a function of time, using *fluorescent assisted capillary electrophoresis* (FACE), a very sensitive quantitative method for determining oligosaccharides of widely different sizes. In the early stages of the reaction [0.2 conversion period, where a conversion period (CP) is the theoretical amount of sucrose that could be converted into dextran for the amount of enzyme present] gave glucose, fructose, leucrose, and isomaltodextrins in small exponentially decreasing amounts from d.p. 2-5, with minuscule amounts of d.p. 6-11; 0.5 CP gave the same compounds, but with exponentially decreasing amounts down to minuscule amounts of d.p. 10-20; 1.00 CP gave the same compounds, with minuscule amounts of d.p. 11-26; and 2.00 CPs gave the same compounds, with minuscule amounts of d.p. 15-26. The number average MWs of the dextrans for these same conversion periods were 172,000 ± 1,500 (d.p. ~1000), 178,000 ± 2,000 (d.p. ~1100), 239,000 ± 3,500 (d.p. ~1475), and 240,000 ± 3,500 Da (d.p. ~1480), respectively. These experiments definitely show that (a) glucose and sucrose are not initiator primers and that (b) the polymerization of dextran does not occur by the addition of glucose from sucrose to the nonreducing ends of isomaltodextrins, as postulated by Moulis et al. (2006). If the polymerization was occurring by this mechanism, just the opposite result should have been observed, namely there should have been exponentially increasing amounts of higher d.p. isomaltodextrins, going up to and including, d.p. 100-1000 or higher.

Moulis et al. (2006) also postulated that dextransucrase has only one active-site that involves the three conserved amino acids (Asp551, Glu589, and Asp662) found in all GH-family 70 enzymes, including glucansucrases, and not two sets of the three conserved amino acids that should have been found for two active-sites. They, therefore, concluded that the two-site insertion mechanism was not valid for the biosynthesis of dextran. Two active-sites, however, was never proposed for the two-site insertion mechanism. What was proposed was two catalytic-groups, with two sucrose binding-sites that were involved in the insertion mechanism for the biosynthesis of dextran at one active-site. Robyt et al. (2008) have now shown how the three conserved amino acids participate in the two catalytic-site, insertion mechanism at one active-site.

Robyt et al. (2008) further show that the molecular size of the dextran is inversely proportional to the concentration of the enzyme, indicating that the elongation of dextran is a highly processive reaction in which glucose is rapidly added to the reducing-end of the covalently linked, growing dextran chain, which is extruded from the active-site until it is released by an acceptor reaction with water or a carbohydrate acceptor, such as glucose, isomaltose, or a dextran chain to give a branch linkage.

From these experiments, Robyt et al. (2008) concluded that the evidence suggests that the most reasonable and logical mechanism for the biosynthesis of dextran is the two catalytic-site, insertion mechanism that occurs at one active-site and not by the one-site nonreducing-end, primer mechanism proposed by Moulis et al. (2006). The mechanism for dextran chain elongation is very similar to that of the biosynthesis of starch chains by starch synthase (see, Fig. 1), except that α-1→6 linkages are synthesized instead of α-1→4 linkages.

MECHANISM FOR THE BIOSYNTHESIS OF *ACETOBACTER XYLINUM* BACTERIAL CELLULOSE AND *XANTHOMONAS CAMPESTRIS* XANTHAN

Cellulose is another homopolysaccharide, consisting of linear chains of glucose linked β-1→4, making up approximately 50% of all plant cell walls. It also is produced by a few species of bacteria that synthesize relatively pure cellulose, as an extracellular product that is extruded from the surface of the cell (Haigler, 1991).

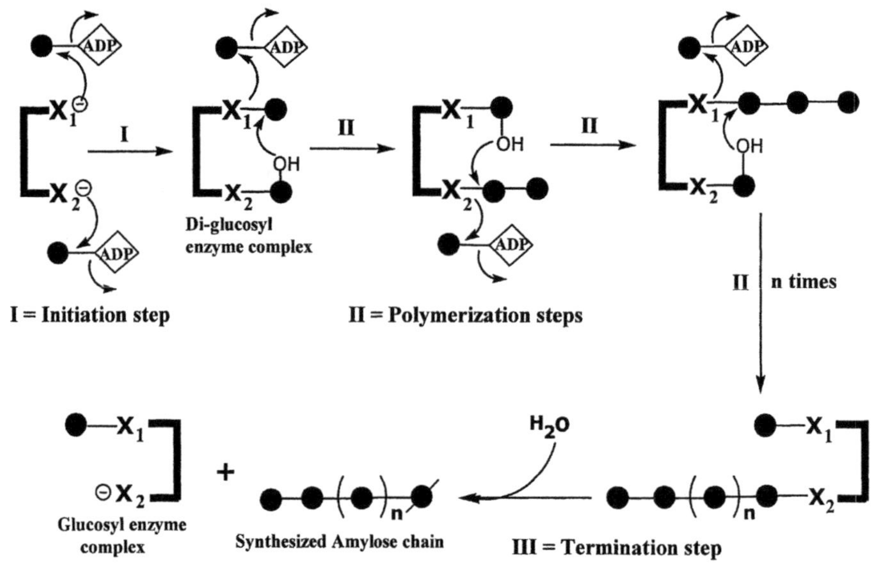

Figure 1 Mechanism for the biosynthesis of starch chain by starch synthase, using ADPGlc as the substrate

The circles represent glucose units. X_1 and X_2 represent nucleophilic catalytic groups at the active-site of the enzyme.

Several strains of *Acetobacter xylinum* synthesize cellulose from UDPGlc by the enzyme, cellulose synthase. Sequence analysis of the enzyme indicates that it is an anchored membrane protein. Efforts to study the biosynthesis of cellulose in plants have not been successful, due to the inability to obtain active cellulose synthase and demonstrate the synthesis *in vitro*. It had been proposed that the cellulose chain is elongated from the reducing-end. This was based on deductions made from a comparative study of the sequence of several different polysaccharide synthesizing and hydrolyzing enzymes; and direct experimental evidence was not presented. A few years later Saxema et al. (1990) proposed that *A. xylinum* cellulose was synthesized by the addition of glucose from UDPGlc to the nonreducing ends of the cellulose chains. This was based on the silver staining of the reducing-ends of the cellulose chains that were extruded from the surface of the bacteria and the microdiffusion-tilting electron crystallographic analysis of the cellulose fibers. The evidence here was very sketchy, indirect, non-quantitative, and arrived at primarily from reasoning by analogy.

To resolve these two opposite positions, the *de novo* synthesis of cellulose by resting *A. xylinum*-cells and *A. xylinum*-membrane preparations was studied, using UDP-[^{14}C]Glc pulse and UDPGlc chase reactions by Han and Robyt (1998). They found that the synthesized cellulose was tightly associated extra-cellularly with the cells and their cell membrane. The cellulose chains could be released from the cells and the membrane preparation by treating them at pH 2, 100 °C for 20 min, which obviously was not by the hydrolysis of the cellulose chain per se. The cellulose chains that were released from the pulse and chase reactions were purified and separated from low molecular weight compounds by gel chromatography on Bio-Gel P4 (fine). The pulsed products from the resting cells, after reduction with sodium borohydride, and acid hydrolysis gave 1799 cpm in ^{14}C-glucitol and 239 cpm in the chased cellulose, indicating that bacterial cellulose was being biosynthesized by the addition of glucose from UDPGlc to the reducing-end of cellulose. These results resolved the conflict, as ^{14}C-glucitol could only be obtained by the addition of ^{14}C-glucose to the reducing-ends of cellulose and non-labeled UDPGlc could only chase it into the cellulose chains, if the synthesis is by the addition of glucose to the reducing-end. Evidence for the involvement of a lipid pyrophosphate in the biosynthesis of cellulose by *A. xylinum* was obtained before this (Colvin, 1959; Garcia et al., 1974; Copper and St. John Manley, 1975; Swissa et al., 1980). The lipid pyrophosphate was found to be an absolutely required component. It was determined to be a polyisoprenyl alcohol (bactoprenol), containing 55 carbons with a pyrophosphate ester linkage to the alcohol group, identical to the lipid phosphate involved in *Salmonella* O-antigen and bacterial cell wall murein-pentapeptide polysaccharides syntheses, previously described here.

Although the elongation of the cellulose chain is by cellulose synthase, the actual mechanism proposed for cellulose biosynthesis by Han and Robyt (1998) involves three enzyme catalyzed reactions: (a) the first reaction is catalyzed by *Lipid pyrophosphate: UDPGlc phosphotransferase* (*LP: UDPGlc-PT*) that transfers Glc-1-P from UDPGlc to bactoprenol phosphate to give bactoprenol pyrophosphate α-glucose); (b) the second reaction is catalyzed by *cellulose synthase* (*CS*) that produces the polymerization of the glucose residues by a two-catalytic site insertion mechanism, releasing bactoprenol pyrophosphates; (c) the third reaction is catalyzed by *lipid pyrophosphate pyrophosphtase* (*LPP*) and gives hydrolysis of the pyrophosphate to give bactoprenol phosphate that can again attack UDPGlc, giving bactoprenol pyrophosphate α-glucose that continues to add glucose, forming a β-linkage to a growing cellulose chain (see Fig. 1B). Initially, it might be thought that the lipid intermediate is not necessarily required for the synthesis and that the glucosyl unit and the growing cellulose chain could be directly attached to the cellulose by the synthase, like dextran is attached to dextransucrase. This kind of attachment, however, would give the glucosyl residue attached to the active-site of the enzyme in a β-configuration. The subsequent reactions of this glucosyl intermediate would then give the addition of the glucosyl intermediate to the growing polymer, but the glycosidic linkage would be alpha to give an α-glucan instead of a β-glucan, cellulose chain. The formation of the lipid pyrophosphate glucosyl intermediate has the glucose attached alpha to the pyrophosphate group because of the way it is formed from UDPGlc (see Fig. 2) and then when it reacts with the lipid-phosphate, the α-configuration is retained and then inverted when added to the growing cellulose chain. The lipid-phosphate and lipid-pyrophosphate also play another role in that they bind to the enzyme/protein at a hydrophobic site at the active-site of cellulose synthase. Glucose is added from UDPGlc to the lipid phosphate inside the cell and then it is enveloped by the lipid and carried thru the lipid by-layer membrane to the outside of the cell, where the lipid unfolds and allows glucose to be added to the growing cellulose chain (see Fig. 3 for the mechanism for cellulose elongation). The proposed reducing-end, insertion mechanism, thus, has no need for preformed oligosaccharide- or polysaccharide-primers, identical to the previous polysaccharides biosynthesized from the reducing-end of a growing polysaccharide by a two catalytic-site, insertion mechanism.

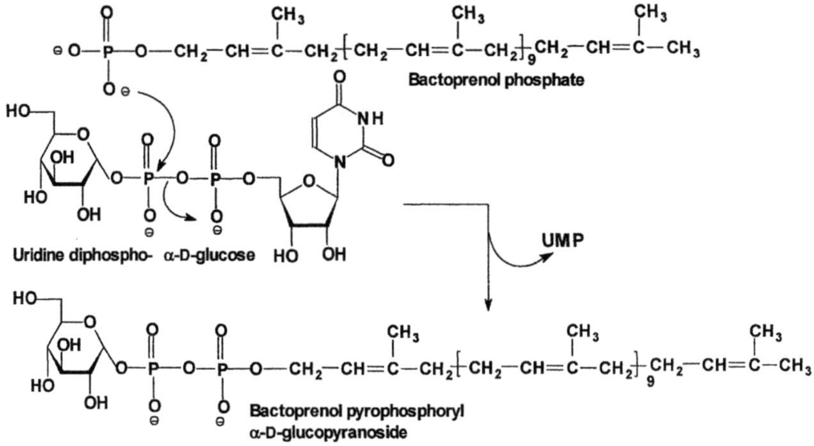

Figure 2 Mechanism for the biosynthesis of *Acetobacter xylinum* bacterial cellulose; Formation of bactoprenol pyrophosphate α-D-glucopyranoside

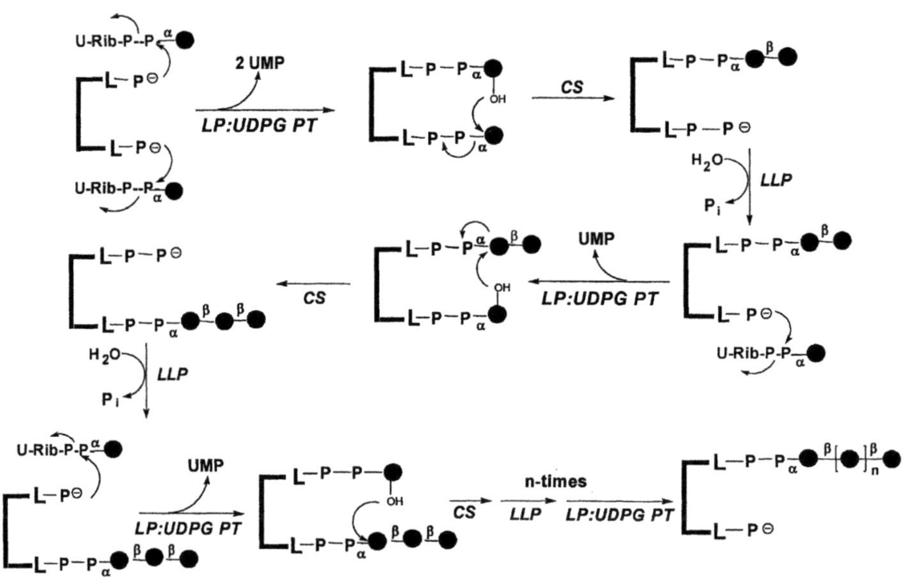

Figure 3 Mechanism for the biosynthesis of *Acetobacter xylinum* bacterial cellulose
Polymerization of cellulose by *cellulose synthase* (CS) and the formation of intermediates by *Lipid pyrophosphate:UDPGlc phosphotransferase* (LP:UDPGlc-PT) and *lipid pyrophosphate pyrophosphatase* (LPP). The circles represent glucose units, L represents lipid bactoprenol.

Xanthan is a water-soluble cellulose analogue that is used as a filler, fiber, and gum in many prepared foods, replacing such polysaccharides as guar and gum arabic. It is produced by *Xanthomonas campestris* and is a hetero-linked, hetero-polysaccharide whose main chain is cellulose (a homo-polysaccharide), with a hetero-linked, hetero-trisaccharide, composed of 4,6-pyruvyl-Man-β-(1→4)-GlcUA-β-(1→2)-Man-linked α-(1→3) to every other D-glucose residue in the cellulose chains. The biosynthetic mechanism for xanthan was shown in by Ilepi et al. (1993) to be the addition of the pyruvyl-cellobiose-trisaccharide bactoprenol pyrophosphate that is inserted into the growing polysaccharide at the reducing-end to form a β-linkage. The mechanism, seen below, is identical to that of *Salmonella* O-antigen polysaccharide (see Section 2), bacterial cell wall, peptido-murein (see Section 3), and cellulose described above in this Section.

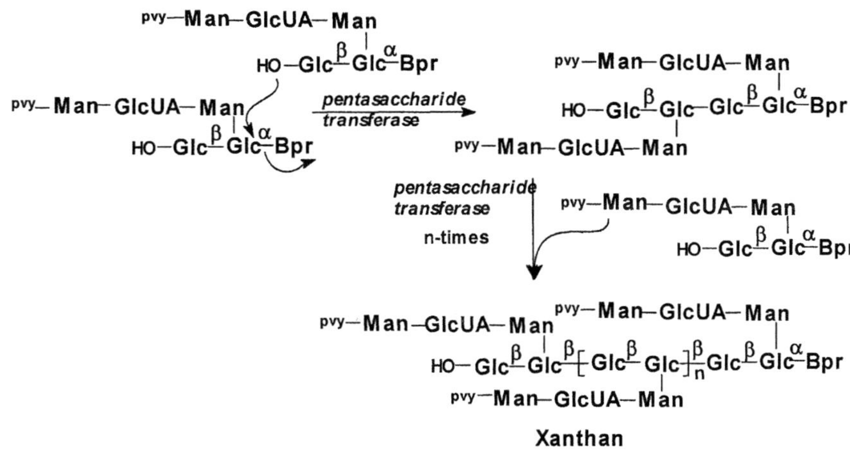

MECHANISM FOR THE BIOSYNTHESIS OF STARCH CHAINS IN STARCH GRANULES

As indicated in introduction, the biosynthesis of starch chains were first postulated to be catalyzed by plant phosphorylase in which glucose is added to the nonreducing-ends of starch primer chains. Phosphorylase, however, is not responsible for the biosynthesis of starch *in vivo*, as the amount of inorganic phosphate to α-G-1-P was too high and phosphorylase exclusively catalyzes a degradative reaction and not a synthetic reaction. De Fekete et al. (1960), Recondo and Leloir (1961), and Leloir et al. (1961) found that the enzyme responsible for starch biosynthesis was *starch synthase* that had been entrapped in starch granules and used ADPGlc as the substrate to synthesize starch. Leloir et al. (1961) assumed that starch synthesis was a primer-dependent reaction in which glucose from ADPGlc was added to the nonreducing-ends of the primers, but this assumption was not necessarily correct in that if the starch chain had been synthesized *de novo* from the reducing-end by ADP-[^{14}C]Glc, it also would have given ^{14}C-maltose when reacted with β-amylase (see introduction). Further, no one had ever synthesized starch chains of significant size through *in vitro* reaction with primers, ADPGlc, and starch synthase (Denyer et al., 1999; Damager et al., 2001). Because of these problems and a lack of definitive experiments on the mechanism of the biosynthesis of starch, Mukerjea and Robyt (2002) reinvestigated the reaction of ADPGlc with eight different varieties of starch granules, using pulse reactions with ADP-[^{14}C]Glc and chase reactions with nonlabeled ADPGlc. After reaction and solubilization of the starch granules, reduction with $NaBH_4$, hydrolysis of the reduced starch with glucoamylase, and descending paper chromatographic separation of glucose and glucitol, a significant amount of ^{14}C-labeled glucitol from the reducing-end and ^{14}C-labeled glucose from the rest of the starch chains was obtained for all eight varieties of starches; the chase reaction also showed a significant decrease in the amount of ^{14}C-glucitol. The formation of ^{14}C-glucitol and its chase indicated that glucose from ADPGlc was being added to the reducing-ends of the growing starch chains and not to the nonreducing-ends of primer chains, as it would have been impossible to obtain any ^{14}C-glucitol if the glucose was being added to the nonreducing-ends of primer chains. The biosynthesis was, thus, identical to what had been found for the biosynthesis of the five previously described polysaccharides. It was also found that a significantly sized starch chain was synthesized (see the following Table 1 for the pulse and chase data and the size of the pulsed synthesized eight starches).

Table 1 Pulse reactions for eight varieties of starches with ADP-[^{14}C]Glc and chase reaction with nonlabeled ADPGlc and the number average d.p. and number average molecular weights of the synthesized starches

Starches	Pulsed ^{14}C-glucitol counts[a]	30 min Chased ^{14}C-glucitol counts[a]	Number Average d.p.	Number Average Molecular Weight
Maize	5240	3140	827	133,992
Waxy maize	3480	2840	436	70,650
Taro	3050	890	462	74,862
Rice	2280	1360	467	75,672
Wheat	1750	1560	476	77,130
Potato	1240	910	524	84,906
Barley	1960	350	127	20,592
Rye	500	160	441	71,460

[a]Samples were counted for 10 minutes in a liquid scintillation counter.

The second set of experiments involved the addition of the putative maltodextrin primers, maltose, maltotriose, and d.p. 12 maltodextrin in increasing amounts. All three of these putative primers, inhibited the biosynthesis of starch chains, with increasing inhibition as the concentrations of the putative primers were increased. This is a result that is just opposite to that expected for a primer, which should stimulate synthesis, if indeed they are required primers for the biosynthesis. The putative primers did give reaction products: glucose units were added to the nonreducing-ends of the putative primers; maltose gave maltotriose as the major product with exponentially decreasing amounts of maltodextrins, d.p. 3 to 9; maltotriose gave maltotetraose and maltopentaose as the major products, with exponentially decreasing amounts of maltodextrins of d.p. 6 to 9. It was concluded from these experiments that the putative primers were acting as *acceptors*, instead of primers, identical to the experiments observed for acceptors when present or added to dextransucrase digests (Robyt and Eklund, 1983; Su and Robyt, 1994; see, Section 4).

From the pulse and chase experiments Mukerjea et al. (2002) and the putative primer experiments, showing inhibition of starch synthesis Mukerjea and Robyt (2005), it was concluded that starch biosynthesis occurs by the addition of glucose to the reducing-end of a growing starch chain by a two catalytic-site insertion mechanism from a single active-site and not by the addition of glucose to the nonreducing-ends of required primers, as had been previously postulated and believed for several decades. The mechanism for starch chain biosynthesis by starch synthase is shown in Fig. 2.

In more recent studies, Mukerjea and Robyt (2007) have isolated, stabilized, and purified starch synthase and starch branching enzyme from potato, with high specific activities, and have found that amylose chains are synthesized from ADPGlc in the absence of any primers. Other purified fractions synthesized α-1$\rightarrow$6 branched starch components, indicating the presence of both starch synthase and starch branching enzymes, and yet another fraction only contained starch branching enzyme. Other starch and carbohydrate metabolizing enzymes, such as phosphorylase, amylase, glucosidase, and debranching enzyme were absent in all of the fractions. Additional studies with the starch synthase and starch branching enzymes are in progress.

SUMMARY AND CONCLUSIONS

B-512F Dextran was the first α-linked homo-polysaccharide that was shown to be biosynthesized by the two catalytic-site, insertion mechanism. The starch chain is the second homo-polysaccharide that is α-linked and shown to be biosynthesized from the reducing-end by the two catalytic-site, insertion mechanism. In both biosyntheses, glucose and the growing polysaccharide form covalent enzyme intermediates with their synthetic enzymes, dextransucrase and starch synthase, respectively. The other homo-polysaccharide, bacterial cellulose, which also is synthesized by the addition of the monomer unit to the reducing-end of the growing polysaccharide chain, has both the monomer unit and the growing polysaccharide chain covalently attached to a lipid pyrophosphate that bind at the active-site of cellulose synthase. For all three polysaccharides, the two covalent intermediates, the monomer unit and the growing polysaccharide, act in concert in which the monomer is inserted between the reducing-end of the growing polysaccharide chain and the enzyme or the lipid pyrophosphate. The insertion is actually a transglycosylation reaction in which the growing chain is transferred to the monomer bactoprenol pyrophosphate by the monomer unit making a nucleophilic attack onto C1 of the growing polysaccharide chain, inverting the configuration of the polysaccharide from β to α and the release of bactoprenol pyrophosphate from the polysaccharide.

The more structurally complex polysaccharides: bacterial cell wall, peptide-murein, *Salmonella* O-antigen polysaccharide, and xanthan are also biosynthesized by the two catalytic-site, insertion mechanism. All three are hetero-linked, hetero-polysaccharides, that is, they have more than one type of glycosidic linkages and two or more monosaccharides in a repeating sequence. The first monosaccharide is enzymatically added to bactoprenol phosphate by the reaction of its nucleotide diphospho derivative to retain the α-linkage of the monosaccharide to bactoprenol pyrophosphate monomer. The repeating sequences are then built-up, by the sequential enzymatic transfer from a nucleotide diphospho monosaccharide to the first monosaccharide that is attached to bactoprenol pyrophosphate. The repeating unit is then transferred to another repeating unit attached to bactoprenol pyrophosphate or to a growing polysaccharide chain that is attached to bactoprenol pyrophosphate by insertion between the repeating unit or the growing polysaccharide, releasing bactoprenol pyrophosphate and giving inversion of the configuration from α to β.

The biosynthesis of the plant cell wall cellulose most probably occurs by a mechanism identical to that of bacterial cellulose biosynthesis, with the possible exception of having a slightly changed lipid pyrophosphate from bactoprenol pyrophosphate to dolichol pyrophosphate or something similar.

There now have been six structurally and functionally diverse polysaccharides that have definitively been shown to be biosynthesized from the reducing-end by the two catalytic-site, insertion mechanism, making this the norm for polysaccharide biosynthesis, rather than the exception. Five of the six polysaccharides are biosynthesized by bacteria, with starch being the only one to date that has been shown to be biosynthesized by eight, eukaryotic plant sources.

REFERENCES

Anderson J S, Matsuhashi M, Haskin M A, and Strominger J L (1965) 'Lipid-phospho *N*-acetyl muramyl-pentapeptide: presumed membrane transport intermediates in cell wall synthesis' *Proc Natl Acad Sci US.*, 53, 881–889.

Ball S G and Morell M K (2003) 'From bacterial glycogen to starch: Understanding the biogenesis of the plant starch granule' *Ann Rev Plant Biol*, 54, 207–233.

Ball S G, Van de Wal H B J M, and Visser R G F (1998) 'Progress in understanding the biosynthesis of amylose' *Trends Plant Sci*, 3, 1360–385.

Bocca S N, Rothschild A and Tandecarz J S (1997) 'Initiation of starch biosynthesis: Purification and characterization of UDP-glucose: protein transglucosylation from potato tubers' *Plant Physiol Biochem*, 35, 205–212.

Bray D and Robbins P W (1967) 'The Direction of chain growth in *Salmonella anatum O*-antigen biosynthesis' *Biochem Biophys Res Commun*, 28: 334–339.

Colvin J R (1959) 'Synthesis of cellulose in ethanol extracts of *Acetobacter xylinum*' *Nature*, 183, 1135–1137.

Copper D and St. John Manley R (1975) 'Evidence for the involvement of a bactoprenol phosphate in bacterial cellulose biosynthesis' *Biochim Biophys Acta*, 381, 78–96.

Cori G T and Cori C F (1939) 'The activating effect of glycogen on the enzymic synthesis of glycogen from glucose-1-phosphate' *J Biol Chem*, 131, 397–398.

Damager I, Denyer K, Motawia M S, Møller B L, and Blennow A (2001) 'The action of starch synthase on 6^{III}-α-maltotriosyl-maltohexaose comprising the branch point of amylopectin' *Eur J Biochem*, 268, 4878–4884.

Dankert M, Wright A, Kelley W S, and Robbins P W (1966) 'Isolation, purification and properties of the lipid-linked intermediates of *O*-antigen biosynthesis' *Arch Biochem Biophys*, 116, 425–435.

De Fekete M A R, Leloir L F, and Cardini, D. E. (1960) 'Mechanism of starch biosynthesis' *Nature*, 187, 918–919.

Denyer K, Waite D, Edwards A, Martin C, and Smith A M (1999) 'Interaction with amylopectin influences the ability of granule-bound starch synthase I to elongate malto-oligosaccharides' *Biochem J*, 342, 647–653.

Ditson S L and Mayer R M (1984) 'Dextransucrase: The direction of chain growth during autopolymerization' *Carbohydr Res*, 126, 170–175.

Ebert K H and Schenk G (1968) 'Mechanisms of biopolymer growth: the Formation of dextran and levan' *Adv Enzymol*, 30, 179–221.

Ewart M H, Siminovitch D, and Briggs D R (1954) 'Possible enzymic processes involved in starch-sucrose interconversions' *Plant Physiol*, 29, 407–413.

Fu D and Robyt J F (1990) 'Acceptor reactions of maltodextrins with *Leuconostoc mesenteroides* B-512FM dextransucrase' *Arch Biochem Biophys*, 283, 379–387.

Fu D and Robyt J F (1991) 'Maltodextrin acceptor reactions with *Streptococcus mutans* 6715 glucosyltransferases' *Carbohydr Res*, 217, 201–211.

Garcia R C, Recondo E, and Dankert M (1974) 'Polysaccharide biosynthesis in *Acetobacter xylinum*. Enzymatic synthesis of lipid diphosphate and monophosphate sugars' *Eur J Biochem*, 43, 93–105.

Haigler C H, (1991) 'Relationship between polymerization and crystallization' in "*Biosynthesis and Biodegradation of Cellulose*" Haigler C H and Weimer P J, Eds., New York, Marcel Dekker, 99–124.

Han N S and Robyt J F (1998) 'The mechanism of *Acetobacter xylinum* cellulose biosynthesis: direction of chain elongation and the role of lipid pyrophosphate intermediates in the cell membrane' *Carbohydr Res*, 313, 125–133.

Hanes, C. S. (1940) 'The reversible formation of starch from glucose-1-phosphate catalyzed by potato phosphorylase' *Proc Roy Soc B*, 129, 174–208.

Ilepi L, Couso R O, and Dankert M (1993) 'Sequential assembly and polymerization of the polyprenol-linked pentasaccharide repeating unit of the xanthan polysaccharide in *Xanthomonas campestris*' *J Bacteriol*, 175, 2490–2500.

Koyama M, Helbert W, Imai T, Sugiyama J, and Henrissat B (1997) 'Parallel-up structure evidences for the molecular directionality during biosynthesis of bacterial cellulose' *Proc Natl Acad Sci US*, 94, 9091–9095.

Leloir L F, De Fekete M A R, and Cardini C E (1961) 'Starch and oligosaccharide synthesis from uridine diphosphate glucose' *J Biol Chem*, 236, 636–641.

Liu T F and Shannon J C (1981) 'Measurement of metabolites associated with nonaqueously isolated starch granules from immature *Zea mays L.* endosperm' *Plant Physiol*, 67, 525–533.

Moulis C, Joucha G, Harrison D, Fabre E, Potocki-Veronese G, Monsan P, and Remaud-Simeon M (2006) 'Understanding the polymerization mechanism of glycoside-hydrolase family 70 glucansucrases' *J Biol Chem*, 281: 31254–67.

Mukerjea Ru and Robyt J F (2005) 'Starch biosynthesis: the primer nonreducing-end mechanism versus the nonprimer reducing-end two-site insertion mechanism' *Carbohydr Res*, 340, 245–255.

Mukerjea Ru and Robyt J F (2007) Unpublished results on the purification and characterization of potato starch synthesizing enzymes.

Mukerjea Ru and Robyt J F (2000) Unpublished results: active starch synthase activities in starch granules stored for ten to forty years at 23°C.

Mukerjea Ru, Yu L, and Robyt J F (2002) 'Starch biosynthesis: mechanism for the elongation of starch chains' *Carbohydr Res*, 337, 1015–1022.

Parnaik V K, Luzio G A, Grahme D A, Ditson S L, and Mayer R M (1983) 'A D-glucosylated form of dextransucrase: Preparation and characteristics' *Carbohydr Res*, 121, 257–268.

Recondo E and Leloir L F (1961) 'Adenosine diphosphate glucose and starch synthesis' *Biochem Biophys Res Commun*, 6, 85–88.

Robbins P W, Bray D, Dankert M, and Wright A (1967) 'Direction of chain growth in polysaccharide synthesis' *Science*, 158, 1536–1542.

Robyt J F and Eklund S H (1983) 'Relative quantitative effects of acceptors in the reaction of *Leuconostoc mesenteroides* B-512F dextransucrase' *Carbohydr Res*, 121, 279–286.

Robyt J F and Martin P J (1983) 'Mechanism of synthesis of glucan by glucosyltransferaeses from *Streptococcus mutans 6715*' *Carbohydr Res*, 113, 301–315.

Robyt J F and Taniguchi H (1976) 'The mechanism of dextransucrase action: II. Biosynthesis of branch linkages by acceptor reactions with dextran' *Carbohydr Res*, 174, 129–137.

Robyt J F and Walseth T F (1978) 'The mechanism of acceptor reactions of *Leuconostoc* mesenteroides B-512F dextransucrase' *Carbohydr Res*, 61, 433–444.

Robyt J F (1995) 'Mechanisms in the glucansucrase synthesis of polysaccharides and oligosaccharides from sucrose' *Adv. Carbohydr. Chem Biochem*, 51, 133–168.

Robyt J F, Kimble B K, and Walseth T F (1974) 'The Mechanism of dextransucrase action: I. Direction of dextran biosynthesis' *Arch Biochem Biophys*, 165, 634–644.

Robyt J F, Yoon S H, and Mukerjea Ru (2008) 'On the mechanism of the synthesis of B-512F dextran by *Leuconostoc mesenteroides* B-512FMC dextransucrase' *Submitted to Carbohydr Res.*

Saxena I M, Brown R M, Fevre M, Geremia R A, and Henrissat B (1995) 'Multidomain architecture of β-glycosyltransferase: implications for mechanism of action' *J Bacteriol*, 177, 1419 –1424.

Saxena I M, Lin F C and Brown R M (1990) 'Cloning and sequencing of the cellulose synthase catalytic subunit gene of *Acetobacter xylinum*' *Plant Mol Biol*, 15, 673–683.

Struve W G and Neuhaus F C (1965) 'Evidence for an initial acceptor of UDP-NAc-muramyl-pentapeptide in the synthesis of bacterial mucopeptide' *Biochem Biophys Res Commun*, 18, 6–12.

Struve W G, Sinha R K and Neuhaus F C (1966) 'On the initial stage in peptidoglycan synthesis: Phospho-N-acetyl-muramyl-pentapeptide' *Biochemistry*, 5, 82–93.

Su D and Robyt J F (1993) 'Control of the synthesis of dextran and acceptor-products by *Leuconostoc mesenteroides* B-512FM dextransucrase' *Carbohydr Res*, 248, 339–348.

Su D and Robyt J F (1994) 'Determination of the number of sucrose and acceptor binding sites for *Leuconostoc mesenteroides* B-512FM dextransucrase and confirmation of the two-site mechanism for dextran synthesis' *Arch Biochem Biophys*, 308, 471–476.

Swanson M A and Cori C F (1948) 'Structure of polysaccharides: III. Relation of structure to activation of phosphorylases' *J Biol Chem*, 172, 815–824.

Swissa M, Aloni Y, Weinhouse H, and Benziman M (1980) 'Intermediary steps in *Acetobacter xylinum* cellulose synthesis: studies with whole cells and cell-free preparations of the wild type and a celluloseless mutant' *J Bacteriol*, 143, 1142–1150.

Tomlinson K and Denyer K (2003) 'Starch Synthesis in Cereal Grains' *Adv Bot Res*, 40, 1–61.

Trevelyan W E, Mann P F E, and Harrison J S (1952) 'The phosphorylase reaction. I. Equilibrium constant: principles and preliminary survey' *Arch Biochem Biophys*, 39, 419–427.

Ward J B and Perkins H R (1973) 'The direction of glycan synthesis in a bacterial peptidoglycan' *Biochem J*, 135, 721–728.

Wright A, Dankert M, Fennessey P, and Robbins P W (1967) 'Characterization of a poly-isoprenoid compound functional in O-antigen Biosynthesis' *Proc Natl Acad Sci US*, 57, 1798–1803.

Yoon S H, Fulton D B, and Robyt J F (2004) 'Enzymatic synthesis of two salicin analogues by reaction of salicyl alcohol with *Bacillus macerans* cyclomaltodextrin glucanyltransferase and *Leuconostoc mesenteroides* B-742CB dextransucrase' *Carbohydr Res*, 339, 1517–1529.

α-AMYLASES. INTERACTION WITH POLYSACCHARIDE SUBSTRATES, PROTEINACEOUS INHIBITORS AND REGULATORY PROTEINS

E. S. Seo, M. M. Nielsen, J. M. Andersen, M. B. Vester-Christensen, J. M. Jensen, C. Christiansen, A. Dilokpimol, M. Abou Hachem, P. Hägglund, K. Maeda1, C. Finnie, A. Blennow, and B. Svensson

ABSTRACT

α-Amylases occur widely in plants, animals, and microorganisms. They often act in synergy with other related and degradative enzymes and may also be regulated by proteinaceous inhibitors. Open questions exist on how α-amylases interact with polysaccharides. Several enzymes possess secondary carbohydrate binding sites situated on the surface at a certain distance of the active site cleft. The functions of such sites were studied in barley α-amylase isozymes by structure-guided mutational analysis and measurement of activity and binding parameters. Two surface sites were assigned distinct roles. One of the sites seems to participate in hydrolysis of polysaccharides by a multiple attack mechanism. Polysaccharide processing enzymes can also contain carbohydrate binding modules, e.g. starch binding domains that assist in the attack on macromolecular substrates and are useful in engineering of enzyme efficiency. The multidomain nature of these enzymes raises questions on the dynamics and structural properties in solution and in substrate complexes.

Key words: barley α-amylase; carbohydrate surface binding sites; starch binding domains; proteinaceous inhibitors; thioredoxin-mediated disulfide reduction

INTRODUCTION

α-amylases and related enzymes hydrolyze polysaccharides with diverse specificity and can also act with synergism resulting in efficient degradation of starch granules to glucose and short maltooligosaccharides. Certain enzymes carry auxiliary tools facilitating contact with supramolecular substrate structures. These include separate starch binding domains (carbohydrate binding modules; http://www.cazy.org/) and secondary carbohydrate binding sites situated on the surface of the integral enzyme structure. A variety of enzyme isoforms can furthermore display distinctly different functional and stability properties. A regulatory point resides in the specificity of proteinaceous inhibitors directed towards individual enzymes. A wide range of approaches including site-directed mutagenesis, domain fusion, and formation of chimeras have been applied to investigate structure/function relationships in starch-degrading enzymes with focus on features engaged in polysaccharide processing.

Enzymes involved in hydrolysis of α-glucosides and α-glucans belong to glycoside hydrolase families 13, 14, 15, 31, 57, 70, and 77 (http://www.cazy.org/). Glycoside hydrolase family 13 (GH13) is by far the largest, both with regard to diversity of specificity and number of sequence entries (currently > 4500). Glycoside hydrolase clan H (GH-H) is formed of GH13 together with the small families GH70 and GH77. GH13 itself has been subdivided according to sequence relationships (Stam et al., 2006) in more than 36 clusters, which provides a grouping according to specificity and taxonomy, but which also reflects that multispecificity is one of the major problems for correct prediction of specificity and biological role from genome data. Members of some of the individual clusters have yet to be characterised. Very recently intracellular enzymes with low hydrolytic activity from GH13_5 were thus linked to the biosynthesis of fungal cell wall α-glucans (van der Kaaij et al., 2007).

Degradation of macromolecular substrates or synthesis of α-glucan polymers mostly involve less well understood albeit essential protein–polysaccharide interactions. One of our main interests is to gain more knowledge on how secondary binding sites at the molecular level assist and participate in enzymatic reactions towards different polysaccharides. One other set of tools are the various starch binding domains at present found in eight families of carbohydrate binding modules (CBMs; http://www.cazy.org/) of which detailed analysis of structure/function relationships was concentrated to just a few. Other yet to be explored facets of the structural basis of the mechanism of action towards polysaccharide substrates include the degree of multiple attack (DMA), enzyme catalysed degradation in the vicinity of branch points and how these and other phenomena also implicate surface binding sites.

Finally, the starch degradative reactions occur in microenvironments with other players in the form of enzymes that attack the substrates or products in conjunction with α-amylases thus conferring a synergistic breakdown, proteinaceous inhibitors acting on specific amylolytic enzymes, and regulation of both inhibitors and enzymes, e.g. by thioredoxin. Some of these questions have been addressed by applying different proteomics approaches (Maeda et al., 2005; Bak-Jensen et al., 2007).

α-AMYLASES AND RELATED ENZYMES IN DIFFERENT LIVING SYSTEMS

Traditionally α-amylases have been studied from germinating seeds, the digestive tract of a variety of animals including insects and mammals, as well as from numerous bacteria and fungi that produce and secrete α-amylases some of which represent an important source of commercial enzymes. In these systems α-amylases very often act in synergy with other degradative hydrolases. The debranching amylolytic enzymes limit dextrinase and isoamylase are even reported to play a role in trimming polysaccharide intermediates in starch biosynthesis. Some enzymes are regulated by endogenous proteinaceous inhibitors as in germinating barley seeds, where the α-amylase isozyme 2 (AMY2) is specifically inhibited by barley α-amylase/subtilisin inhibitor (BASI) (Mundy et al., 1983) and limit dextrinase (LD) is inhibited by limit dextrinase inhibitor (LDI) belonging to the CM-proteins (MacGregor et al., 2004). Other amylase inhibitors participate in defence against pathogens and pests (Svensson et al., 2004). Remarkably, mapping of α-amylase forms during barley seed germination using proteomics techniques revealed that the two isozyme families containing four (family 1; AMY1) and six (family 2; AMY2) genes, gave rise to products which as identified at the protein level originated from only one and two genes, respectively. Moreover of the two AMY2 members only one was found as numerous degradation products by 2D gel electrophoresis and immunoblotting using antibodies recognising both isozymes and identification by mass spectrometry; this suggests that the germinating seed system shows isozyme-specific variation in biological stability (Bak-Jensen et al., 2007).

CARBOHYDRATE BINDING SURFACE SITES

Several carbohydrate-active enzymes possess binding sites situated at a certain distance of the active site cleft. While it is easy to imagine that there can be an advantage of such sites in interaction with polysaccharide substrates including the very large starch granules, there is limited insight into the various ways by which these sites operate in the action on polysaccharides as well as of their actual functional importance.

The first surface site ever reported in GH13 was from barley α-amylase (AMY2) and identified by a differential chemical modification strategy in which tryptophan residues were subject to oxidation by N-bromosuccinimide in the presence and absence of β-cyclodextrin (β-CD), respectively (Gibson and Svensson, 1987). Two adjacent tryptophans, situated on the surface of the catalytic domain and later seen to bind acarbose in the crystal structure of AMY2 (Kadziola et al., 1998), were in this way found to be protected by β-CD against the oxidation together with a more weakly protected tryptophan localised to substrate binding subsite +2 by crystallography (Kadziola et al., 1998). Only the 80% sequence identical isozyme AMY1 was produced recombinantly and site-directed mutagenesis of the equivalent residues confirmed a role of the surface site in binding of β-CD and starch (Søgaard et al., 1993). Very recently, thorough site-directed mutagenesis at this site (Trp^{278} and Trp^{279}, AMY1 numbering) using a more efficient expression system, indicated its dominating role in adsorption of AMY1 onto starch granules (Nielsen et al., unpublished).

(a)

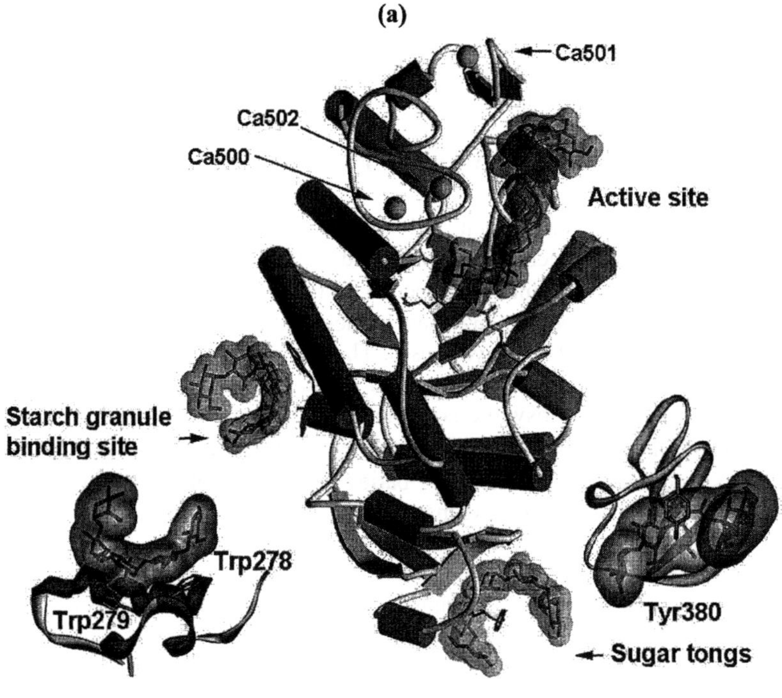

(b)

```
AMY1   347  GITATSALKILMHEGDAYVAEIDGKVVVKIGSRYDVG.AVIPAG
AMY2   345  GIHNESKLOIIEADADLYLAEIDGKVIVKLGPRYDVG.NLIPGG
                                            ↑ ↑
```

Figure 1 Comparison of barely α-amylase 1 and 2

(a) Surface binding sites in barley α-amylase 1 (AMY1). The D180A inactive catalytic nucleophile mutant in complex with maltoheptaose. Three calcium ions are found both in AMY1 and AMY2 (Ca500, Ca501, Ca502). (b) Sequence comparison of a C-terminal domain segment in AMY1 and AMY2.

Some α-amylases, but not AMY1 and AMY2, are inhibited by certain cyclodextrins (α-, β-, γ-), and α-cyclodextrin is seen to bind to several surface sites e.g. in porcine pancreatic α-amylase (Larson et al., 1994). Other α-amylases, but not AMY1 or AMY2, hydrolysed β- and γ-CDs.

The inactive catalytic nucleophile D180A AMY1 mutant binds the substrate maltoheptaose both at the active site and at two secondary sites – one containing the two adjacent tryptophan residues described above, the other being at a longer distance situated on the non-catalytic C-terminal domain and involving Tyr^{380} (Fig. 1a; Robert et al., 2005). This latter site is called "a pair of sugar tongs" because Tyr^{380} swings 3 Å to grasp the sugar ligand. The site on the catalytic $(\beta/\alpha)_8$-barrel shows carbohydrate stacking to the adjacent Trp^{278} and Trp^{279} matching neighboring sugar rings' geometry (Robert et al., 2003, 2005). This site in AMY1 also binds acarbose similarly to in AMY2 (Kadziola et al., 1998; Robert et al., 2005). Remarkably, the orientation of the three bound maltoheptaose molecules in D180A AMY1 is such that no connection between them can be proposed (Robert et al., 2005).

Thorough analysis of AMY1 surface sites indicated that they had somewhat different roles in the interaction with polysaccharides and starch granules (Bozonnet et al., 2007; Nielsen et al., 2008). We used surface plasmon resonance (SPR) to monitor binding of the small starch mimic β-CD and found K_D to increase from 0.2 mM of wild-type AMY1 to 1.4 mM for Y380A (Table 1). Y380A AMY1 also had 13-fold reduced affinity and ~90% reduced catalytic efficiency towards starch granules as compared to wild-type. Alanine substitution in AMY1 of $Trp^{278}Trp^{279}$ allowed specific roles to be deduced for these residues in interaction with starch granules and poly- and oligosaccharide substrates and also indicated a synergistic effect with the "sugar tongs" site.

In contrast to AMY1 (Robert et al., 2003, 2005), oligosaccharide binding at the "sugar tongs" was not observed in the structure of the 80% sequence identical AMY2 (Kadziola et al., 1998), although Tyr^{378} was conserved (corresponding to AMY1 Tyr^{380}) (Fig. 1b). This isozyme difference is investigated by aid of site-directed mutagenesis.

Table 1 Binding properties and activity of "sugar tongs" mutants

Enzyme	β-CD K_d mM	Starch granules K_d mg ml^{-1}	Insoluble Blue Starch U mg^{-1}
Y380A AMY1[a]	1.4	5.9	1400
S378P AMY1[a]	0.25	0.57	2695
Wild-type AMY1[a]	0.2	0.47	2900
Wild-type AMY2	0.24	3.5	5000
M6	0.24	3.2	4925
P376S M6	0.22	2.1	4600

[a]Bozonnet et al., 2007

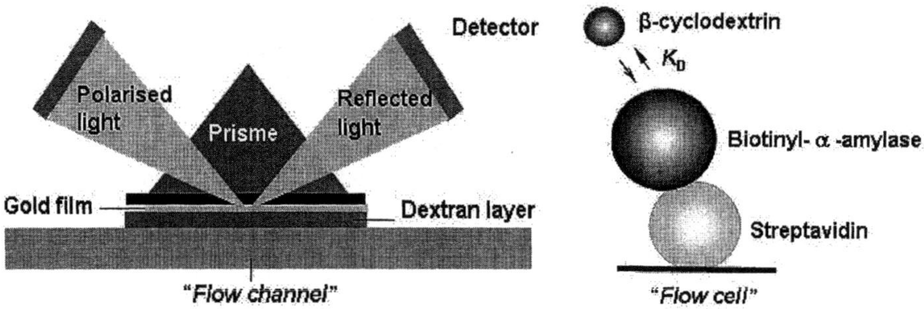

Figure 2 Surface Plasmon Resonance (SPR): β-Cyclodextrin binding

Unfortunately, this cannot be performed using AMY2 itself as parent as this isozyme is produced in low yields in the otherwise efficient heterologous host *Pichia pastoris* (Juge et al., 1996). However, A42P AMY2 (M6), a single mutant obtained using degenerate oligonucleotide gene shuffling in a combinatorial screen involving the 10 positional differences between AMY1 and AMY2 in the N-terminal segment, which was proposed to cause the low expression, increased the yield by 15–60 fold (Fukuda et al., 2005). M6 was an excellent mimic as it shared enzymatic properties and stability characteristics with AMY2, including recognition of the proteinaceous barley α-amylase/subtilisin inhibitor BASI (Fukuda et al., 2005). The P376S M6 mutant addressing the characteristic sequence difference between AMY1 and AMY2 at the "sugar tongs" examined the suggestion that Pro^{376} in AMY2 (AMY1 Ser^{378}) would prevent the conformational change seen for Tyr^{380} in ligand binding to AMY1 due to backbone rigidity. P376S M6 (Fig. 1b), however, showed slightly improved, but still weaker affinity than AMY1 (Table 1) (Seo et al., unpublished). Preliminary data suggest that Tyr^{378} in M6 has a role in binding onto starch granules, whereas no effect was observed by mutation of this residue in binding of β-CD (Seo et al., unpublished).

In another series dual site mutants involving Tyr^{105} at subsite –6, which has the highest substrate affinity at the active site (Kandra et al., 2006) and Tyr^{380} (Nielsen et al., 2008) were constructed to study cooperation between the active site and the surface site. In this way it was indicated that Tyr^{380} at the "sugar tongs" dominated in degradation of amylose over Tyr^{105} by contributing to the multiple attack and by advancing hydrolysis of an insoluble starch substrate. Experiments are furthermore in progress on dual surface site mutants involving Trp^{278}, Trp^{279} and Tyr^{380} to identify the main functional roles of these secondary sites and their possible cooperation. This also involves surface plasmon resonance binding analysis (Fig. 2) of the oligosaccharide substrate maltoheptaose to the corresponding inactive variants in which the catalytic nucleophile D180A is introduced together with the different single and multiple surface site mutants.

THE PROCESSIVE MECHANISM

Certain depolymerases apply a multiple attack mechanism in which the substrate is cleaved several times by the enzyme in a single enzyme-substrate encounter. AMY1 thus hydrolysed amylose an in release on average of two oligosaccharide/ maltodextrin molecules following the initial endo-cleavage of the substrate chain, which corresponds to a degree of multiple attack (DMA) of 2 (Kramhøft et al., 2005).

Table 2 DMA of wild-type and "sugar tongs" mutants

Enzyme	R_t [a]	R_s [a]	R_p [a]	DMA [b]
		(s^{-1})		[(R_t/R_p)-1]
Y380A AMY1[c]	53	25	28	1.0
Y380M AMY1[c]	90	60	30	2.0
S378P AMY1[c]	152	105	47	2.2
Wild-type AMY1[d]	138	90	48	1.9
Wild-type AMY2	248	163	85	0.5
M6	269	189	80	0.4

[a]Amylose DP400 (1 mg/ml) was used as substrate (Kramhøft et al., 2005).
R_t is the total reducing power of reaction mixture.
R_p is the reducing power of the polysaccharide fraction.
R_s is the reducing power of soluble fraction and is calculated as R_t-R_p.
[b]DMA values are means calculated from the linear rates of reducing value formation in each individual experiment.
[c]Bozonnet et al., 2007; [d]Kramhøft et al., 2005

The AMY1 "sugar tongs" mutant Y380A has the degree of multiple attack reduced from two to one and the "sugar tongs" (Bozonnet et al., 2007) is proposed to constitute a point of attachment of the polysaccharide on the surface of the enzyme such that the substrate chain maintains sufficient flexibility to reorganise itself for several cleavages at the active site without loosing contact to the enzyme (Table 2). This result supports our hypothesis that a distant polysaccharide binding site is involved in the processive degradation of amylose (Kramhøft et al., 2005). The "sugar tongs" may also be related to an earlier identified allosteric regulatory site, where oligosaccharide binding enhanced hydrolytic activity (Oudjeriouat et al., 2003) as the enzymatic activity of Y380A AMY1 towards an oligosaccharide substrate was reduced, even though the "sugar tongs" is situated at a distance of >40Å from the active site cleft (Bozonnet et al., 2007). We are currently analysing data on proposed multiple attack on amylopectin. Moreover, analysis of the DMA is in progress for AMY2 and mutants of Tyr^{378} in the "sugar tongs", which have different binding properties compared to the "sugar tongs" of AMY1. AMY2 thus showed lower DMA compared to AMY1 (Table 2) even though the isozymes have 80% sequence identity (Seo et al., unpublished).

STARCH BINDING DOMAINS

Polysaccharide active enzymes commonly contain carbohydrate binding modules (CBMs) which can assist in degradation of insoluble substrates in various ways (Janeček et al., 2003; Boraston et al., 2004; Machovic and Janeček, 2006). Thus for starch binding domains (SBDs) pioneering work was done on various family CBM20 members, with emphasis on SBDs from *Aspergillus niger* glucoamylase and bacterial cyclodextrin glucanotransferases (CGTases), respectively. Today a total of eight different SBD families have been reported. Moreover, polypeptide chain segments in GH31 from plants were also found to provide binding to granular starch (Nakai et al., 2008).

It was proposed that the SBD increased susceptibility of granular starches to the hydrolase by disentangling the starch α-glucan double helix (Southall et al., 1999). Furthermore, in several cases fusion proteins have been demonstrated to show enhanced activity towards starch granules (Ohdan et al., 2000; Juge et al., 2006). The SBD from *A. niger* glucoamylase by itself induced supramolecular structures with amylose as demonstrated by atomic force microscopy (Giardina et al., 2001; Morris et al., 2005) supporting the proposed role in disentanglement. The actual enhanced interaction with the starch granule surface for the fusion protein between AMY1 and this SBD increased the rate of release of soluble oligosaccharides from the granules by a factor of 15 (Juge et al., 2006). The affinity for β-CD of the two non-identical binding sites was substantially higher than found for the "sugar tongs" site in AMY1 (Giardina et al., 2001; Bozonnet et al., 2007).

Bioinformatics analysis on the relation between CBM20 and CBM21 gave an evolutionary tree based on a common alignment of sequences of both modules (Machovic et al., 2005) which confirms an early sequence alignment of SBD from *Rhizopus oryzae* glucoamylase with the above mentioned enzymes (Svensson et al., 1989). CBM21 SBDs from α-amylases and glucoamylases are the closest relatives to the CBM20, with the CBM20 from GH13 amylopullulanases being possible candidates for an intermediate between the two CBM families (Machovic al., 2005). A dimer of two cross-linked CBM21 of *Rh. oryzae* glucoamylase has been obtained in which β-CD interacts with one of the two binding sites present in each SBD (Liu et al., 2007).

The mechanism of action and dynamics of the multidomain enzymes is envisaged to depend on the architecture and differ for those enzymes where an extended polypeptide linker connects the SBD with the catalytic or another domain, as for example in the glucoamylase, and those where the SBD domain is intimately interacting with the rest of the structure and has a well defined interface, exemplified by CGTases (Janeček et al., 2003). Whereas previous experiments using a double headed inhibitor targeted to the catalytic site by an acarbose moiety and to the starch binding sites in SBD by β-CD suggested that the SBD and the catalytic domain would approach each other in a unimolecular complex (Sigurskjold et al., 1998; Payre et al., 1999), we have demonstrated very recently by solution studies using small angle x-ray scattering of *A. niger* glucoamylase wild-type, its SBD truncated form, and a variant with a shortened and non-glycosylated linker, that the two-domain molecule is dumbbell shaped and appears rigid with low flexibility (Jørgensen et al., 2008). Addition of the double headed synthetic oligosaccharide inhibitor mentioned above elicited dimerisation in which two inhibitor molecules bound the domains together head-to-tail in two molecules of glucoamylase (Jørgensen et al., 2008).

BINDING TO STARCHES

A variety of starch metabolising enzymes have the ability to bind to starch granules. In certain cases this happens via starch binding domains (http://www.cazy.org/). We have focused on different SBDs of CBM20 and found that a CBM20 of plant origin from the N-terminal region of a glucan, water dikinase 3 (GWD3) is able to bind onto starch granules as shown after fluorophore labeling of the recombinant domain by using confocal laser scanning microscopy to monitor binding (Christiansen et al., unpublished). Glucan, water dikinase is targeted to the plastid and catalyses starch phosphorylation (Blennow et al., 2002) which results in increased degradability of the granule *in vivo*. The CBM20 indeed localizes the enzyme on the starch molecule. The low affinity for starch of this domain as compared to other CBM20 family members emphasizes the importance and possibility of organisms to modulate starch affinity in order to permit dynamic partitioning of enzymes to the granule surface. This domain is further characterized with respect to carbohydrate ligand affinity (Christiansen et al., unpublished). An SBD belonging to CBM45 from GWD1 was similarly shown previously to be involved in binding onto starch granules in connection with phosphorylation (Mikkelsen et al., 2006). Along the same lines, the surface sites on AMY1 were implicated in binding to starch granules, as their mutation resulted in varying degree of loss of affinity for starch granules as determined using Langmuir binding analysis to the solid substrate and also illustrated by confocal laser scanning microscopy (Nielsen et al., unpublished).

STARCH DEGRADING ENZYMES AND PROTEINACEOUS INHIBITORS

Proteinaceous inhibitors present in the mature barley seed and available during germination have the ability to complex with and suppress the activity of AMY2 – but not AMY1 – and the debranching enzyme, limit dextrinase (LD) (Mundy et al., 1983; Vallée et al., 1998; Nielsen et al., 2003; Svensson et al., 2004; MacGregor, 2004; Bønsager et al., 2005). These are both examples of regulation of endogenous enzymes, however very many α-amylase inhibitors of plant origin are directed against enzymes in pests and pathogens and hence considered to be part of the plant defence system (Svensson et al., 2004). The BASI-AMY2 complex is relatively well understood and has high stability with K_d in the sub-nanomolar range (Nielsen et al., 2003; Bønsager et al., 2005). A number of residues were assigned functional roles for the complex formation from both the enzyme and the inhibitor by using site-directed mutagenesis, crystallography, surface plasmon resonance, and activity inhibition analyses (Vallée et al., 1998; Rodenburg et al., 2000; Nielsen et al., 2003; Bønsager et al., 2005). It was also demonstrated that the ability of BASI mutants to form complex with AMY2 was sensitive to pH and ionic strength. In fact, a single mutant in BASI modestly weakened the complex, however, in a way that allowed manipulating the inhibitor affinity by rather subtly adjusting pH and ionic strength (Rodenburg et al., 2000; Bønsager et al., 2005).

The LDI-LD complex on the other hand has only been subject to an initial analysis (MacGregor et al., 1994, 2003). Both proteins have been problematical to produce in recombinant form, but are now obtained by heterologous expression in *P. pastoris* (Vester-Christensen et al., unpublished). A system for monitoring the complex formation between LD and LDI has been established using surface plasmon resonance, which in initial experiments indicated sub-nanomolar K_d values. Using different reaction conditions it was furthermore concluded that hydrophobic interactions were important for the complex formation (Jensen et al., unpublished). Mutational analysis of the LDI-LD complex formation is in progress.

PROTEIN DISULFIDE REDUCTION BY THIOREDOXIN

In an indirect manner, the metabolism of starch is anticipated to be under the influence also of the protein disulfide reductase thioredoxin as this was earlier proposed to act on disulfide bonds of relevant enzymes and inhibitors in barley seeds (Cho et al., 1999) accompanied by modification of the functional and physico-chemical properties. Barley has two thioredoxin h isoforms (h1 and h2) and also two isoforms of NADPH dependent thioredoxin reductase (NTR1 and NTR2) that reduce the disulfide formed in the thioredoxin active site motif CXXC by reduction of a target protein disulfide bond

(Maeda et al., 2003; Shahpiri et al., 2008). A proteomics-based procedure was developed for global identification of target disulfides in protein extracts which provides a wealth of information including both mere identification of protein targets and specific identification of which disulfide is reduced (Maeda et al., 2005; Hägglund et al., unpublished). Use of isotope coded alkylating reagents coupled to a cleavable biotin affinity tag enabled purification of peptides containing thiol groups originating from thioredoxin target disulfides, which were subsequently affinity purified and subjected to mass spectrometric identification as well as relative quantification of the extent of reduction (Hägglund et al., unpublished). To further understand what makes a disulfide bond a target for thioredoxin, we determined the crystal structure of barley thioredoxin h in complex with BASI, which enabled identification of structural deteminants for protein recognition (Maeda et al., 2006).

Another point to elaborate is a clear distinction of the *in vivo* roles of thioredoxin h1 and h2 and the thioredoxin reductase NTR1 and NTR2. *In vitro*, however, one specific pair is found to be as much as three times more efficient than the least efficient pair; noticeably the most efficient pair also predominates in the aleurone layer (Shahpiri et al., 2008). It is certainly possible that different spatio-temporal occurrence of isoforms is a key in efficient recycling of oxidised thioredoxin. The impact of thioredoxin is currently analysed in dissected embryo, aleurone and endosperm tissues from germinating barley seeds by the developed quantitative proteomics procedure that ranks the identified target disulfides according to their degree of susceptibility to thioredoxin (Hägglund et al., unpublished). Indeed the germinating seed is a highly dynamic biological system that undergoes numerous metabolic as well as morphological changes and the emerging analysis of the proteomics of germinating seeds will be accompanied by transcriptomics and preferably also by metabolomic data.

GH13 AND GH31. RELATED STRUCTURES OF α-GLUCOSIDE ACTIVE ENZYMES

Although the catalytic machinery in GH13 and GH31 is different, some sequence similarity can still be seen of GH31 to clan GH-H at β3, β4, β7, and β8 of the catalytic $(\beta/\alpha)_8$-barrel; the resemblance is closest with GH77 members (Janeček et al., 2007). Thanks to massive efforts, crystal structures were solved of a few α-glycosidases from GH31. The first solved structure was for an enzyme (YicI) encoded by *Escherichia coli* that turned out to be an α-xylosidase, a "new" specificity in GH31 (Kitamura et al., 2005; Lovering et al., 2005). Guided by the structure, YicI was engineered to an α-glucosidase (Okuyama et al., 2006), confirming the close relationship between these two GH31 enzymes. The *Sulfolobus solfataricus* α-glucosidase MalA was solved shortly after (Ernst et al., 2006). The structures will be complemented by kinetics analyses which may be useful also in explanation of the difference in reaction mechanism between the GH31 starch lyases (Yu et al., 1999) as compared to GH31 hydrolases (Lee et al., 2003). A GH31 α-glucosidase from barley has both been purified from germinated barley seeds and produced recombinantly (Frandsen et al., 2000; Naested et al., 2006). This enzyme shows highest specificity for maltose and maltotriose and decreasing values (k_{cat}/K_m) towards higher maltooligosaccharides, although these are still substrates. Various nitrogen containing sugar analogues were good inhibitors of the barley GH31 α-glucosidase (Naested et al., unpublished).

SPECIFICITY ENGINEERING OF α-AMYLASES AND RELATED ENZYMES

A very large number of sequences are available for clan GH-H which combined with three-dimensional structures covering a broad variety of enzyme specificities (http://www.cazy.org/; MacGregor et al., 2001; Stam et al., 2006) can guide engineering of enzymatic properties by applying semi-rational approaches, such as modeling and also 1D/3D comparison. Across clan GH-H structural similarities and differences help characterize key determinants in specificity and other properties, which thus advances understanding of structure/function relationships and facilitates rational engineering. GH70 enzymes have a circularly permuted catalytic $(\beta/\alpha)_8$-barrel domain (MacGregor et al., 1996) and the crystallization and solving of the first structure in this multidomain family of *Lactobacillus reuterii* 180 glucansucrase (Pijning et al., 2008) unveiled how segments from distant parts of the polypeptide conform with clan GH-H active site sequence motifs, as already exploited to manipulate bond-type specificity of reuteransucrase in GH70 to be mainly that of an α-1,6-synthesizing enzyme at the expense of the α-1,4-glucosidic linkage specificity (Kralj et al., 2005). As the relation between active site motifs and product specificity is known for a large number of GH70 enzymes there is potential possibility for engineering polymer product properties (Fabre et al., 2005; Kralj et al., 2006; van Leeuwen et al., 2008).

More classically, specificity engineering modified product composition for cyclodextrin glucanotransferases, neopullulanases, and maltogenic α-amylase using the conserved sequence motifs at the four active site β-strands (Kuriki et al., 1996; Beier et al., 2000; Park et al., 2000, 2008; Kim et al., 2001; MacGregor al., 2001; Leemhuis et al., 2003).

Site-directed mutagenesis of a single residue in barley α-amylase at one of the substrate binding subsites could dramatically shift – while maintaining wild-type activity level – the preference for starch over oligosaccharide or *vice versa* (Gottschalk et al., 2001; Mori et al., 2001; Bak-Jensen et al., 2004). Some of these mutants located at outer subsites −6 (Y105A) or +4 (T212W) also elicited changes of the subsite affinity profile (Kandra et al., 2006) and Y105A at subsite −6 surprisingly enhanced activity on insoluble starch and highly reduced activity on oligosaccharides (Bak-Jensen et al., 2004). Double mutation at subsites −6 and +4 resulted in stronger binding energy for subsite +2 than in wild-type or any of the corresponding single mutants (Kandra et al., 2006). Correlation between mutational manipulation of the subsite structure and the affinity profile eventually provides a basis for engineering the product profile.

α-AMYLASES AND CALCIUM IONS

α-amylases almost as the rule have bound Ca^{2+} which is related to stability and activity; some other GH-H members, however, do not have structural metal ions. One highly conserved Ca^{2+} is seen in the three-dimensional structure of α-amylases as shown for Ca500 in AMY1 next to the catalytic site (Fig. 1a); several structures have more Ca^{2+} or different metal ions (Na^{+}, Zn^{+2}). The stability of AMY1 and AMY2 depends differently on $[Ca^{2+}]$ and differential scanning calorimetry of site-directed mutants showed AMY2 as the more sensitive to removal of Ca^{2+} by EDTA especially at lower pH values, while at higher $[Ca^{2+}]$ and pH the isozymes both displayed high and essentially the same thermostability (Abou Hachem et al., unpublished). Quite remarkably, substitution of certain side chains that bind to or are near structural Ca^{2+} could result in either weakening or strengthening the conformational stability.

CLOSING REMARKS

Crystal structures have greatly improved insight into the relationship between structure and function of starch- and related α-glucan-active enzymes. Recently, we focused on secondary binding sites that participate in multi-site substrate interactions with polysaccharides on the enzyme surface at a certain distance of the active site. Binding analysis of oligosaccharides and starch granules indicated different carbohydrate ligand preferences for two surface sites described in AMY1 from barley. The open question on why the other isozyme AMY2 has different carbohydrate binding properties is here approached by mutational analysis in M6 a mimic of AMY2. This indicated differences between surface sites properties in AMY1 and AMY2. Insight into and understanding of the concerted action of the catalytic and these remote substrate binding sites is highly limited and questions remain on the mechanism of action and role of such sites in enzymatic conversions and utilization of sugars, as well as on how individual domains interact during catalysis.

The modular nature of amylolytic enzymes can be further developed by engineering through combining functionalities. Starch binding domains are also characterized including plant CBM20s and others. We use confocal laser scanning microscopy to monitor whether a protein binds or not to starch granules. This was also used to analyse AMY1 mutants and obviously mutation at both surface sites eliminates ability to bind onto starch granules. Currently we gather information on affinity for starch granules of different botanical or genetic origin. This has relevance for action on e.g. recalcitrant substrates. In fact an earlier suggestion that multi-domain protein glucoamylase in which a peptide linker of approx. 100 residues flexibly connects the catalytic and the starch binding domains, respectively, has been disproven by a recent small angle x-ray scattering structure determination that shows a rigid dumbbell structure, which however can dimerise or perhaps oligomerise by multi-dentate ligand binding (Jørgensen et al., 2008).

ACKNOWLEDGEMENTS

The expert technical assistance of Susanne Blume and Karina Rasmussen is gratefully acknowledged. The work has been supported by the Danish Natural Science Research Council, the Danish Research Council for Technology and Production Sciences and the Carlsberg Foundation. ESS held a Korea Research Foundation Grant funded by the Korean Government (MOEHRD) (KRF-2005-214-D00275) and a H.C. Ørsted postdoctoral fellowship from DTU. MMN and MBVC received Ph.D. stipends from DTU. CC holds a Ph.D. stipend from the FOBI graduate school. JMA received a Novo student scholarship.

REFERENCES

Bak-Jensen K S, André G, Gottschalk T E, Paës G, Tran V, and Svensson B (2004), 'Tyrosine 105 and threonine 212 at outermost substrate binding subsites −6 and +4 control substrate specificity, oligosaccharide cleavage patterns, and multiple binding modes of barley α-amylase 1', *J Biol Chem*, 279, 10093–10102.

Bak-Jensen K S, Laugesen S, Østergaard O, Finnie C, Roepstorff P, and Svensson B (2007), 'Spatio-temporal profiling and degradation of α-amylase isozymes during barley seed germination', *FEBS J*, 274, 2552–2565.

Beier L, Svendsen A, Andersen C, Frandsen T P, Borchert T V, and Cherry J R (2000), 'Conversion of the maltogenic α-amylase Novamyl into a CGTase', *Protein Eng*, 13, 509–513.

Blennow A, Engelsen S B, Nielsen T H, Baunsgaard L, and Mikkelsen R (2002), 'Starch phosphorylation — a new front line in starch research', *Trends in Plant Sci*, 7, 445–450.

Boraston A B, Bolam D N, Gilbert H J, and Davies G J (2004), 'Carbohydrate-binding modules: fine-tuning polysaccharide recognition', *Biochem J*, 382, 769–781.

Bozonnet S, Jensen M T, Nielsen M M, Aghajari N, Jensen M H, Kramhøft B, Willemoës M, Tranier S, Haser R, and Svensson B (2007), 'The "pair of sugar tongs" site on the non-catalytic domain C of barley α-amylase participates in substrate binding and activity', *FEBS J*, 274, 5055–5067.

Bønsager B C, Nielsen P K, Abou Hachem M, Fukuda K, Prætorius-Ibba M, and Svensson B (2005), 'Mutational analysis of target enzyme recognition of the β-trefoil fold barley α-amylase/subtilisin inhibitor', *J Biol Chem*, 280, 14855–14864.

Cho M J, Wong J H, Marx C, Jiang W, Lemaux P G, and Buchanan B B (1999), 'Overexpression of thioredoxin *h* leads to enhanced activity of starch debranching enzyme (pullulanase) in barley grain', *Proc Natl Acad Sci USA*, 96, 14641–14646.

Ernst H A, Lo Leggio L, Willemoës M, Leonard G, Blum P, and Larsen S (2006), 'Structure of the *Sulfolobus solfataricus* α-glucosidase: implications for domain conservation and substrate recognition in GH31', *J Mol Biol*, 358, 1106–1124.

Fabre E, Bozonnet S, Arcache A, Willemot R M, Vignon M, Monsan P, and Remaud-Simeon M (2005), 'Role of the two catalytic domains of DSR-E dextransucrase and their involvement in the formation of highly α-1,2 branched dextran', *J Bacteriol*, 187, 296–303.

Frandsen T P, Lok F, Mirgorodskaya E, Roepstorff P, and Svensson B (2000), 'Purification, enzymatic characterization, and nucleotide sequence of a high-isoelectric-point α-glucosidase from barley malt', *Plant Physiol*, 123, 275–286.

Fukuda K, Jensen M H, Haser R, Aghajari N, and Svensson B (2005), 'Biased mutagenesis in the N-terminal region by degenerate oligonucleotide gene shuffling enhances secretory expression of barley α-amylase 2 in yeast', *Protein Eng Des Sel*, 18, 515–526.

Giardina T, Gunning A P, Juge N, Faulds C B, Furniss C S M, Svensson B, Morris V J, and Williamson G (2001), 'Both binding sites of the starch-binding domain of *Aspergillus niger* glucoamylase are essential for inducing a conformational change in amylose', *J Mol Biol*, 313, 1149–1159.

Gibson R M and Svensson B (1987), 'Identification of tryptophanyl residues involved in binding of carbohydrate ligands to barley α-amylase 2', *Carlsberg Res Commun*, 52, 373–379.

Gottschalk T E, Tull D, Aghajari N, Haser R, and Svensson B (2001), 'Specificity modulation of barley α-amylase 1 by biased random mutation of a tripeptide in β→α loop 7 of the catalytic $(\beta/\alpha)_8$-domain', *Biochemistry*, 40, 12844–12854.

Janeček Š, Svensson B, and MacGregor E A (2003), 'Relation between domain evolution, specificity, and taxonomy of the α-amylase family members containing a C-terminal starch-binding domain', *Eur J Biochem*, 270, 635–645.

Janeček Š, Svensson B, and MacGregor E A (2007), 'A remote but significant sequence homology between glycoside hydrolase clan GH-H and family GH31', *FEBS Lett*, 581, 1261–1268.

Jørgensen A D, Nøhr J, Kastrup J S, Gajhede M, Sigurskjold B W, Sauer J, Svergun D I, Svensson B, and Vestergaard B (2008), 'Small angle X-ray studies reveal that *Aspergillus niger* glucoamylase has a defined extended conformation and can form dimers in solution', *J Biol Chem*, 283, 14772–14780.

Juge N, Andersen J S, Tull D, Roepstorff P, and Svensson B (1996), 'Overexpression, purification, and characterization of recombinant barley α-amylases 1 and 2 secreted by the methylotrophic yeast *Pichia pastoris*', *Protein Express Purif*, 8, 204–214.

Juge N, Nøhr J, Le Gal-Coëffet M-F, Kramhøft B, Furniss C S M, Planchot V, Archer D B, Williamson G, and Svensson B (2006), 'The activity of barley α-amylase on starch granules is enhanced by fusion of a starch binding domain from *Aspergillus niger* glucoamylase', *Biochim Biophys Acta*, 1764, 275–284.

Kadziola A, Søgaard M, Svensson B, and Haser R (1998), 'Molecular structure of a barley α-amylase-inhibitor complex: implications for starch binding and catalysis', *J Mol Biol*, 279, 205–217.

Kandra L, Abou Hachem M, Gyémant G, Kramhøft B, and Svensson B (2006), 'Mapping of barley α-amylase and outer subsite mutants reveal high-affinity subsites and barriers in the long substrate binding cleft', *FEBS Lett*, 580, 5049–5053.

Kim T J, Nguyen V D, Lee H S, Kim M J, Cho H Y, Kim Y W, Moon T W, Park C S, Kim J W, Oh B H, Lee S B, Svensson B, and Park K H (2001), 'Modulation of the multisubstrate specificity of Thermus maltogenic amylase by truncation of the N-terminal domain and by a salt-induced shift of the monomer/dimer equilibrium', *Biochemistry*, 40, 14182–14190.

Kitamura M, Ose T, Okuyama M, Watanabe H, Yao M, Mori H, Kimura A, and Tanaka I (2005), 'Crystallization and preliminary X-ray analysis of α-xylosidase from *Escherichia coli*', *Acta Cryst Sect F*, 61, 178–179.

Kralj S, van Geel-Schutten I G, Faber E J, van der Maarel M J, and Dijkhuizen L (2005), 'Rational transformation of *Lactobacillus reuteri* 121 reuteransucrase into a dextransucrase', *Biochemistry*, 44, 9206–9216.

Kralj S, Eeuwema W, Eckhardt T H, and Dijkhuizen L (2006), 'Role of asparagine 1134 in glucosidic bond and transglycosylation specificity of reuteransucrase from *Lactobacillus reuteri* 121', *FEBS J*, 273, 3735–3742.

Kramhøft B, Bak-Jensen K S, Mori H, Juge N, Nøhr J, and Svensson B (2005), 'Involvement of individual subsites and secondary substrate binding sites in multiple attack on amylose by barley α-amylase', *Biochemistry*, 44, 1824–1832.

Kuriki T, Kaneko H, Yanase M, Takata H, Shimada J, Handa S, Takada T, Umeyama H, and Okada S (1996), 'Controlling substrate preference and transglycosylation activity of neopullulanase by manipulating steric constraint and hydrophobicity in active center', *J Biol Chem*, 271, 17321–17329.

Larson S B, Greenwood A, Cascio D, Day J, and McPherson A (1994), 'Refined molecular structure of pig pancreatic α-amylase at 2.1 A resolution', *J Mol Biol*, 235, 1560–1584.

Lee S S, Yu S, and Withers S G (2003), 'Detailed dissection of a new mechanism for glycoside cleavage: α-1,4-glucan lyase', *Biochemistry*, 42, 13081–13090.

Leemhuis H, Kragh K M, Dijkstra B W, and Dijkhuizen L (2003), 'Engineering cyclodextrin glycosyltransferase into a starch hydrolase with a high exo-specificity', *J Biotechnol*, 103, 203–212.

Liu Y N, Lai Y T, Chou W I, Chang M D, and Lyu P C (2007), 'Solution structure of family 21 carbohydrate-binding module from *Rhizopus oryzae* glucoamylase', *Biochem J*, 403, 21–30.

Lovering A L, Lee S S, Kim Y W, Withers S G, and Strynadka N C (2005), 'Mechanistic and structural analysis of a family 31 α-glycosidase and its glycosyl-enzyme intermediate', *J Biol Chem*, 280, 2105–2115.

MacGregor A W, Macri L J, Schroeder S W, and Bazin S L (1994), 'Purification and characterisation of limit dextrinase inhibitors from barley' *J Cereal Sci*, 20, 33–41.

MacGregor A W, Donald L J, MacGregor E A, and Duckworth H W (2003), 'Stoichiometry of the complex formed by barley limit dextrinase with its endogenous inhibitor. Determination by electrospray time-of-flight mass spectrometry' *J Cereal Sci*, 37, 357–362.

MacGregor E A, Jespersen H M, and Svensson B (1996), 'A circularly permuted α-amylase-type α/β-barrel in glucan-synthesizing glucosyltransferases', *FEBS Lett*, 378, 263–266.

MacGregor E A, Janeček Š, and Svensson B (2001), 'Relationship of sequence and structure to specificity in the α-amylase family of enzymes', *Biochim Biophys Acta*, 1546, 1–20.

MacGregor E A (2004), 'The proteinaceous inhibitor of limit dextrinase in barley and malt', *Biochim Biophys Acta*, 1696, 165–170.

Machovic M and Janeček Š (2006), 'Starch-binding domains in the post-genome era', *Cell Mol Life Sci*, 63, 2710–2724.

Machovic M, Svensson B, MacGregor E A, and Janeček Š (2005), 'A new clan of CBM families based on bioinformatics of starch-binding domains from families CBM20 and CBM21', *FEBS J*, 272, 5497–5513.

Maeda K, Finnie C, Østergaard O, and Svensson B (2003), 'Identification, cloning and characterisation of two thioredoxin h isoforms, HvTrx1 and HvTrx2, from the barley seed proteome', *Eur J Biochem*, 270, 2633–2643.

Maeda K, Finnie C, and Svensson B (2005), 'Identification of thioredoxin h-reducible disulphides in proteomes by differential labeling of cysteines: Insight into recognition of proteins in barley seeds by thioredoxin h', *Proteomics*, 5, 1634–1644.

Maeda K, Hägglund P, Finnie C, Svensson B, and Henriksen A (2006), 'Structural basis for target protein recognition by the protein disulfide reductase thioredoxin. *Structure*, 14, 1701–1710.

Mikkelsen R, Suszkiewicz K, and Blennow A (2006), 'A novel type carbohydrate-binding module identified in α-glucan, water dikinases is specific for regulated plastidial starch metabolism' *Biochemistry*, 45, 4674–4682.

Mori H, Bak-Jensen K S, Gottschalk T E, Motawia M S, Damager I, Møller B L, and Svensson B (2001), 'Modulation of activity and substrate binding modes by single and double subsites +1/+2 and –5/–6 mutation of barley α-amylase 1', *Eur J Biochem*, 268, 6545–6558.

Morris V J, Gunning A P, Faulds C B, Williamson G, and Svensson B (2005), 'AFM images of complexes between amylose and *Aspergillus niger* glucoamylase mutants, native and mutant starch binding domains: a model for the action of glucoamylase', *Starch*, 57, 1–7.

Mundy J, Svendsen I, and Hejgaard J 1983, 'Barley α-amylase/subtilisin inhibitor. Isolation and characterization', *Carlsberg Res Commun*, 48, 81–91.

Naested H, Kramhøft B, Lok F, Bojsen K, Yu S, and Svensson B (2006), 'Production of enzymatically active recombinant full-length barley high pI α-glucosidase of glycoside family 31 by high cell-density fermentation of *Pichia pastoris* and affinity purification', *Protein Express Purif*, 46, 56–63.

Nakai H, Tanizawa S, Ito T, Kamiya K, Kim Y-M, Yamamoto T, Matsubara K, Sakai M, Sato H, Imbe T, Okuyama M, Mori H, Chiba S, Sano Y, and Kimura A (2008), 'Rice α-glucosidase isozymes and isoforms showing different starch granules-binding and -degrading ability', *Biocatal Biotransf*, 26, 104–110.

Nielsen M M, Seo E S, Bozonnet S, Aghajari N, Robert X, Haser R, and Svensson B (in press), 'Multi-site substrate binding and interplay in barley α-amylase 1', *FEBS Lett*

Nielsen P K, Bønsager B C, Berland C R, Sigurskjold B W, and Svensson B (2003), 'Kinetics and energetics of the binding between barley α-amylase/subtilisin inhibitor and barley α-amylase 2 studied by surface plasmon resonance and isothermal titration calorimetry', *Biochemistry*, 42, 1478–1487.

Ohdan K, Kuriki T, Takata H, Kaneko H, and Okada S (2000), 'Introduction of raw starch-binding domains into *Bacillus subtilis* α-amylase by fusion with the starch-binding domain of *Bacillus* cyclomaltodextrin glucanotransferase', *Appl Environ Microbiol*, 66, 3058–3064.

Okuyama M, Kaneko A, Mori H, Chiba S, and Kimura A (2006), 'Structural elements to convert *Escherichia coli* α-xylosidase (YicI) into α-glucosidase', *FEBS Lett*, 580, 2707–2711.

Oudjeriouat N, Moreau Y, Santimone M, Svensson B, Marchis-Mouren G, and Desseaux V (2003), 'On the mechanism of α-amylase', *Eur J Biochem*, 270, 3871–3879.

Park J T, Park H S, Kang H K, Hong J S, Cha H, Woo E J, Kim J W, Kim M J, Boos W, Lee S, and Park K H (2008), 'Oligomeric and functional properties of a debranching enzyme (TreX) from the archaeon *Sulfolobus solfataricus* P2', *Biocatal Biotransf*, 26, 76–85.

Park K H, Kim T J, Cheong T K, Kim J W, Oh B H, and Svensson B (2000), 'Structure, specificity and function of cyclomaltodextrinase, a multispecific enzyme of the α-amylase family', *Biochim Biophys Acta*, 1478, 165–185.

Payre N, Cottaz S, Boisset C, Borsali R, Svensson B, Henrissat B, and Driguez H (1999), 'Dynamic light scattering evidence of a ligand-induced motion between the two domains of glucoamylase G1 of *Aspergillus niger* with heterobivalent substrate analogues', *Angew Chem*, 38, 974–977.

Pijning T, Vujičič-Žagar A, Kralj S, Eeuwema W, Dijkhuizen L, and Dijkstra B W (2008), 'Biochemical and crystallographic characterization of a glucansucrase from *Lactobacillus reuteri* 180', *Biocatal Biotransf*, 26, 12–17.

Robert X, Haser R, Gottschalk T E, Ratajczek F, Driguez H, Svensson B, and Aghajari N (2003), 'The structure of barley α-amylase isozyme 1 reveals a novel role of domain C in substrate recognition and binding: "a pair of sugar tongs"', *Structure*, 11, 973–984.

Robert X, Haser R, Mori H, Svensson B, and Aghajari N (2005), 'Oligosaccharide binding to barley α-amylase', *J Biol Chem*, 280, 32968–32978.

Rodenburg K W, Vallée F, Juge N, Aghajari N, Guo X J, Haser R, and Svensson B (2000), 'Specific inhibition of barley α-amylase 2 by barley α-amylase/subtilisin inhibitor depends on charge interactions and can be conferred isozyme 1 by mutation', *Eur J Biochem*, 267, 1019–1029.

Shahpiri A, Svensson B, and Finnie C (2008), 'The NADPH-dependent thioredoxin reductase/thioredoxin system in germinating barley seeds: gene expression, protein profiles, and interaction between isoforms of thioredoxin *h* and thioredoxin reductase', *Plant Physiol*, 146, 789–799.

Sigurskjold B W, Christensen T, Payre N, Cottaz S, Driguez H, and Svensson B (1998), 'Thermodynamics of binding of heterobidentate ligands consisting of spacer-connected acarbose and beta-cyclodextrin to the catalytic and starch-binding domains of glucoamylase from *Aspergillus niger* shows that the catalytic and starch-binding sites are in close proximity in space', *Biochemistry*, 37, 10446–10452.

Southall S M, Simpson P J, Gilbert H J, Williamson G, and Williamson M P (1999), 'The starch binding domain from glucoamylase disrupts the structure of starch', *FEBS Lett*, 447, 58–60.

Stam M R, Danchin E G, Rancurel C, Coutinho P M, and Henrissat B (2006), 'Dividing the large glycoside hydrolase family 13 into subfamilies: towards improved functional annotations of α-amylase-related proteins', *Protein Eng Des Sel*, 19, 555–562.

Svensson B, Fukuda K, Nielsen P K, and Bønsager B C (2004), 'Proteinaceous α-amylase inhibitors', *Biochim Biophys Acta*, 1696, 145–156.

Svensson B, Jespersen H, Sierks M R, and MacGregor E A (1989), 'Sequence homology between putative raw-starch binding domains from different starch-degrading enzymes', *Biochem J*, 264, 309–311.

Søgaard M, Kadziola A, Haser R, and Svensson B (1993), 'Site-directed mutagenesis of histidine 93, aspartic acid 180, glutamic acid 205, histidine 290, and aspartic acid 291 at the active site and tryptophan 279 at the raw starch binding site in barley α-amylase 1', *J Biol Chem*, 268, 22480–22484.

Vallée F, Kadziola A, Bourne Y, Juy M, Rodenburg K W, Svensson B, and Haser R (1998), 'Crystal structure of barley α-amylase complexed with the endogenous protein inhibitor BASI at 1.9 Å resolution', *Structure*, 6, 649–659.

van der Kaaij R M, Janeček Š, van der Maarel M J E C, and Dijkhuizen L (2007), 'Phylogenetic and biochemical characterisation of a novel cluster of intracellular fungal α-amylase enzymes', *Microbiology*, 153, 4003–4015.

van Leeuwen S S, Krajl S, Geel-Shutten I H, Gerwig G J, Dijkhuizen L, and Kamerling J P (2008), 'Structural analysis of the α-D-glucan (EPS180) produced by the *Lactobacillus strain* 180 glucansucrase GTF180 enzyme', *Carbohydr Res*, 343, 1237–1250.

Yu S, Bojsen K, Svensson B, and Marcussen J (1999), 'α-1,4-Glucan lyases producing 1,5-anhydrofructose from starch and glycogen have sequence similarity to α-glucosidases', *Biochim Biophys Acta*, 1433, 1–15.

WHY COULD ISOPULLULANASE, AN ODD PULLULAN-HYDROLYZING ENZYME, BE DISCOVERED?

Yoshiyuki Sakano

ABSTRACT

I write a tiny story on the discovery, cloning, expression, crystallization, and 3D structure of *Aspergillus niger* isopullulanase (EC 3.2.1.57), that hydrolyzes pullulan to produce isopanose (Glc(α1$\rightarrow$4)Glc(α1$\rightarrow$6)Glc).

Key words: isopullulanase; *Aspergillus niger*; isoamylase; pullulan-hydrolyzing enzyme; TVA

INTRODUCTION

A novel and odd pullulan-hydrolyzing enzyme, isopullulanase, was incidentally discovered at a small laboratory in 1970. At first I researched on yeast isoamylase (that is amylopectin 6-glucanohydrolase (EC 3.2.1.9)) on the master course at graduate school of Tokyo University of Agriculture and Technology in 1966-1967, because yeast isoamylase was given to me as the thesis of master course. At that time my boss was Professor Tsueno Kobayashi, who had discovered the enzyme with Dr. Bunji Maruo in 1951 (Maruo and Kobayashi, 1951). The enzyme had been believed to be a synthetase getting longer chains to branch of amylopectin before their discovery, but they showed that the enzyme was a typical hydrolase attacking α-1,6 glucosidic linkages of starch and glycogen to produce shorter linear maltooligosaccharides. After passing through the master course, I prepared a paper, "Purification and Substrate Specificity of Yeast Isoamylase", which was published in *Agri. Biol. Chem.* in 1969 (Sakano et al., 1969). But this enzyme was very unstable on heat and solvent treatments, so it was very hard to continue this thesis moreover.

ENDEAVOR TO A NEW THESIS

Before 1970 isoamylase was never discovered from any organisms besides yeast, plants and bacteria. In 1961 for the first time pullulanase was discovered from *Klebsiella pneumoniae* (traditional name, *Aerobacter aerogenes*) as a hydrolase cleaving α-1,6 glucosidic linkages of pullulan, and also hydrolyzed α-1,6 linkages of starch like isoamylase (Bender and Wallenfels, 1961). Some bacteria (e.g., *Escherichia intermedia* (Ueda and Nanri, 1967) and *Streptococcus mitis* (Walker, 1968)) were reported to produce pullulanase till 1970. At those days amylases (Florkin and Stotz, 1964) were classified to α-amylase, β-amylase, glucoamylase and debranching amylase (i.e., debranching enzyme); 1) α-amylase (EC 3.2.1.1) hydrolyzes exclusively endo-wise α-1,4 linkages of starch to produce α-anomer products, mainly α-maltose, 2) β-amylase (EC 3.2.1.2) does exo-wise α-1,4 linkages of starch from nonreducing end to produce β-maltose only, 3) glucoamylase (EC 3.2.1.3) does α-1,4 and α-1,6 linkages of starch from nonreducing end to produce β-glucose only, and 4) debranching amylase (isoamylase (EC 3.2.1.9) and pullulanase (3.2.1.41)) does α-1,6 linkages of starch to produce linear maltooligosaccharides (Fig.1).

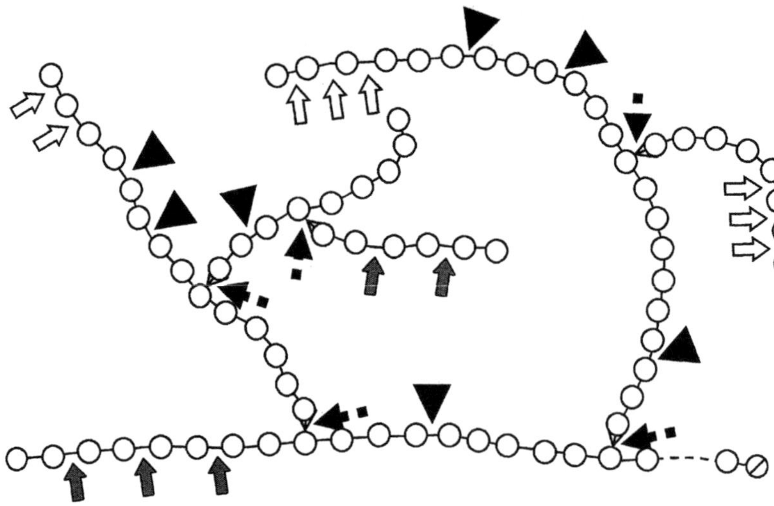

Figure 1 Schematic action pattern of starch-hydrolyzing enzymes
Symbols: Circle, glucose; Circle with slash mark; glucose with reducing end; —, α-(1→4)-glucosidic linkage; ↓, α-(1→6)-glucosidic linkage. Enzymes: White arrow, glucoamylase; Gray arrow, β-amylase; Black arrow, α-amylase; Dashed arrow, debranching amylase (isoamylase or pullulanase).

Therefore I decided to plan screening molds producing a debranching amylase (preferably isoamylase). But there is a big trouble in progressing this new thesis, because in general mold produces a lot of α-amylase and glucoamylase hydrolyzing starch (amylose and amylopectin). At first my colleague and I thought that nobody can detect molds producing isoamylase by using isoamylase activity of hydrolyzing α-1,6 linkages in amylopectin to produce linear oligosaccharides, that is the activity changing the color of iodo-starch reaction from brown to blue, even if there are molds producing

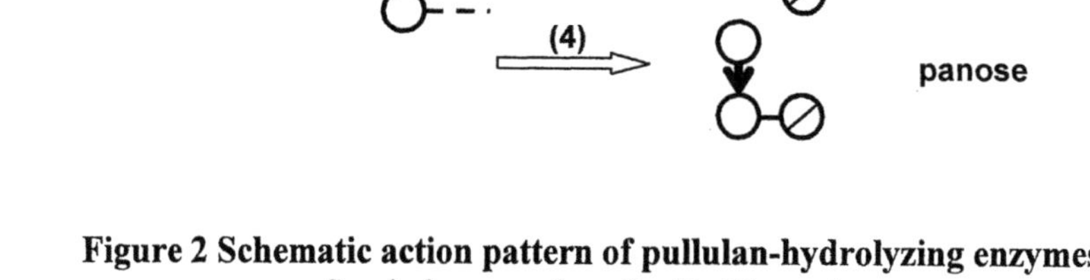

Figure 2 Schematic action pattern of pullulan-hydrolyzing enzymes
Symbols are as described in Figure 1.

isoamylase or isoamylase activity. Therefore we chose pullulan instead of amylopectin as the substrate. We expected that molds producing debranching amylase could be detected if we had used pullulan-hydrolyzing activity, that is the activity arising reducing sugar from pullulan, though they produced α-amylase and glucoamylase with it. Then this new thesis (big evolution) started from the screening experiment.

In 1970 only pullulanase and glucoamylase were known as a kind of enzyme hydrolyzing pullulan. But we considered that there were the possibility of four types of pullulan-hydrolyzing enzymes (Fig. 2; Sakano and Kobayashi, 1971); 1) glucoamylase-type enzyme cleaving pullulan from nonreducing end to produce glucose, 2) pullulanase-type enzyme doing α-1,6 linkages in pullulan to produce maltotriose, 3) third type enzyme doing α-1,4 linkages in pullulan to produce isopanose (Glc(α1→4)Glc(α 1→6)Glc), and 4) fourth type enzyme doing another kind of α-1,4 linkages in pullulan to produce panose (Glc(α1→6)Glc(α1→4)Glc). Our target was the discovery and acquirement of pullulanase-type enzyme. But anytime we were thinking (watching) that we might be able to get a new type (i.e., third or fourth type in Fig.2) of pullulan-hydrolyzing enzyme, while we were really looking for pullulanase-type (debranching amylase). Presently in 1971 we discovered the third type of pullulan-hydrolyzing enzyme (Sakano et al., 1971), named to isopullulanase (pullulan 4-glucano-hydrolase, EC 3.2.1.57; Sakano et al., 1972). Actually long afterwards *Thermoactinomyces vulgaris* α-amylase (EC 3.2.1.1, abbreviated to TVA; Shimizu et al., 1978), *Bacillus stearothermophilus* neopullulanase (EC 3.2.1.35; Kuriki et al., 1988) and *Bacillus licheniformis* maltogenic amylase (BLMA; Kim et al., 1992) were reported as the fourth type in 1978, 1988, and 1992, respectively.

DISCOVER OF A NOVEL PULLULAN-HYDROLYZING ENZYME, ISOPULLULANASE FROM *ASPERGILLUS NIGER*

Wheat bran medium or rice koji medium were used for screening of the molds producing pullulan-hydrolyzing activity, then *Aspergillus niger* ATCC 9642 strain was selected as the best mold producing the activity. Throughout this research, we were attending enzymes that contaminated with the object enzyme and disturbed its enzyme reaction, checking maltose-hydrolyzing activity except pullulan-hydrolyzing activity. This concept steered us to victory; we could discover "isopullulanse" and publish its paper in 1971 (Sakano et al., 1971). *Aspergillus niger* ATCC 9642 isopullulanase were purified from the water extract of wheat bran culture using acetone precipitation and chromatographies of p-cellulose, DEAE-cellulose and Sephadex G-150. Its molecular weight was estimated to be 74k by gel filtration. Purified enzyme, perfectly removed maltose-hydrolyzing activity, hydrolyzed α-1,4 links of pullulan to produce isopanose, so this enzyme was indicated to be the third type of pullulan-hydrolyzing enzyme in Figure 2. We named this new enzyme to isopullulanase (EC 3.2.1.57 pullulan 4-glucanohydrolase; abbreviated to IPU; Sakano et al., 1972) according to the enzyme nomenclature of the Enzyme Committee, the International Union of Biochemistry. The substrate specificity of IPU is summarized to Figure 3 (Akeboshi et al., 2003). Looking back to those days in 1971-2 it was fortunate for us that we had had none of PAGE apparatus. If we had had it and done its PAGE experiments, we would be much confused to judge the purity of the purified enzyme from the results of native and SDS-PAGE and couldn't advance more this research.

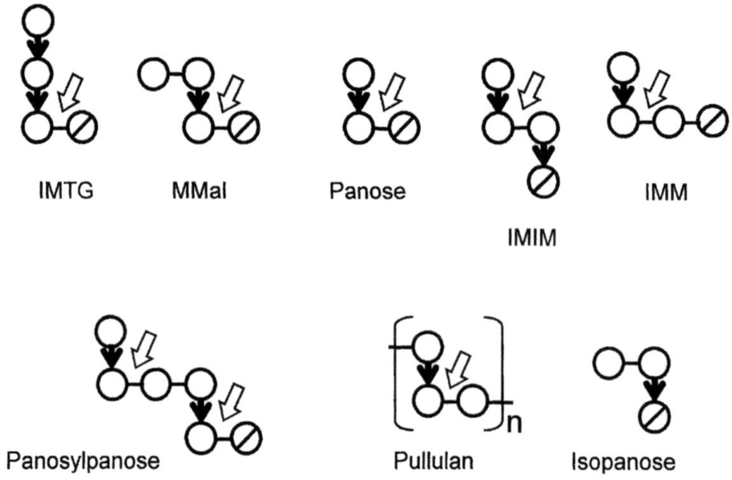

Figure 3 Schematic structure of substrates with the panose motif and isopanose
Bold arrows, enzymatic cleaving points. Other symbols are as described in Figure 1.

CLONING, EXPRESSION, CRYSTALLIZATION, AND 3D STRUCTURE OF ISOPULLULANSE

Molecular cloning and expression of the isopullulanase (IPU) gene of *A. niger* ATCC 9642 were done in 1997 (Aoki et al., 1997) after researches (Sakano et al., 1973; Sakano et al., 1990; Aoki et al., 1996) on production of IPU in solid culture and submerged culture, purification and substrate specificity of extracellular and cell-bound IPU, and carbohydrate content of IPU. Really twenty six years had passed since discovery of IPU. It was apparent at last that the IPU gene encodes an open reading frame of 1696 bp (564 amino acids). IPU contained a signal sequence of 19 amino acids, the molecular weight of the mature form was calculated to be 59 k. Contrary to our beginning expectation, the primary structure of IPU is completely different from those of pullulanase, TVAs I (Tonozuka et al., 1993) and II (Tonozuka et al., 1995), and α-amylases (GH family 13; Matsuura, 1995), but is highly similar to those of the *Penicillium* and *Arthrobacter* dextranases (Aoki and Sakano, 1997) classified into GH family 49. It contains 15 potential N-glycosylation sites, Asn-X-Ser/Thr. Recombinant IPUs expressed in *Aspergillus oryzae* M-2-3 and in *Pichia pastoris* (named to IPU-AO and IPU-PP respectively) had higher carbohydrate contents than that of native IPU; their carbohydrate contents were 34% (Padomajanti et al., 2000), 41% (Akeboshi et al., 2003) and 12-15% (Aoki et al., 1996), respectively; much contents of carbohydrate are characteristic of this enzyme.

IPU-PP was treated with Endo H_f and was purified with HiLoad Q-Sepharose 16/10 HP, then the purified protein was used for crystallization. The crystals of IPU (Fig.4) were grown at 20°C using the hanging-drop vapor-diffusion method (Mizuno et al.,

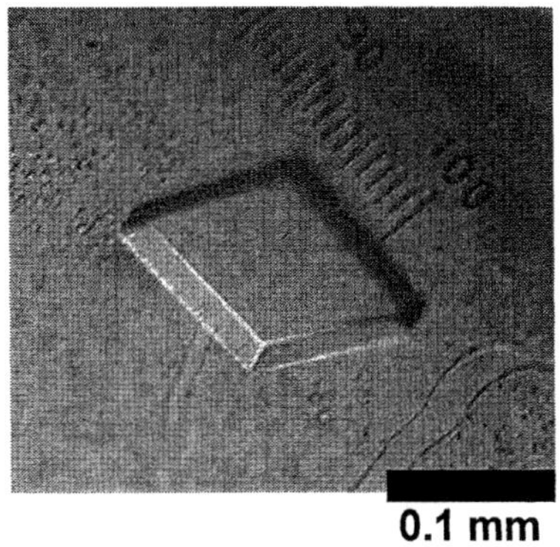

Figure 4 A crystal of isopullulanase
Scale bar represents 0.1 mm.

2008). Its crystal structure has been determined at 1.7 Å resolution (Fig. 5). IPU consists of two domains, domain N (residues 20-182) and domain C (residues 197-564), jointed by a short linker (residues 183-196). Domain N is made up of 13 β-strands and 9 of them form a β-sandwich structure. Domain C is folded into a large right-handed cylinder (termed parallel β-helix), and its structure is composed of 10 complete coils and 3 incomplete coils. The former contain three parallel β-sheets (PB1, PB2 and PB3), connected by three turns (T1, T2, and T3) (Fig. 5). This structure is quite different from those of TVAs I (Kamitori et al., 2002) and II (Kamitori et al., 1999) with typical $(\beta/\alpha)_8$ barrel structure, N-terminal domain (antiparallel β-strands) and C-terminal domain (β-sandwich structure).

Many succession of accidents, chances, or lucks at the moments when I met professors, students, and researchers have made me and the colleagues to discover IPU and develop its continuous research. Progress of the study on IPU has been dependent on combination of partners, time, scientific current, and experimental technology. The enzyme discovered from mold was different from our original target enzyme, but was quite a novel one, predicted one. We pursued the research by following the results anytime, and we declared acquirement of the enzyme gene, construction of enzyme expression system, and crystallization and 3D structure of the enzyme, unable to image before 37 years. But it has been unclear what the proper substrate to IPU is. I expect somebody to resolve this pending question.

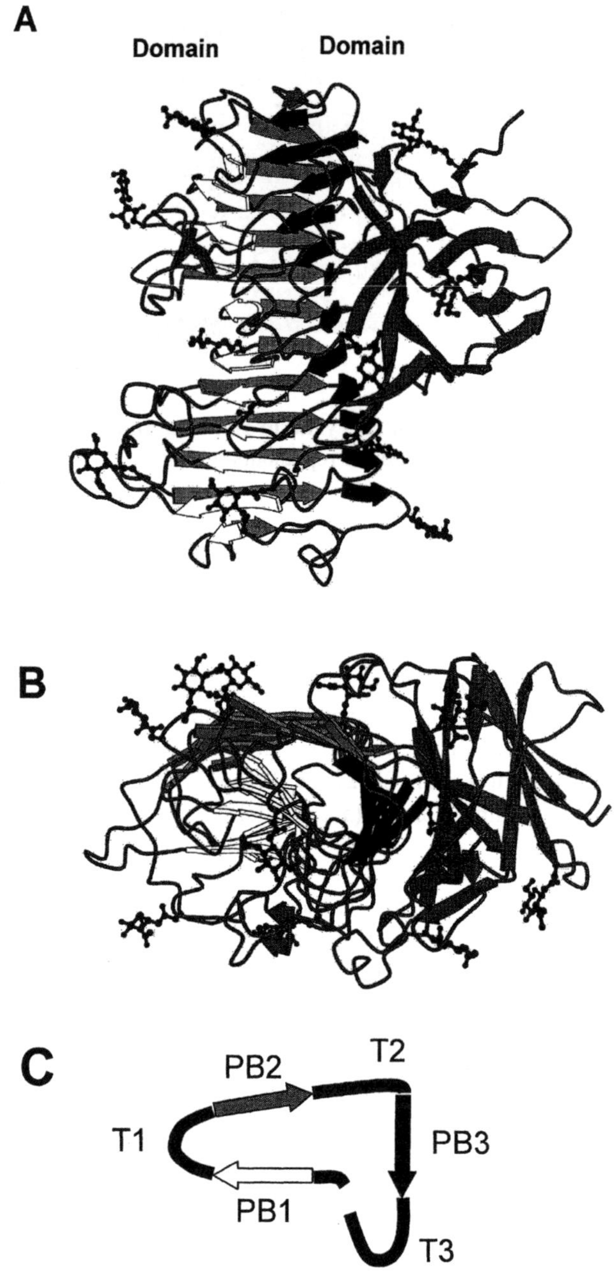

Figure 5 Three-dimensional structures of isopullulanase

(A) Overall structure of isopullulanase. Domain N and three β-strands (PB1, PB2 and PB3) are shown by different grey scales. *N*-acetylglucosamine residues are indicated as ball-and- stick model. (B) A top view of isopullulanase. The orientation is rotated through 90° from that of (A). (C) Schematic representation of three β-strands (PB1, PB2 and PB3) and three turns (T1, T2 and T3) forming the β-helix fold of domain C.

REFERENCE

Akeboshi H, Kashiwagi Y, Aoki H, Tonozuka T, Nishikawa A, and Sakano Yoshiyuki (2003), 'Construction of a efficient expression system for *Aspergillus* isopullulanse in *Pichia pastoris*, and a simple purification method', *Biosci Biotechnol Biochem*, 67, 1149-53.

Aoki H, Yopi, Padomajanti A, and Sakano Y (1996), 'Two components of cell-bound isopullulanase from *Aspergillus niger* ATCC 9642 --- Their purification and enzymatic properties', *Biosci Biotechnol Biochem*, 60, 1795-98.

Aoki H and Sakano Y (1997), 'A classification of dextran-hydrolyzing enzymes based on amino-acid-sequence similarities', *Biochem J*, 323, 859-861.

Aoki H, Yopi, and Sakano Y (1997), 'Molecular cloning and heterologous expression of the isopullulanse gene from *Aspergillus niger* A.T.C.C. 9642', *Biochem J*, 323, 757-64.

Bender H and Wallenfels K (1961), 'Untersuchungen an pullulan', *Biochem Z*, 334, 180-187.

Florkin M and Stotz E H (Ed) (1964), Enzyme Nomenclature, Elsevier Publishing Company, Amsterdam.

Kamitori S, Kondo S, Okuyama K, Yokota T, Shimura Y, Tonozuka T, and Sakano Y (1999), 'Crystal structure of *Thermoactinomyces vulgaris* R-47 α-amylase II (TVA II) hydrolyzing cyclodextrins and pullulan at 2.6 Å resolution', *J Mol Biol*, 287, 907-921.

Kamitori S, Abe A, Ohtaki A, Kaji A, Tonozuka T, and Sakano Y (2002), 'Crystal structure and structural comparison of *Thermoactinomyces vulgaris* α-amylase 1 (TVA I) at 1.6 Å resolution and α-amylase 2 (TVA II) at 2.3 Å resolution', *J Mol Biol*, 318, 443-453.

Kim I C, Cha J H, Kim J R, Jang S Y, Seo B C, Cheong T K, Lee D S, Choi Y D, and Park K H (1992), Catalytic properties of the cloned amylase from *Bacillus licheniformis*, *J Biol Chem*, 267, 22108-14.

Kuriki T, Okada S, and Imanaka T (1988), 'New type of pullulanase from *Bacillus stearothermophilus* and molecular cloning and expression of the gene in *Bacillus subtilis*', *J Bacteriol*, 170, 1554-59.

Maruo B and Kobayashi T (1951), 'Enzymic scission of the branch links of amylopectin', *Nature*, 167, 606-7.

Matsuura Y (1995), 'Crystal structure of α-amylases and their catalytic implications', in *Enzyme Chemistry and Molecular Biology of Amylases and Related Enzymes*, The Amylase Research Society of Japan, CRC Press, 137-145.

Mizuno M, Koide A, Yamamura A, Akeboshi H, Yoshida H, Kamitori S, Sakano Y Nishikawa A, and Tonozuka T (2008), 'Crystal structure of *Aspergillus niger* isopullulanase, a member of glycoside hydrolase family 49', *J Mol Biol*, 376, 210-220.

Padomajanti A, Tonozuka T, and Sakano Y (2000), 'Deglycosylated isopullulanse retains enzymatic activity', *J Appl Glycosci*, 47, 287-292.

Sakano Y, Kobayashi T, and Kosugi Y (1969), 'Purification and substrate specificity of yeast isoamylase', *Agric Biol Chem*, 33, 1535-1540.

Sakano Y and Kobayashi T (1971), 'On a novel pullulan-hydrolyzing enzyme', *Proceedings of Amylase Symposium*, 6, 25-30 (in Japanese).

Sakano Y, Masuda N, and Kobayashi T (1971), 'Hydrolysis of pullulan by a novel enzyme from *Aspergillus niger*', *Agric Biol Chem*, 35, 971-973.

Sakano Y, Higuchi M, and Kobayashi T (1972), 'Pullulan 4-glucanohydrolase from *Aspergillus niger*', *Arch Biochem Biophys*, 153, 180-187.

Sakano Y, Higuchi M, Masuda N, and Kobayashi T (1973), 'Production of pullulan 4-glucanohydrolase by *Aspergillus niger*', *J Ferment Technol*, 51, 726-733.

Sakano Y, Taguchi A, Hisamatsu R, Kobayashi S, Fujimoto D, and Kobayashi T (1990), 'Comparison of cell-bound and extracellular isopullulanse from *Aspergillus niger*', *Denpun Kagaku*, 37, 39-41.

Shimizu M, Tamura M, and Suekane M (1978), 'Purification and some properties of a novel α-amylase produced by a strain of *Thermoactinomyces vulgaris*', *Agric Biol Chem*, 42, 1681-1688.

Tonozuka T, Ohtsuka M, Mogi S, Sakai H, Ohta T, and Sakano Y (1993), 'A neopullulanase-type α-amylase gene from *Thermoactinomyces vulgaris* R-47', *Biosci Biotechnol Biochem*, 57, 395-401.

Tonozuka T, Mogi S, Shimura Y, Ibuka A, Sakai H, Matsuzawa H, Sakano Y, and Ohta T (1995), 'Comparison of primary structures and substrate specificities of two pullulan-hydrolyzing α-amylases, TVA I and TVA II, from *Thermoactinomyces vulgaris* R-47', *Biochim Biohpys Acta*, 1252, 35-42.

Ueda S and Nanri N (1967), 'Production of isoamylase by *Escherichia intermedia*', *Appl Microbiol*, 15, 492-496.

Walker G (1968), 'Metabolism of the reserve polysaccharide of *Streptococcus mitis*. Some properties of a pullulanase', *Biochem J*, 108 33-40.

SEQUENCE FINGERPRINTS IN THE EVOLUTION OF THE α-AMYLASE FAMILY

Štefan Janeček

ABSTRACT

The α-amylase family constitutes the clan GH-H consisting of the three glycoside hydrolase (GH) families GH13, GH70 and GH77. The entire α-amylase family counts ~5,000 known amino acid sequences that exhibit almost 30 different enzyme specificities. Besides the other basic characteristics of the family, such as hydrolysis and/or transglycosylation of several α-glucosidic linkages and the employment of the catalytic triad in the enzymatic reaction, the α-amylase family members should contain from four up-to seven conserved sequence regions. These regions may be considered to be the 'sequence fingerprints' of the individual enzyme specificities and/or taxonomic sources. This article is therefore focused on showing the importance of the conserved sequence regions as the sequence fingerprints in special cases of: (i) the GH13 α-amylases from plants and archaeons; (ii) the oligo-1,6-glucosidase and neopullulanase GH13 subfamilies; and (iii) the unique GH77 amylomaltases from borreliae.

Key words: α-amylase family; glycoside hydrolase clan GH-H; conserved sequence regions; sequence fingerprints; evolutionary relationships

INTRODUCTION

In the amino acid sequence of a protein, i.e. in its primary structure, there is only a very small percentage of residues that have to be conserved in order to save the function of the protein. A substantial part of the sequence can tolerate the substitutions of amino acid residues, because the tertiary structure that is able to accommodate the sequence changes, is more conserved than primary structure. In the case the residues essential for the function of the protein are conserved, no changes (or no any dramatic changes) in the function and activity of the protein have to be seen. The essential residues usually form the so-called conserved sequence regions that cover the isolated sequence stretches belonging to the active site. If proteins and/or enzymes form the families then it is very probable that the conserved sequence regions contain the residues responsible for often very subtle differences among the family members. The conserved sequence regions can thus become the 'sequence fingerprints' characteristic of the individual members of the protein/enzyme family.

The α-amylase family represents just such a family (Janecek, 2002). It consists of almost 30 different enzyme specificities (MacGregor et al., 2001; Kuriki et al., 2006) covering hydrolases (EC 3), transferases (EC 2) and even isomerases (EC 5). Remarkably, some representatives of the heteromeric amino acid transporter proteins, rBAT protein and 4F2 heavy-chain antigen, may be added to the family (Table 1) from structure and evolutionary points of view (Janecek et al., 1997; Fort et al., 2007). At present the α-amylase family is known as the clan GH-H, consisting of the three families of glycoside hydrolases (GH), GH13, GH70 and GH77, classified among the other GHs within the well-accepted CAZy server (Coutinho and Henrissat, 1999). The family counts ~5,000 known amino acid sequences so that it is a very good model for studies focused on elucidating the relationship between sequences, structures, function,

Table 1 The members of the α-amylase family (clan GH-H)

Enzyme class	Enzyme	EC	GH
Hydrolases	α-Amylase	3.2.1.1	13
	Oligo-1,6-glucosidase	3.2.1.10	13
	α-Glucosidase	3.2.1.20	13
	Pullulanase	3.2.1.41	13
	Amylopullulanase	3.2.1.1/41	13
	Cyclomaltodextrinase	3.2.1.54	13
	Maltotetraohydrolase	3.2.1.60	13
	Isoamylase	3.2.1.68	13
	Dextran glucosidase	3.2.1.70	13
	Trehalose-6-phosphate hydrolase	3.2.1.93	13
	Maltohexaohydrolase	3.2.1.98	13
	Maltotriohydrolase	3.2.1.116	13
	Maltogenic α-amylase	3.2.1.133	13
	Maltogenic amylase	3.2.1.133	13
	Neopullulanase	3.2.1.135	13
	Maltooligosyltrehalose hydrolase	3.2.1.141	13
	Maltopentaohydrolase	3.2.1.-	13
	Sucrose hydrolase	3.2.1.-	13
Transferases	Amylosucrase	2.4.1.4	13
	Glucansucrase	2.4.1.5	70
	Sucrose phosphorylase	2.4.1.7	13
	Glucan branching enzyme	2.4.1.18	13
	Cyclodextrin glucanotransferase	2.4.1.19	13
	4-α-Glucanotransferase (Amylomaltase)	2.4.1.25	13, 77
	Glucan debranching enzyme	2.4.1.25/3.2.1.33	13
	Alternansucrase	2.4.1.140	70
	Maltosyltransferase	2.4.1.-	13
Isomerases	Isomaltulose synthase	5.4.99.11	13
	Maltooligosyltrehalose synthase	5.4.99.15	13
	Trehalose synthase	5.4.99.16	13
HATs[1)]	rBAT protein	---	13
	4F2hc antigen	---	13

[1)] HATs means the heteromeric amino acid transporter proteins.

specificity, stability and evolution. The family GH13 is the main and also the largest GH family of the clan GH-H (Table 1). The members are multidomain proteins that adopt the structure of catalytic $(\beta/\alpha)_8$-barrel domain (MacGregor et al., 2001) first identified in Taka-amylase A, i.e. the α-amylase from *Aspergillus oryzae* (Matsuura et al., 1984). The members of the family GH70 were predicted to have a circularly permuted version of the catalytic $(\beta/\alpha)_8$-barrel (MacGregor et al., 1996) and this has recently been really confirmed (Dijkstra et al., 2007). The main feature that discriminates the family GH77 from GH13 is the lack of domain C (Przylas et al., 2000) that is a characteristic domain succeeding the catalytic $(\beta/\alpha)_8$-barrel in all GH13 members (MacGregor et al., 2001).

In general, four requirements were postulated for an enzyme to be a member of the α-amylase family (Takata et al., 1992; Kuriki and Imanaka, 1999): (i) the enzyme should be active towards the α-1,4- and α-1,6-glucosidic bonds; (ii) it should hydrolyze those α-glucosidic bonds or form them by transglycosylation; (iii) it should contain 4 conserved sequence regions in the primary structure; and (iv) it should possess the catalytic triad of residues corresponding with the Asp^{206} (catalytic nucleophile), Glu^{230} (proton donor) and Asp^{297} (transition-state stabilizer) of Taka-amylase A. In the light of the present-day knowledge - mainly due to a lot of new sequences and specificities - it has become clear that also the α-1,1-, α-1,2- and α-1,3-glucosidic linkages should be considered, and isomerization should be added to hydrolysis and transglycosylation as an eventual enzymatic reaction (MacGregor et al., 2001). With regard to the number of conserved sequence regions, it was clearly documented that, in addition to the four best known regions I, II, III and IV (Nakajima et al., 1986), also the three other regions V, VI and VII (Janecek, 1992; 1994a; 1994b; 1995) exhibit unambiguously a conservative nature and contain the sequence features specifically ascribable to the individual enzyme specificities (Fig. 1). Concerning the number of the amino acid residues that are conserved invariantly throughout the α-amylase family, it was originally thought that ~10% of residues are conserved among the α-amylases (Nakajima et al., 1986; MacGregor, 1988). It is worth mentioning, however, that as the number of sequences and specificities has dramatically increased to several thousands and almost 30, respectively, the number of the invariant residues has decreased to 8-10 in 1990s (Janecek, 1994b; Svensson, 1994). At the beginning of 2000s, it seemed that only 4 residues may be totally invariantly conserved throughout the family (Janecek, 2002): the three involved in the catalytic reaction plus the arginine (Arg^{204} in Taka-amylase A) positioned two residues before the catalytic nucleophile (Asp^{206} in Taka-amylase A). At present it has become evident, using the GH77 amylomaltase from *Borrelia burgdorferi*, that a real and functional member of the α-amylase family may contain only the catalytic triad (Fig. 1) conserved invariantly (Machovic and Janecek, 2003; Godany et al., 2008) although the otherwise invariant arginine (Arg^{204} in Taka-amylase A) is in the above-mentioned GH77 amylomaltase substituted conservatively by lysine. From the evolutionary point of view, the α-amylase family is a diverged family in which the individual enzyme specificities do not form their own and independent hydrolase, transferase and isomerase clusters (Stam et al., 2006; Janecek et al., 2007).

SEQUENCE FINGERPRINTS OF THE α-AMYLASE FAMILY

Any study focused on the structure/function and evolutionary relationships within the α-amylase family has to take into account the existence of the conserved sequence regions (Janecek, 2002). The catalytic triad and the almost invariant arginine mentioned above together with two histidines (His^{122} and His^{296}; Fig. 1) were postulated to be crucial for Taka-amylase A (Matsuura et al., 1984), and are considered as being essential for almost every specificity from the α-amylase family (MacGregor et al., 2001). Nevertheless, the two histidines are not totally invariantly conserved (Fig. 1). All these residues form the basis of the four best known conserved sequence regions (the regions I, II, III and IV in Fig. 1) that are located at or near the C-termini of the strands β3, β4, β5 and β7 of the catalytic $(\beta/\alpha)_8$-barrel domain. The additional three conserved sequence regions (the regions V, VI and VII in Fig. 1) are positioned near the C-terminus of domain B around

EC	Year	Enzyme/Protein	VI β2	I β3	V loop3	II β4 *	III β5 *	IV β7 *	VII β8
Family GH13:									
3.2.1.1	1984	Alpha-amylase	56_GFTAIWITP	117_DVVANH	173_LPDLD	202_GLRIDTVKH	226_YCIGEVLD	292_FVENHD	323_GIPIIYAGQ
2.4.1.19	1991	Cyclodextrin glucanotransferase	70_GVTALWISQ	135_DFAPNH	197_LADFN	225_GIRVDAVKH	253_FTFGEWFL	323_FIDNHD	354_GVPAIYYGT
3.2.1.10	1993	Oligo-1,6-glucosidase	44_GIDVIWLSP	98_DLVVNH	167_QPDLN	195_GFRMDVINF	251_MTVGEMPG	324_YWNNHD	360_GTPYIYQGE
3.2.1.60	1997	Maltotetraohydrolase	50_GFSAIWMPV	112_DVVPNH	160_DADLN	189_GFRFDFVRG	215_FCVGELWK	288_FVDNHD	326_GTPVVYWSH
3.2.1.68	1998	Isoamylase	218_GVTAVEFLP	292_DVVYNH	342_GANFN	371_GFRFDLASV	431_DLFAEPWA	505_FIDVHD	574_GTPLMQGGD
3.2.1.133	1999	Maltogenic amylase	189_GITGIYLTP	242_DAVFNH	295_MPKLN	324_GWRLDVANE	353_YILGEIWH	419_LLGSHD	451_GSPCIYYGD
3.2.1.133	1999	Maltogenic alpha-amylase	65_GVTTIWLSP	127_DFVPNH	196_LADLS	221_GLRIDAVKH	249_FLVGEWYG	321_FIDNHD	352_VRPPIYYGT
3.2.1.141	2000	Maltooligosyltrehalose hydrolase	132_GITAIEIMP	187_DVVYNH	220_NFDDA	248_GFRLDAVHA	279_IVIAESDL	372_YIQNHD	409_YIPMIFMGE
2.4.1.4	2001	Amylosucrase	134_GLTYLHLMP	190_DFIFNH	262_QWDLN	290_ILRMDAVAF	332_FFKSEAIV	396_YVRSHD	488_GLPLIYLGD
2.4.1.-	2001	Maltosyltransferase	133_GADAIYLLP	201_DFIPRT	-----	381_GARLDMGHA	410_VMIAEELD	463_SVETPD	495_SIPYVNTGQ
2.4.1.18	2002	Glucan branching enzyme	280_GFTHLELLP	335_DWVPGH	-----	401_ALRVDAVAS	454_VTMAEEST	521_LPLSHD	562_GKKLLFMGN
2.4.1.25	2002	4-Alpha-glucanotransferase	36_GIDFVWLMP	89_DLPIHH	154_SPDLN	182_GFRFDAAKH	212_IFLAEIWA	272_FTSNHD	306_GVPLVFYGD
3.2.1.54	2002	Cyclomaltodextrinase	185_GVNAVYFTP	238_DAVFNH	292_MPKLN	321_GWRLDVANE	350_YILGEVWH	416_LLDSHD	448_GTPCIYYGD
3.2.1.135	2003	Neopullulanase	189_GITGIYLTP	242_DAVFNH	295_MPKLN	324_GWRLDVANE	353_YILGEIWH	419_LLGSHD	451_GSPCIYYGD
5.4.99.11	2003	Isomaltulose synthase	86_GIDAIWINP	140_DVVINH	209_QPDLN	237_GMRFDTVAT	291_ATAGEIFG	364_FLDNHD	391_ATPFIYQGS
5.4.99.15	2003	Maltooligosyltrehalose synthase	30_GVSHLYLSP	85_DIVPNH	-----	224_GYRIDHIDG	251_IIIVEKIL	438_ATSTHD	579_GIPDFYQGT
3.2.1.98	2004	Maltohexaohydrolase	38_GITAVWIPP	102_DVVMNH	203_YADID	232_GFRIDAVKH	262_FAVAEFWK	328_FVDNHD	362_GYPSVFYGD
2.4.1.7	2004	Sucrose phosphorylase	33_VYGDVHILP	83_DAIVNH	160_QIDID	188_YIRLDAVGY	228_EILIEVHS	285_VLDTHD	373_GVPQVYYVG
3.2.1.41	2006	Pullulanase	458_GVTHVELLP	590_DVVYNH	632_CSDSA	661_GFRFDLMGY	690_YFFGEGWD	817_YVSKHD	859_GIAFDQQGS
3.2.1.70	2008	Dextranglucosidase	44_GVMAIWLSP	98_DLVVNH	162_QPDLN	190_GFRMDVIDM	232_LTVGETWG	308_FWNNHD	344_GTPYIYQGE
3.2.1.-	2008	Sucrose hydrolase	119_GVRYLHLLP	175_DFVLNH	248_QWDLN	276_AFRLDSTAY	318_VMKAEAIV	387_YVRCHD	486_GVPLIYMGD
3.2.1.20		Alpha-glucosidase	52_GVDAIWVCP	106_DLVINH	181_QVDLN	210_GFRIDTAGL	272_MRVGEVAH	344_YIENHD	381_GTLYVYQGQ
3.2.1.1/41		Amylopullulanase	434_GVSVIYLNP	487_DGVFNH	562_WANFI	590_GWRLDVENE	619_PMIAENWG	695_LLGSHD	742_GMADIYYGD
3.2.1.93		Trehalose-6-phosphate hydrolase	47_QVDVLWLTP	101_DLVVNH	171_QADLN	198_GFRLDVINL	249_MTVGEMSS	323_FWCNHD	359_GTPYIYQGE
3.2.1.116		Maltotriohydrolase	35_GVSAIWIPQ	119_DIVLNH	172_LLDLP	200_GLRIDAAAH	228_WRVGEVWD	292_FVQNHD	325_GMPMLYRGG
3.2.1.-		Maltopentaohydrolase	31_GFAAVQISP	87_DAVINH	153_LQDLN	181_GLRVDAAKH	212_FWFLEVIG	280_FVDNHD	318_GYPALMSAT
2.4.1.25/3.2.1.33		Glucan debranching enzyme	137_GYNMIHFTP	198_DVVYNH	-----	505_GVRLDNCHS	534_YVVAELFT	605_MDITHD	642_GYDELVPHQ
5.4.99.16		Trehalose synthase	54_GVDCLWVPP	108_DFVMNH	178_QPDLN	206_GFRLDAVPY	248_VLLYEANQ	322_FLRNHD	385_GSPVLYYGD
Family GH70:									
2.4.1.5	2007	Glucansucrase	828_GITDFEMAP	894_DWVPDQ	378_ANDVD	411_SIRVDAVDN	429_VSIVEAWS	521_FARAHD	591_SIPRVYYGD
2.4.1.141		Alternansucrase	1092_GITSFELAP	1169_DWVPDQ	593_ANDID	631_GIRVDAVDN	669_LSILEDWN	762_FVRAHD	834_TVPRVYYGD
Family GH77:									
2.4.1.25	2000	4-Alpha-glucanotransferase	40_GGRYWQVLP	213_DMPIFV	262_LYRWD	289_LVRIDHFRG	336_PVLAEDLG	390_YTGTHD	442_SVARLAVYP
HATs:									
---	2007	rBAT protein	156_NIKTVWITS	210_DFIPNH	282_QPDLN	310_GFSLDAVKF	370_QYSTEPGR	444_MIGGPD	474_GTPITYYGE
---		4F2hc antigen	286_KVKGLVLGP	338_DLTPNY	-----	375_GFQVRDIEN	405_RLLIAGTN	465_SLSQ--	493_GTPVFSYGD

Figure 1 Sequence fingerprints of the α-amylase family members

One representative of each enzyme specificity is presented. The catalytic triad is highlighted in black-and-white inversion and signified by asterisks. The residues conserved in at least 50% of sequences are colored with grey background. The representatives of heteromeric amino acid transporter proteins (HATs) are also shown. The 'Year' denotes the year of three-dimensional structure determination (if any). Adapted from Janecek (2002).

the calcium-binding aspartate (Asp^{175} in Taka-amylase A; region V) and at or near the C-termini of the strands β2 (region VI) and β8 (region VII) of the catalytic $(\beta/\alpha)_8$-barrel domain. It should be pointed out that while the first four regions do contain most of the residues responsible for catalytic action and function of α-amylase family members, the additional three regions may cover the sequence features characteristic of certain enzyme specificities from the family (Janecek, 2002).

The chronology and history of discovering the individual conserved sequence regions were described in detail by Janecek (2002). Although the regions I, II, III and IV were defined for α-amylases by Nakajima et al. (1986) and for the entire family by Takata et al. (1992), the pioneering works by Toda et al. (1982), Friedberg (1983) and Rogers (1985) offered the original ideas on conserved sequence segments among the bacterial, fungal, plant and animal α-amylases. With regard to the three additional conserved sequence regions V, VI and VII - namely the region V for α-amylases (Janecek, 1992) and for the entire family (Janecek, 1995) as well as the regions VI and VII for α-amylases (Janecek, 1994a) and for the entire family (Janecek, 1994b) - these were proposed also thanks to the influence by the excellent fundamental studies by MacGregor (1988), Svensson (1988), MacGregor and Svensson (1989) and Jespersen et al. (1991; 1993). The conserved sequence regions in the circularly permuted GH70 members of the α-amylase family were identified in the prediction study by MacGregor et al. (1996). The overall philosophy of conserved sequence regions would not be possible to develop irrespective of the created classification system of all GHs into the GH families by Henrissat (1991), now available within the CAZy server (Coutinho and Henrissat, 1999).

It should thus be understandable that the α-amylase family is to be considered with as many conserved sequence regions as possible (Janecek, 2002). In other words, every novel member of the family (or even a novel sequence of already known family member) should be characterized by all available conserved sequence regions, i.e. by its sequence fingerprints. Even an absence of one of the regions in the amino acid sequence (Fig. 1) may be utilized as a sequence feature typical of a given enzyme specificity (Janecek, 2000).

The unambiguous importance of the power of the appropriate utilization of conserved sequence regions was first demonstrated on a description (Janecek et al., 1995) how to distinguish between the α-amylases and cyclodextrin glucanotransferases (CGTases), the two very closely related enzyme specificities from the α-amylase family. Previously, due to close interplay between the hydrolysis and transglycosylation brought about a CGTase (Uitdehaag et al., 1999), some CGTases were originally erroneously described as α-amylases, e.g., the sequences from *Bacillus* sp. strain B1018 (Itkor et al., 1990) and *Thermoanaerobacter thermosulfurogenes* (Bahl et al., 1991). The main sequence features that may help to discriminate the CGTases from α-amylases and are present in conserved sequence regions were identified as follows (Janecek et al., 1995): (i) Ala^{137}-Pro^{138} (or Thr-Pro) (the numbering is from the CGTase from *Bacillus circulans* strain 8; Nitschke et al., 1990) preceding the Asn^{139}-His^{140} and Phe^{136} (or tyrosine) succeeding the Asp^{135} in region I (strand β3); (ii) Trp^{258}-Phe^{259} succeeding the catalytic proton donor (Glu^{257}) in region III (strand β5); and (iii) a glutamine (Gln^{78}) preceding the proline at the end of the region VI (strand β2) resulting in the fact that there is an octapeptide in CGTases in comparison with heptapeptide in α-amylases between the conserved glycine and proline positioned at the beginning and the end of the region, respectively (Fig. 1). It is worth mentioning that the enzyme from *T.*

thermosulfurogenes was later confirmed to be the CGTase also experimentally (Wind et al., 1995; Knegtel et al., 1996).

α-AMYLASES FROM ARCHAEA AND PLANTS

One of the remarkable findings observed in the α-amylase family using the conserved sequence regions as sequence fingerprints was that concerning the close relatedness between the α-amylases from Archaea and plants (Janecek et al., 1999). At the beginning of 1990s when no sequence of an archaeal α-amylase was available, plant α-amylases were found to occupy the position in the evolutionary tree on a common cluster with bacterial liquefying and intracellular α-amylases represented by bacilli and enterobacteria, respectively (Janecek, 1994a).

The situation was dramatically changed by involving the first sequences of hyperthermostable α-amylases from Archaea (Janecek et al., 1999; Jones et al., 1999) that were unexpectedly found on a common branch with their 'mesostable' plant counterparts. The sequence features exclusive for the α-amylases from hyperthermophilic archaeons (Fig. 2A) are as follows (Janecek et al., 1999): (i) Ile^{107} (the numbering is from the α-amylase from *Thermococcus hydrothermalis*; Leveque et al., 2000) succeeding the conserved aspartate in region I (strand β3); (ii) (Ala^{194})-Trp^{195} at the beginning, Tyr^{199} in the middle and Gly^{202} at the end of the region II (strand β4); (iii) Ala^{219} succeeding the conserved tryptophane and Tyr^{223}-Trp^{224} succeeding the catalytic proton donor (Glu^{222}) in the region III (strand β5); (iv) Ala^{286} in the region IV (strand β7); (v) Ile^{165} in the region V (located within the loop3, i.e. domain B); (vi) Ile^{42} succeeding the conserved glycine at the beginning and dipeptide Pro^{48}-Pro^{49} at the end of the region VI (strand β2); and (vii) Gln^{309} succeeding the conserved glycine at the beginning, tripeptide Ile^{312}-Phe^{313}-Tyr^{314} in the middle and Asp^{316} at the end of the region VII (strand β8). It is evident that most of these archaeal features are present exclusively in the α-amylases originated from plants (Fig. 2A). Some of these residues have already been shown to play an important functional role, for example, the glycine from the region II (Gly^{202} of the archaeal α-amylase) and the tryptophane from the region III (Trp^{224} of the archaeal α-amylase) were found to provide a specific ligand for calcium ion and form a stacking interaction with one of the acarbose rings bound in the active site, respectively, in the structure of barley α-amylase-acarbose complex (Kadziola et al. 1998). The same is true for the analogous structure of the α-amylase from *Pyrococcus woesei* (Linden et al., 2003).

It is worth mentioning that the evolutionary clustering of the α-amylases from Archaea and plants seen for a limited sample of taxonomical representatives (Fig. 2B) has been retained also in the recent evolutionary trees constructed for a wide spectrum of organisms including novel groups of α-amylases from bacteria (Da Lage et al., 2004) and fungi (van der Kaaij et al., 2007). Within the recently created GH13 subfamilies at CAZy, these plant and archaeal α-amylases belong to the subfamilies GH13_6 and GH13_7, respectively (Stam et al., 2006). In any case, the bacterial liquefying and intracellular α-amylases represented for example by the enzymes from *Bacillus licheniformis* (Yuuki et al., 1985) and *Escherichia coli* (Raha et al., 1992), respectively, should still be considered as most closely related groups to plant and archaeal α-amylases (Fig. 2B). It only seems that the α-amylases from non-hyperthermohilic halophilic archaeons, such as those from *Natronococcus* sp. strain Ah-36 (Kobayashi et al., 1994) and *Haloarcula hispanica* (Hutcheon et al., 2005), could cluster among the bacterial counterparts (Fig. 2B), probably as a group of halophilic prokaryotic α-

(A)

Source	VI β2	I β3	V loop3	II β4 *	III β5 *	IV β7 *	VII β8
Bacteria:							
Aeromonas hydrophila	25_GYKQVLISP	81_DVVLNH	163_LPDLD	190_GFRVDAVKH	217_HVFGEVIT	288_FAITHD	323_GSPLVYSDH
Alteromonas haloplanktis	28_GYAAVQVSP	84_DTLINH	142_LADLD	170_GFRFDASKH	196_VVFQEVID	259_FVDNHD	297_GYPKVMSSY
Bacillus licheniformis	36_GITAVWIPP	100_DVVINH	198_YADID	227_GFRLDAVKH	257_FTVAEYWQ	323_FVDNHD	357_GYPQVFYGD
Bacillus subtilis	33_GYTAIQTSP	97_DAVINH	144_LYDWN	172_GFRFDAAKH	204_FQYGEILD	264_WVESHD	299_STPLFFSRP
Escherichia coli	35_GINMVWLPP	100_DVVVNH	202_GENID	231_GFRIDAVKH	261_FIVAEYWS	327_LVANHD	361_GVPSVFYPD
Lactobacillus amylovorus	45_GYTAVQTSP	109_DATLND	156_FYDWN	184_GFRYDAATH	218_FQYGEVLQ	278_WVESHD	313_SVPLFFDRP
Streptomyces albidoflavus	32_GYGYVQVSP	88_DSVINH	145_LADLD	173_GFRIDAAKH	200_YWKQEAIH	263_FVDNHD	297_GSPDVHSGY
Thermotoga maritima	70_GVDAVWFMP	123_DLVINH	186_MPDLN	214_GFRIDAAKH	254_ILVGEVFS	305_FLENHD	349_GSPVIYYGG
Archaea:							
Pyrococcus furiosus	40_GISAIWLPP	105_DVVINH	161_FPDIC	193_GWRFDYVKG	217_WAVGEYWD	283_FVANHD	307_GQPVIFYRD
Pyrococcus woesei	41_GISAIWLPP	106_DVVINH	162_FPDIC	194_GWRFDYVKG	218_WAVGEYWD	284_FVANHD	308_GQPVIFYRD
Thermococcus hydrothermalis	41_GISAIWIPP	106_DIVINH	162_YPDIC	194_AWRFDYVKG	218_WAVGEYWD	284_FVANHD	308_GQPAIFYRD
Thermococcus kodakaraensis	41_GISAIWIPP	106_DIVINH	162_FPDIA	194_AWRFDYVKG	218_WAVGEYWD	284_FVANHD	308_GQPVIFYRD
Thermococcus profundus	41_GISAIWIPP	106_DIVINH	162_FPDIA	194_AWRFDYVKG	218_WAVGEYWD	284_FVANHD	308_GQPVIFYRD
Thermococcus sp. strain Rt3	41_GISAIWIPP	106_DIVINH	162_FPDIA	194_AWRFDYVKG	217_LAVGEYWD	283_FVANHD	307_GQPVIFYRD
Haloarcula hispanica	28_GYDAIQVPP	91_DAVINH	145_LKDLK	173_GIRWDAAKH	201_WTVGEVLD	263_FVSNHD	287_GYPRVYSNR
Plants:							
Apple	56_GFTSAWLPP	108_DIVINH	160_VPNID	189_DFRFDFARG	214_FSVGEYWD	295_FLDNHD	327_GIPTVFYDH
Banana	33_GVTHVWLPP	87_DIVINH	146_APDID	175_GWRLDFAKG	200_FVVAEIWS	282_FVDNHD	314_GVPSIFYDH
Barley low-pI isozyme	34_GVTHVWLPP	88_DIVINH	147_APDID	176_AWRLDFARG	201_LAVAEVWD	286_FVDNHD	318_GIPCIFYDH
Barley high-pI isozyme	33_GITHVWLPP	87_DIVINH	146_APDID	175_GWRFDFAKG	200_FAVAEIWT	284_FVDNHD	316_GTPCIFYDH
Kidney bean	32_GITHVWLPP	86_DIVINH	145_APDID	174_GWRFDFVKG	199_FAVGEKWD	280_FIDNHD	312_GTPSIFYDH
Maize	61_GATHVWLPP	115_DVVINH	174_APDID	203_GWRLDFAKG	228_FVVAEIWS	310_FVDNHD	342_GTPCIFYDH
Potato	32_GFTTAWLPP	84_DIVINH	136_VPNID	165_DFRFDFAKG	190_FAVGEYWD	271_FIDNHD	303_GIPSVFFDH
Rice	33_GITHVWLPP	87_DIVINH	145_APDID	174_AWRLDFAKG	199_FAVAEIWT	284_FVDNHD	316_GNPCIFYDH
Other Eucarya:							
Aspergillus oryzae	56_GFTAIWITP	117_DVVANH	173_LPDLD	202_GLRIDTVKH	226_YCIGEVLD	292_FVENHD	323_GIPI YAGQ
Cryptococcus sp. strain S-2	58_GFTAIWISP	124_DVVVNH	186_LVDLR	215_GLRIDSLQQ	240_YMVGEVFN	307_FLENQD	338_GIPITYYGQ
Fruit fly	36_GYAGVQVSP	94_DVVFNH	154_LRDLN	182_GFRVDAAKH	219_YIVQEVID	283_FVDNHD	322_GTPRVMSSF
Shrimp	35_GFAGVQVSP	97_DAVINH	165_LNDLN	193_GFRIDASKH	229_FIFQEVID	293_FIDNHD	332_GYTRVMSSY
Chicken	36_GFGGVQVSP	96_DAVVNH	165_LLDLA	193_GFRIDAAKH	229_FIYQEVID	295_FVDNHD	334_GFTRVMSSY
Pig	36_GFGGVQVSP	96_DAVINH	165_LLDLA	193_GFRLDASKH	229_FIFQEVID	295_FVDNHD	334_GFTRVMSSY
Human	36_GFGGVQVSP	96_DAVINH	165_LLDLA	193_GFRIDASKH	229_FIYQEVID	295_FVDNHD	334_GFTRVMSSY

Figure 2 Sequence fingerprints of the α-amylases representing the individual taxonomic sources focused on Archaea and plants (A) and the corresponding evolutionary tree (B)

The sequence features characteristic of the archaeal α-amylases are highlighted in black-and-white inversion. The catalytic triad is signified by asterisks. The tree was calculated using the program CLUSTAL-W (Thompson et al., 1994) and displayed with the program Tree-View (Page, 1996). Adapted from Janecek et al. (1999).

(B)

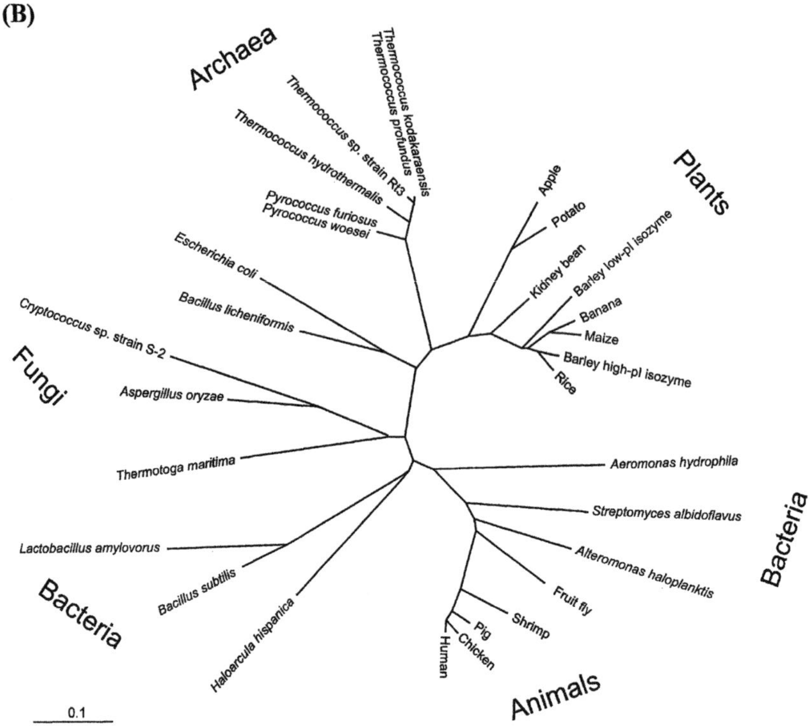

Figure 2 Continued

amylases. It should, however, be taken into account that the enzyme from *Natronococcus* sp. strain Ah-36 (not included here) is a maltotriohydrolase (Kobayashi et al., 1992), i.e. EC 3.2.1.116 (Fig. 1), so that this may also be the case of the 'α-amylase' from *H. hispanica.* Consequently all true archaeal α-amylases (i.e. EC 3.2.1.1) may be most related to plant α-amylases as seen in Figure 2B.

OLIGO-1,6-GLUCOSIDASE AND NEOPULLULANASE SUBFAMILIES

The establishment of the oligo-1,6-glucosidase and neopullulanase subfamilies was based on identifying the fifth conserved sequence region, first defined in α-amylases (Janecek, 1992), also in other enzyme specificities from the α-amylase family (Janecek, 1995). Already in that time it was recognized that the region 173_LPDLD (Taka-amylase A numbering; Fig. 1) is in the group formed by oligo-1,6-glucosidase, α-glucosidase, dextran glucosidase and trehalose-6-phosphate hydrolase conserved mostly as QPDLN, whereas in the group of neopullulanase and cyclomaltodextrinase as MPKLN, with some cases of special 'α-amylases' having MPDLN (Fig. 3A). Later when the heteromeric amino acid transporter proteins, rBAT and 4F2hc, were confirmed to belong to the α-amylase family, both were interestingly revealed to be most closely related to the large oligo-1,6-glucosidase group including in that time also the neopullulanase and cyclomaltodextrinase (Janecek et al., 1997). The idea on the existence of the two independent GH13 subfamilies of oligo-1,6-glucosidase and neopullulanase was developed by Oslancova and Janecek (2002) who described the

(A)

Source	VI β2	I β3	V loop3	II β4 *	III β5 *	IV β7 *	VII β8
Oligo-1,6-glucosidase subfamily:							
Bacillus cereus OGLU	44_GIDVIWLSP	98_DLVVNH	167_QPDLN	195_GFRMDVINF	251_MTVGEMPG	324_YWNNHD	359_GTPYIYQGE
Bacillus thermoglucosidasius OGLU	44_GVDVVWLSP	98_DLVVNH	167_QPDLN	195_GFRMDVINM	252_MTVGETPG	325_YLNNHD	361_GTPYIYQGE
Erwinia rhapontici OGLU	45_GITLLWLSP	99_DLVVNH	169_QPDLN	197_GFRMDVIDL	239_VTVGETWS	316_FWSNHD	352_GTPYIYQGE
Honey bee AGLU	65_GVDMFWLSP	119_DFVPNH	191_QPDLN	219_GFRVDALPY	282_HMLIEAYT	343_VPGNHD	373_GVAVNYYGD
Bacillus stearothermophilus AGLU	44_GVDIVWICP	98_DLVINH	167_QPDLN	195_GFRIDAISH	252_MTVGEANG	321_FLENHD	357_GTPFIYQGQ
Saccharomyces cerevisiae AGLU	52_GVDAIWVCP	106_DLVINH	181_QVDLN	210_GFRIDTAGL	272_MRVGEVAH	344_YIENHD	381_GTLYVYQGQ
Aspergillus parasiticus DGLU	55_GIDLVWLSP	109_DLVVNH	183_QPDLN	211_GFRMDVINM	256_ARAGEGSE	327_YLENHD	364_GTPFVYQGQ
Streptococcus mutans DGLU	44_GVMAIWLSP	98_DLVVNH	162_QPDLN	190_GFRMDVIDM	232_LTVGETWG	308_FWNNHD	344_GTPYIYQGE
Bacillus subtilis T6PH	47_QVDVLWLTP	101_DLVVNH	170_QADLN	198_GFRLDVINL	249_MTVGEMSS	323_FWCNHD	359_GTPYIYQGE
Escherichia coli T6PH	46_GVDAIWLTP	100_DMVFNH	168_QADLN	196_GLRLDVVNL	247_MTVGEMSS	320_FWCNHD	356_GTPYIYQGE
Neisseria polysaccharea ASU	134_GLTYLHLMP	190_DFIFNH	262_QWDLN	290_ILRMDAVAF	332_FFKSEAIV	396_YVRSHD	487_GLPLIYLGD
Leuconostoc mesenteroides SPH	34_AIGGVHLLP	82_DFMINH	164_QIDID	192_LIRLDAFAY	233_EILPEIHE	290_TLDTHD	369_GIPQIYYVG
Pseudomonas saccharophila SPH	40_VFGGVHLLP	90_DVIVNH	168_QIDIA	196_MVRLDAVGY	236_EVLVEIHA	293_VLDTHD	382_GVPQVYYVG
Erwinia rhapontici ISY	86_GIDAIWINP	140_DIVINH	209_QPDLN	237_GLRFDTVAT	291_ATAGEIFG	364_FLDNHD	391_ATPFIYQGS
Klebsiella sp. LX3 ISY	86_GIDAIWINP	140_DVVINH	209_QPDLN	237_GMRFDTVAT	291_ATAGEIFG	364_FLDNHD	391_ATPFIYQGS
Pimelobacter sp. R48 TSY	54_GVDCLWVPP	108_DFVMNH	178_QPDLN	206_GFRLDAVPY	248_VLLYEANQ	322_FLRNHD	385_GSPVLYYGD
Streptomyces coelicolor TSY	62_GVDCLWLPP	116_DFVMNH	186_QPDLN	214_GFRLDAVPY	256_VVLAEANQ	325_FLRNHD	388_GSPILYYGD
Intermediary group:							
Bacillus megaterium	83_QVNGIWMMP	136_DLVVNH	201_MPDLN	229_GFRLDAALH	259_YLTGEVWD	335_FLTNHD	367_GNPYIYYGE
Clostridium acetobutylicum	78_GYKAVQVSP	139_DVVMNH	190_MPDLN	218_GFRFDAAKH	258_YIYGEVLQ	316_MLETHD	349_SVPLFFDRP
Dictyoglomus thermophilum (AmyC)	70_NITALWLMP	123_DLVVNH	181_MPDLN	209_GFRLDAAKH	244_YLVGEVWD	309_FLRNHD	341_GNTFIYYGE
Thermotoga maritima	91_GVDAVWFMP	144_DLVINH	207_MPDLN	235_GFRIDAAKH	275_ILVGEVFS	326_FLENHD	370_GSPVIYYGG
Xanthomonas campestris	79_GVSGIWLMP	132_DLVINH	195_MPDLN	223_GFRLDAARH	268_YLVGEVSA	336_FLSNHD	368_GRPYLYYGE
Bacillus polymyxa	70_QVSGLWLMP	123_DLVINH	189_MPDLN	217_GFRLDAAMH	257_YLAGEVWD	323_FLSNHD	355_GQPFLYYGE
Neopullulanase subfamily:							
Alicyclobacillus acidocaldarius CMD	180_GVNLMYLTP	233_DAVFNH	286_MPKLN	315_GWRFDVANE	345_LLIGEIWH	410_LLGSHD	442_GIPMVYYGD
Bacillus sp. I-5 CMD	185_GVNAVYFTP	238_DAVFNH	292_MPKLN	321_GWRLDVANE	350_YILGEVWH	416_LLDSHD	448_GTPCIYYGD
Bacillus sphaericus CMD	187_GVNALYFNP	240_DAVFNH	294_MPKLN	323_GWRLDVANE	352_YILGEIMH	418_LLGSHD	450_GTPCIYYGD
Bacillus stearothermophilus MGA	189_GVNGIYFTP	242_DAVFNH	295_MPKLN	324_GWRLDVANE	353_YILGEIWH	419_LLGSHD	451_GTPCIYYGD
Bacillus subtilis MGA	192_GVNGIYLTP	245_DAVFNH	295_MPKLN	324_GWRLDVANE	353_YILGEIWH	419_LLDSHD	451_GSPCIYYGT
Thermus sp. IM6501 MGA	189_GITGIYLTP	242_DAVFNH	295_MPKLN	324_GWRLDVANE	353_YILGEIWH	419_LLGSHD	451_GSPCIYYGD
Bacillus sp. KSM-1876 NPU	188_GINGIYLTP	241_DAVFNH	294_MPKLN	323_GWRLDVANE	352_YILGEVWH	418_LLGSHD	450_GTPCIYYGD
Bacillus stearothermophilus NPU	189_GITGIYLTP	242_DAVFNH	295_MPKLN	324_GWRLDVANE	353_YILGEIWH	419_LLGSHD	451_GSPCIYYGD
Thermoactinomyces vulgaris NPU	186_GVTALYFTP	239_DAVFNH	293_MPKLR	321_GWRLDVANE	350_LIVGEIWH	416_LLDSHD	448_GTPLIYYGD
HATs proteins:							
Human rBAT	156_NIKTVWITS	210_DFIPNH	282_QPDLN	310_GFSLDAVKF	370_QYSTEPGR	444_MIGGPD	474_GTPITYYGE
Dog rBAT	171_NIKTIWITS	225_DFIPNH	297_QPDLN	325_GFSFDAVKF	385_QYSREPGR	459_MIGGPD	489_GTPITYYGE
Opossum rBAT	154_NIKTVWITS	208_DFIPNH	280_QPDLN	308_GFSFDAVKF	368_QYSREPGR	443_MIGGPD	473_GTPITYYGE
Zebrafish rBAT	146_NIKAIWISP	200_DYIPNH	268_QPDLN	296_GFRMDAVKH	356_TYSREPGR	432_VVGNHD	462_GTPTTYYGE
Human 4F2	286_KVKGLVLGP	338_DLTPNY	-----	375_GFQVRDIEN	405_RLLIAGTN	465_SLSQ--	493_GTPVFSYGD
Dog 4F2	193_KVKGFVLGP	245_DLTPNY	-----	283_GFQVRNVEN	314_RLLIAGTE	374_SLSQ--	401_GTPVFSYGD
Opossum 4F2	212_KVKALVLGP	264_DLTPNY	-----	302_GIQVLDVQN	333_NVLIVGTE	394_TYSQ--	421_GTPIFSYGD
Zebrafish 4F2	138_KVKGLVLGP	190_DLTPNY	-----	226_GIFLSDIND	257_KKAIMGTA	317_GLGQTA	345_GTPVFSAGD

Figure 3 Figure caption is shown in next page

(B)

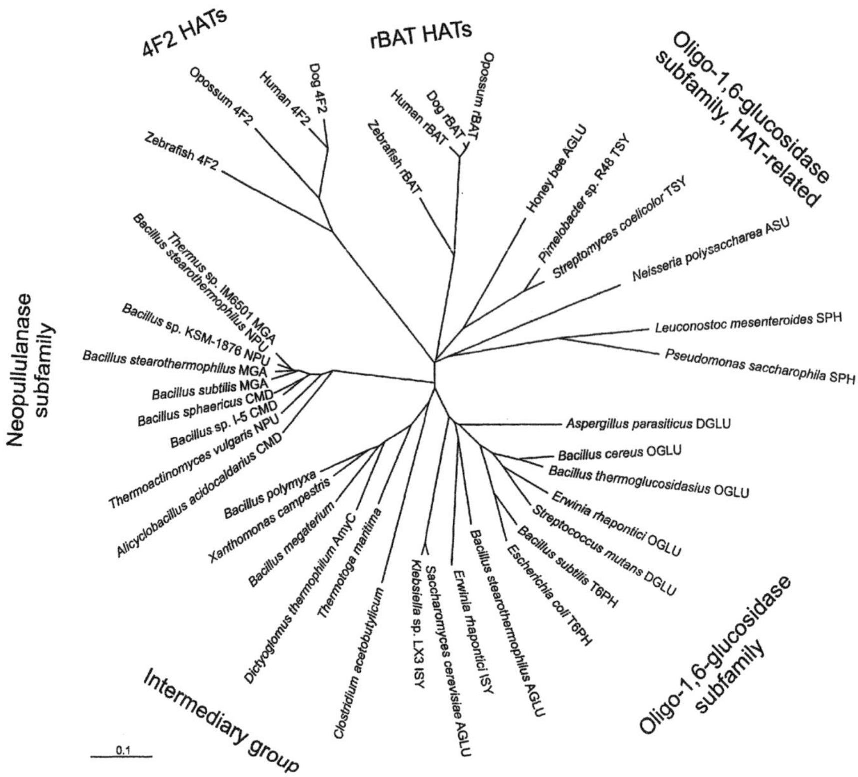

Figure 3 Sequence fingerprints of members of the oligo-1,6-glucosidase and neopullulanase GH13 subfamilies (A) and the corresponding evolutionary tree (B) The sequence features characteristic of the oligo-1,6-glucosidase and neopullulanase subfamilies are highlighted in black-and-white inversion and grey background, respectively. The catalytic triad is signified by asterisks. Abbreviations: OGLU, oligo-1,6-glucosidase; AGLU, α-glucosidase; DGLU, dextran glucosidase; T6PH, trehalose-6-phosphate hydrolase; ASU, amylosucrase; SPH, sucrose phosphorylase; ISY, isomaltulose synthase; TSY, trehalose synthase; CMD, cyclomaltodextrinase; MGA, maltogenic amylase; NPU, neopullulanase; HAT, heteromeric amino acid transporter proteins. The tree was calculated using the program CLUSTAL-W (Thompson et al., 1994) and displayed with the program Tree-View (Page, 1996). Adapted from Oslancova and Janecek (2002).

detailed features characteristic of the two subfamilies present in the conserved sequence regions (Fig. 3A). The members of the oligo-1,6-glucosidase subfamily are, in addition to the four above-mentioned oligo-1,6-glucosidase, α-glucosidase, dextran glucosidase and trehalose-6-phosphate hydrolase, also the isomaltulose synthase and trehalose synthase and in a wider sense also the amylosucrase with sucrose phosphorylase. The member of the neopullulanase subfamily is, in addition to the two above-mentioned neopullulanase and cyclomaltodextrinase, also the maltogenic amylase. It is worth mentioning that while the sequences, structures and specificities of the oligo-1,6-glucosidase subfamily members are more variable, the individual neopullulanase subfamily members are almost indistinguishable from each other (Park et al., 2000; Lee et al., 2002; Oh, 2003). This is clearly reflected also in the evolutionary tree (Fig. 3B) in which the neopullulanase subfamily forms a compact cluster separated from the rest by a quite long branch.

The specific sequence features of the two subfamilies that are present in the conserved sequence regions (Fig. 3A) are as follows (Oslancova and Janecek, 2002): (1) for the oligo-1,6-glucosidase subfamily: (i) a hydrophobic residue, e.g., Leu^{99} (the numbering is from the oligo-1,6-glucosidase from *Bacillus cereus*; Watanabe et al., 1990) succeeding the conserved aspartate in the region I (strand β3); (ii) Gln^{167} and Asp^{169} in the region V (in the loop3, i.e. domain B); (iii) Trp^{49} in region VI (strand β2); and (iv) Gln^{365} near and Glu^{367} at the end of the region VII (strand β8); and (2) for the neopullulanase subfamily: (i) Ala^{243} (the numbering is from the neopullulanase from *Bacillus stearothermophilus*; Kuriki and Imanaka, 1989) succeeding the conserved aspartate in the region I (strand β3); (ii) Trp^{325} succeeding the conserved glycine at the beginning and the tetrapeptide Val^{329}-Ala^{330}-Asn^{331}-Glu^{332} at the end of the region II (strand β4); (iii) His^{360} at the end of the region III (strand β5); (iv) Ser^{422} preceding the conserved functional histidine in the region IV (strand β7); (v) Met^{295} and Lys^{297} in the region V (in the loop3, i.e. domain B); (vi) Tyr^{194} in region VI (strand β2); and (vii) Tyr^{457} near and Asp^{459} at the end of the region VII (strand β8).

The most important feature that served as the basis for definition of the two subfamilies was the specific sequence in the fifth conserved sequence region, i.e. QpDln for the oligo-1,6-glucosidase subfamily and MPKln for the neopullulanase subfamily with the 'intermediary' sequence MPDLN that is typical for the so-called intermediary group possessing a mixed enzyme specificity of α-amylase, neopullulanase and cyclomaltodextrinase (Oslancova and Janecek, 2002). It should be pointed out that the intermediary group exhibits more features of the oligo-1,6-glucosidase subfamily (Fig. 3A). The same is true also for both representatives of heteromeric amino acid transporter proteins although it is clear that the 4F2hc antigens do not contain the α-amylase-type domain B (Fort et al., 2007) and thus might represent a protein group more distantly related to enzymes from the α-amylase family (Janecek et al., 1997). With regard to the recently established division of the family GH13 into the subfamilies at CAZy (Stam et al., 2006), the oligo-1,6-glucosidase subfamily covers the subfamilies 16, 17, 23, 29, 30, 31, 34 and 35 (plus 4 and 18 for transferases), whereas the neopullulanase subfamily forms the CAZy subfamily GH13_20, the intermediary group being assigned the number GH13_36.

GH77 AMYLOMALTASES FROM BORRELIAE

The importance of the conserved sequence regions (Fig. 1) as sequence fingerprints in the α-amylase family, i.e. the clan GH-H, can also be illustrated using the family GH77. Solving the three-dimensional structure of the maltosyltransferase from *Thermotoga maritima* (Roukeinikova et al., 2001) confirmed that in this enzyme no one of the two otherwise in the family conserved functional histidines (one located in strand β3 in the region I and the other positioned in strand β7 in the region IV; Fig. 1) is present. The former histidine was found to be also absent in both GH70 and GH77 representatives (Fig. 1) so that it was generally accepted (Janecek, 2002) that there are only four residues conserved absolutely invariantly throughout the α-amylase family. These are the catalytic triad and the arginine in position *i-2* with respect to the catalytic nucleophile located in the strand β4 (Fig. 1).

This well-established opinion was first shaken by Machovic and Janecek (2003) who pointed out that the putative GH77 amylomaltase from *Borrelia burgdorferi* may contain the otherwise invariant above-mentioned arginine substituted by a lysine. That observation was based on searching the completely sequenced genome of the Lyme disease spirochete *B. burgdorferi* (Fraser et al., 1997). This was recently confirmed by Godany et al. (2008) who not only verified the sequence, but also documented that the natural 'mutant' protein exhibits a typical amylomaltase enzyme activity, i.e. the enzyme catalyzes both the hydrolysis of maltooligosaccharides and formation of their transglycosylation products (Terada et al., 1999; Kaper et al., 2005; Park et al., 2007).

It is worth mentioning that the GH77 amylomaltase from *B. burgdorferi* contains more exclusive sequence features in addition to the unique natural Arg-to-Lys substitution (Godany et al., 2008). It seems, moreover, that this phenomenon could be extended to all GH77 members from borreliae, although there are perhaps two GH77 borrelial groups (Fig. 4A): one with unique features including the Arg-to-Lys mutation (*B. burgdorferi*, *B. afzelii*, *B. garinii* and *B. valaisiana*) and the other with an intermediary character keeping the original arginine (*B. hermsii* and *B. turicatae*).

The specific sequence features of the unique borrelial GH77 amylomaltases that are present in the conserved sequence regions (Fig. 4A) are as follows (Machovic and Janecek, 2003): (i) Asn^{228} (the numbering is from the GH77 amylomaltase from *B. burgdorferi*; Godany et al., 2008) instead of otherwise conserved aspartate at the beginning of the region I (strand β3); (ii) Lys306 instead of otherwise totally invariant arginine two residues before the catalytic nucleophile (Asp^{308}) in the region II (strand β4); (iii) Trp^{353}-Val^{354} preceding the catalytic proton donor (Glu^{355}) and Phe^{357}-Gln^{358} (or Phe-Glu) instead of Leu-Gly at the end of the region III (strand β5); (iv) a glycine (Gly^{407}) or serine or asparagine replacing the conserved functional histidine preceding the transition-state stabilizer (Asp^{408}) in the region IV (strand β7); and (v) Ser^{50} instead of glycine at the beginning and Phe^{57}-Ala^{58} instead of Leu-Pro at the end of the region VI (strand β2). Interestingly, both borrelial representatives with the intermediary features (*B. hermsii* and *B. turicatae*) seem to have the third residue of the catalytic triad, the transition-state stabilizer (aspartate from the strand β7), replaced by a glutamate (Fig. 4A). The eventual functional effects of this feature have, however, to be verified experimentally.

The above-mentioned sequence characteristics of all known borrelial GH77 amylomaltases and/or amylomaltase-like proteins are clearly reflected also in the evolutionary tree (Fig. 4B) that shows also the representatives of the other two α-amylase GH families GH13 and GH70. Each GH family retains its own independence, the unique GH77 members from borreliae being found on a quite long distance from the rest of their family and also from the borrelial counterparts possessing the intermediary signatures (Fig. 4B).

CONCLUSIONS

The aim of the present article was to summarize the knowledge on the conserved sequence regions of the α-amylase family and to show their power as the sequence fingerprints to distinguish between the individual enzyme specificities and taxonomic sources that are often very closely related to each other. The GH13 α-amylases from plants and Archaea, the GH13 oligo-1,6-glucosidase and neopullulanase subfamilies and

(A)

Source	VI β2	I β3	II β4 *	III β5 *	IV β7 *
GH77 borreliae:					
Borrelia afzelii	49_SQSYWQMFA	227_NVPLFI	303_VIKIDHFRG	350_KIWVEDFE	402_YTGISD
Borrelia burgdorferi	50_SQSYWQMFA	228_NVPFFI	304_IIKIDHFRG	351_KIWVEDFQ	403_YTGSGD
Borrelia garinii	49_SQSYWQMFA	227_NLPLFI	303_VIKIDHFRG	350_KIWVEDFE	402_YTGIGD
Borrelia hermsii	38_SQSYWQILP	216_DIPIFV	292_IIRIDYFRG	339_EIWVEDLV	391_YTGSHE
Borrelia turicatae	38_SQGYWQILP	216_DIPIFV	292_IVRIDYFRG	339_EIWVEDLV	391_YTGTHE
Borrelia valaisiana	49_SQSYWQMFA	227_NVPLFI	303_VIKIDHFRG	350_KIWVEDFE	402_YTGIND
Rest of GH77:					
Aquifex aeolicus	37_GFSLWQVLP	206_DLPMYP	282_FLRLDHFRG	329_PFIAEDLG	380_YTSTHD
Clostridium butyricum	30_GQKYWQILP	204_DIPIYI	281_ILRIDHFRG	328_EIIAEDLG	380_YTGTHD
Escherichia coli	178_GGSFIGLNP	374_DLAVGV	444_ALRIDHVMS	492_MVIGEDLG	542_VAATHD
Thermus aquaticus	40_GGRYWQVLP	213_DMPIFV	289_LVRIDHFRG	336_PVLAEDLG	390_YTGTHD
Thermus thermophilus	40_GGRYWQVLP	213_DMPIFV	289_LVRIDHFRG	336_PVLAEDLG	390_YTGTHD
Pyrobaculum aerophilum	38_EQTYWQTLP	191_DLPFYP	267_YVRLDHFRG	314_RLIAEDLG	366_YTGTHD
Chlamydomonas reinhardtii	119_GMQCWQLLP	297_DMPIYV	373_ECRIDHFRG	420_PILAEDLG	472_YPGTHD
Solanum tuberosum DE1	61_GCSLWQVLP	241_DMPIYV	317_EFRIDHFRG	364_NIIAEDLG	416_YTGTHD
Solanum tuberosum DE2	294_GFHLVQLLP	478_DLPIGV	548_AYRIDHILG	743_LACGEDLG	794_APSCHD
Triticum aestivum	111_GCTLWQVLP	291_DMPIYV	367_EFRIDHFRG	414_NIIAEDLG	466_YTGTHD
GH13 representatives:					
Alpha-amylase	56_GFTAIWITP	117_DVVANH	202_GLRIDTVKH	226_YCIGEVLD	292_FVENHD
Cyclodextrin glucanotransferase	70_GVTALWISQ	135_DFAPNH	225_GIRVDAVKH	253_FTFGEWFL	323_FIDNHD
Oligo-1,6-glucosidase	44_GIDVIWLSP	98_DLVVNH	195_GFRMDVINF	251_MTVGEMPG	324_YWNNHD
Neopullulanase	189_GITGIYLTP	242_DAVFNH	324_GWRLDVANE	353_YILGEIWH	419_LLGSHD
Isomaltulose synthase	86_GIDAIWINP	140_DVVINH	237_GMRFDTVAT	291_ATAGEIFG	364_FLDNHD
GH70 members:					
Glucansucrase	828_GITDFEMAP	894_DWVPDQ	411_SIRVDAVDN	429_VSIVEAWS	521_FARAHD
Alternansucrase	1092_GITSFELAP	1169_DWVPDQ	631_GIRVDAVDN	669_LSILEDWN	762_FVRAHD

Figure 4 Sequence fingerprints of GH77 amylomaltases from borreliae and other taxonomic sources as well as of selected remaining representatives of the α-amylase family (A) and the corresponding evolutionary tree (B)

The sequence features characteristic of the unique borrelial amylomaltases and the rest of GH77 members are highlighted in black-and-white inversion and grey background, respectively. The catalytic triad is signified by asterisks. The tree was calculated using the program CLUSTAL-W (Thompson et al., 1994) and displayed with the program Tree-View (Page, 1996). Adapted from Godany et al. (2008).

(B)

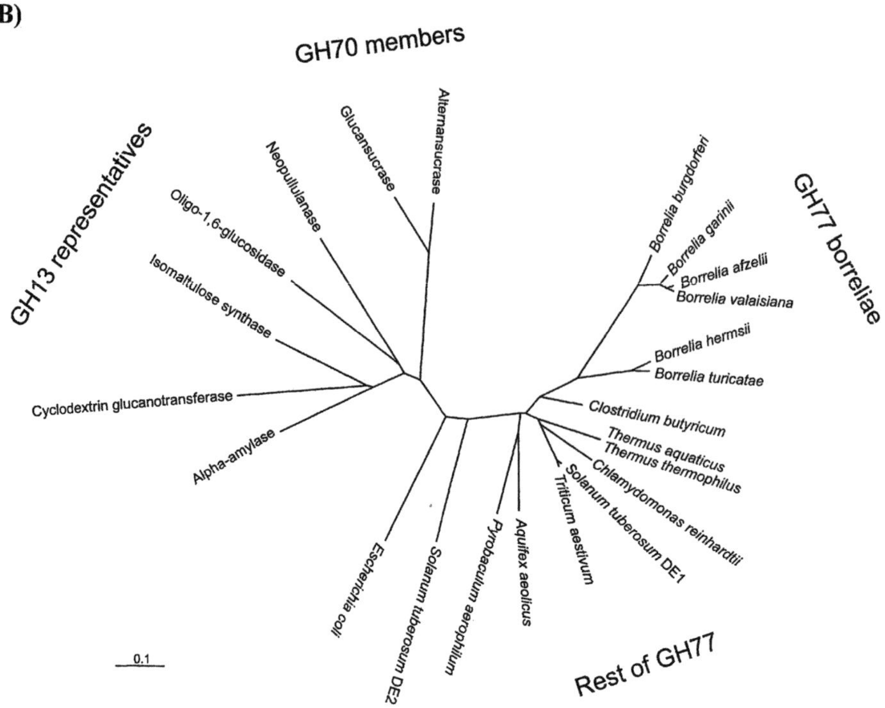

Figure 4 Continued

the GH77 amylomaltases from borreliae were used as specific examples illustrating the usability of the conserved sequence regions in studying the structure/function and evolution of the α-amylase family. The impact of the conserved sequence regions goes beyond the family itself since it has already been indicated (Janecek et al., 2007) that they can be used also in revealing the remote homologies between the α-amylase clan GH-H and family GH31.

ACKNOWLEDGEMENTS

This study was supported by the grant No. 2/0114/08 from the Slovak grant agency VEGA and the project AV-4/2023/08 from the Ministry of Education of the Slovak Republic.

REFERENCES

Bahl H, Burchhardt G, Spreinat A, Haeckel K, Wienecke A, Schmidt B, and Antranikian G (1991), 'α-Amylase of *Clostridium thermosulfurogenes* EM1: nucleotide sequence of the gene, processing of the enzyme, and comparison of other α-amylases', *Appl Environ Microbiol*, 57, 1554-1559.

Coutinho P M and Henrissat B (1999), 'Carbohydrate-active enzymes: an integrated database approach', in Gilbert H J, Davies G, Henrissat B and Svensson B (eds), *Recent*

Advances in Carbohydrate Bioengineering, Cambridge, The Royal Society of Chemistry, 3-12; http://www.cazy.org/.

Da Lage J L, Feller G, and Janecek S (2004), 'Horizontal gene transfer from Eukarya to bacteria and domain shuffling: the α-amylase model', *Cell Mol Life Sci*, 61, 97-109.

Dijkstra B W, Vujicic-Zagar A, Pijning T, Kralj S, and Dijkhuizen L (2007), 'Structural and functional characterization of a glucansucrase (ΔN-GTF180) from *Lactobacillus reuteri* 180, a GH family 70 enzyme', in Janecek S (ed), *3rd Symposium on the Alpha-Amylase Family*, Programme and Abstracts, Sep 23-27, 2007, Bratislava, Asco Arts & Sicence, 15.

Fort J, de la Ballina L R, Burghardt H E, Ferrer-Costa C, Turnay J, Ferrer-Orta C, Uson I, Zorzano A, Fernandez-Recio J, Orozco M, Lizarbe M A, Fita I, and Palacin M (2007), 'The structure of human 4F2hc ectodomain provides a model for homodimerization and electrostatic interaction with plasma membrane', *J Biol Chem*, 282, 31444-31452.

Fraser C M, Casjens S, Huang W M, Sutton G G, Clayton R, Lathigra R, White O, Ketchum K A, Dodson R, Hickey E K, Gwinn M, Dougherty B, Tomb J F, Fleischmann R D, Richardson D, Peterson J, Kerlavage A R, Quackenbush J, Salzberg S, Hanson M, van Vugt R, Palmer N, Adams M D, Gocayne J, Weidman J, Utterback T, Watthey L, McDonald L, Artiach P, Bowman C, Garland S, Fuji C, Cotton M D, Horst K, Roberts K, Hatch B, Smith H O, and Venter JC (1997), 'Genomic sequence of a Lyme disease spirochaete, *Borrelia burgdorferi*', *Nature*, 390, 580-586.

Friedberg F (1983), 'On the primary structure of amylases', *FEBS Lett*, 152, 139-140.

Godany A, Vidova B, and Janecek S (2008), 'The unique glycoside hydrolase family 77 amylomaltase from *Borrelia burgdorferi* with only catalytic triad conserved', *FEMS Microbiol Lett*, 284, 84-91.

Henrissat B (1991), 'A classification of glycosyl hydrolases based on amino acid sequence similarities', *Biochem J*, 280, 309-316.

Hutcheon G W, Vasisht N, and Bolhuis A (2005), 'Characterisation of a highly stable α-amylase from the halophilic archaeon *Haloarcula hispanica*', *Extremophiles*, 9, 487-495.

Itkor P, Tsukagoshi N, and Udaka S (1990), 'Nucleotide sequence of the raw-starch-digesting amylase gene from *Bacillus* sp. B1018 and its strong homology to the cyclodextrin glucanotransferase genes', *Biochem Biophys Res Commun*, 166, 630-636.

Janecek S (1992), 'New conserved amino acid region of α-amylases in the third loop of their $(\beta/\alpha)_8$-barrel domains', *Biochem J*, 288, 1069-1070.

Janecek S (1994a), 'Sequence similarities and evolutionary relationships of microbial, plant and animal α-amylases', *Eur J Biochem*, 224, 519-524.

Janecek S (1994b), 'Parallel β/α-barrels of α-amylase, cyclodextrin glycosyltransferase and oligo-1,6-glucosidase versus the barrel of β-amylase: evolutionary distance is a reflection of unrelated sequences', *FEBS Lett*, 353, 119-123.

Janecek S (1995), 'Close evolutionary relatedness among functionally distantly related members of the $(\alpha/\beta)_8$-barrel glycosyl hydrolases suggested by the similarity of their fifth conserved sequence region', *FEBS Lett*, 377, 6-8.

Janecek S (2000), 'Structural features and evolutionary relationships in the α-amylase family,' in Ohnishi M, Hayashi T, Ishijima S and Kuriki T (eds), *Glycoenzymes*, Tokyo, Basel, Japan Scientific Societies Press, Karger Publishers, 19-54.

Janecek S (2002), 'How many conserved sequence regions are there in the α-amylase family?', *Biologia*, 57 (Suppl 11), 29-41.

Janecek S, Leveque E, Belarbi A, and Haye B (1999), 'Close evolutionary relatedness of α-amylases from Archaea and plants', *J Mol Evol*, 48, 421-426.

Janecek S, MacGregor E A, and Svensson B (1995), 'Characteristic differences in the primary structure allow discrimination of cyclodextrin glucanotransferases from α-amylases', *Biochem J*, 305, 685-686.

Janecek S, Svensson B, and Henrissat B (1997), 'Domain evolution in the α-amylase family', *J Mol Evol*, 45, 322-331.

Janecek S, Svensson B, and MacGregor E A (2007), 'A remote but significant sequence homology between glycoside hydrolase clan GH-H and family GH31', *FEBS Lett*, 581, 1261-1268.

Jespersen H M, MacGregor E A, Henrissat B, Sierks M R, and Svensson B (1993), 'Starch- and glycogen-debranching and branching enzymes: prediction of structural features of the catalytic $(\beta/\alpha)_8$-barrel domain and evolutionary relationship to other amylolytic enzymes', *J Protein Chem*, 12, 791-805.

Jespersen H M, MacGregor E A, Sierks M R, and Svensson B (1991), 'Comparison of the domain-level organization of starch hydrolases and related enzymes', *Biochem J*, 280, 51-55.

Jones R A, Jermiin L S, Easteal S, Patel B K, and Beacham I R (1999), 'Amylase and 16S rRNA genes from a hyperthermophilic archaebacterium', *J Appl Microbiol*, 86, 93-107.

Kadziola A, Sogaard M, Svensson B, and Haser R (1998), 'Molecular structure of a barley α-amylase-inhibitor complex: implications for starch binding and catalysis', *J Mol Biol*, 278, 205-217.

Kaper T, Talik B, Ettema T J, Bos H, van der Maarel M J, and Dijkhuizen L (2005), 'Amylomaltase of *Pyrobaculum aerophilum* IM2 produces thermoreversible starch gels', *Appl Environ Microbiol*, 71, 5098-5106.

Knegtel R M, Wind R D, Rozeboom H J, Kalk K H, Buitelaar R M, Dijkhuizen L, and Dijkstra B W (1996), 'Crystal structure at 2.3 Å resolution and revised nucleotide sequence of the thermostable cylodextrin glycosyltransferase from *Themoanaerobacterium thermosulfurigenes* EM1', *J Mol Biol*, 256, 611-622.

Kobayashi T, Kanai H, Aono R, Horikoshi K, and Kudo T (1994), 'Cloning, expression, and nucleotide sequence of the α-amylase gene from the haloalkaliphilic archaeon *Natronococcus* sp. strain Ah-36', *J Bacteriol*, 176, 5131-5134.

Kobayashi T, Kanai H, Hayashi T, Akiba T, Akaboshi R, and Horikoshi K (1992), 'Haloalkaliphilic maltotriose-forming α-amylase from the archaebacterium *Natronococcus* sp. strain Ah-36', *J Bacteriol*, 174, 3439-3444.

Kuriki T and Imanaka T (1989), 'Nucleotide sequence of the neopullulanase gene from *Bacillus stearothermophilus*', *J Gen Microbiol*, 135, 1521-1528.

Kuriki T and Imanaka T (1999), 'The concept of the α-amylase family: structural similarity and common catalytic mechanism', *J Biosci Bioeng*, 87, 5575-5565.

Kuriki T, Takata H, Yanase M, Ohdan K, Fujii K, Terada Y, Takaha T, Hondoh H, Matsuura Y, and Imanaka T (2006), 'The concept of the α-amylase family: a rational tool for interconverting glucanohydrolases/glucanotransferases, and their specificities', *J Appl Glycosci*, 53, 155-161.

Lee H S, Kim M S, Cho H S, Kim J I, Kim T J, Choi J H, Park C, Lee H S, Oh B H, and Park K H (2002), 'Cyclomaltodextrinase, neopullulanase, and maltogenic amylase are nearly indistinguishable from each other', *J Biol Chem*, 277, 21891-21897.

Leveque E, Haye B, and Belarbi A (2000), 'Cloning and expression of an α-amylase encoding gene from the hyperthermophilic archaebacterium *Thermococcus hydrothermalis* and biochemical characterisation of the recombinant enzyme. *FEMS Microbiol Lett*, 186, 67-71.

Linden A, Mayans O, Meyer-Klaucke W, Antranikian G, and Wilmanns M (2003), 'Differential regulation of a hyperthermophilic α-amylase with a novel (Ca,Zn) two-metal center by zinc', *J Biol Chem*, 278, 9875-9884.

MacGregor E A (1988), 'α-Amylase structure and activity', *J Protein Chem*, 7, 399-415.

MacGregor E A, Janecek S, and Svensson B (2001), 'Relationship of sequence and structure to specificity in the α-amylase family of enzymes', *Biochim Biophys Acta*, 1546, 1-20.

MacGregor E A, Jespersen H M, and Svensson B (1996), 'A circularly permuted α-amylase-type α/β-barrel structure in glucan-synthesizing glucosyltransferases', *FEBS Lett*, 378, 263-266.

MacGregor E A and Svensson B (1989), 'A supersecondary structure predicted to be common to several α-1,4-D-glucan-cleaving enzymes', *Biochem J*, 259, 145-152.

Machovic M and Janecek S (2003), 'The invariant residues in the α-amylase family: just the catalytic triad', *Biologia*, 58, 1127-1132.

Matsuura Y, Kusunoki M, Harada W, and Kakudo M (1984), 'Structure and possible catalytic residues of Taka-amylase A', *J Biochem*, 95, 697-702.

Nakajima R, Imanaka T, and Aiba S (1986), Comparison of amino acid sequences of eleven different α-amylases', *Appl Microbiol Biotechnol*, 23, 355-360.

Nitschke L, Heeger K, Bender H, and Schulz G E (1990), 'Molecular cloning, nucleotide sequence and expression in *Escherichia coli* of the β-cyclodextrin glycosyltransferase gene from *Bacillus circulans* strain no. 8', *Appl Microbiol Biotechnol*, 33, 542-546.

Oh B H (2003), 'The same group of enzymes with different names: cyclomaltodextrinases, neopullulanases, and maltogenic amylases', *Biologia*, 58, 299-305.

Oslancova A and Janecek S (2002), 'Oligo-1,6-glucosidase and neopullulanase enzyme subfamilies from the α-amylase family defined by the fifth conserved sequence region', *Cell Mol Life Sci*, 59, 1945-1959.

Page R D (1996), 'TreeView: an application to display phylogenetic trees on personal computers', *Comput Applic Biosci*, 12, 357-358.

Park J H, Kim H J, Kim Y H, Cha H, Kim Y W, Kim T J, Kim Y R, and Park K H (2007), 'The action mode of *Thermus aquaticus* YT-1 4-α-glucanotransferase and its chimeric enzymes introduced with starch-binding domain on amylose and amylopectin', *Carbohydr Polym*, 67, 164-173.

Park K H, Kim T J, Cheong T K, Kim J W, Oh B H, and Svensson B (2000), 'Structure, specificity and function of cyclomaltodextrinase, a multispecific enzyme of the α-amylase family', *Biochim Biophys Acta*, 1478, 165-185.

Przylas I, Tomoo K, Terada Y, Takaha T, Fujii K, Saenger W, and Straeter N (2000), 'Crystal structure of amylomaltase from *Thermus aquaticus*, a glycosyltransferase catalysing the production of large cyclic glucans', *J Mol Biol*, 296, 873-886.

Raha M, Kawagishi I, Mueller V, Kihara M, and Macnab R M (1992), '*Escherichia coli* produces a cytoplasmic α-amylase, *AmyA*', *J Bacteriol*, 174, 6644-6652.

Rogers J C (1985), 'Conserved amino acid sequence domains in α-amylases from plants, mammals, and bacteria', *Biochem Biophys Res Commun*, 128, 470-476.

Roujeinikova A, Raasch C, Burke J, Baker P J, Liebl W, and Rice D W (2001), 'The crystal structure of *Thermotoga maritima* maltosyltransferase and its implications for the molecular basis of the novel transfer specificity', *J Mol Biol*, 312, 119-131.

Stam M R, Danchin E G, Rancurel C, Coutinho P M, and Henrissat B (2006), 'Dividing the large glycoside hydrolase family 13 into subfamilies: towards improved functional annotations of α-amylase-related proteins', *Protein Eng Des Sel*, 19, 555-562.

Svensson B (1988), Regional distant sequence homology between amylases, α-glucosidases and transglucanosylases', *FEBS Lett*, 230, 72-76.

Svensson B (1994), 'Protein engineering in the α-amylase family: catalytic mechanism, substrate specificity, and stability', *Plant Mol Biol*, 25, 141-157.

Takata H, Kuriki T, Okada S, Takesada Y, Iizuka M, Minamiura N, and Imanaka T (1992), 'Action of neopullulanase. Neopullulanase catalyzes both hydrolysis and transglycosylation at α-(1->4)- and alpha-(1->6)-glucosidic linkages', *J Biol Chem*, 267, 18447-18452.

Terada Y, Fujii K, Takaha T, and Okada S (1999), '*Thermus aquaticus* ATCC 33923 amylomaltase gene cloning and expression and enzyme characterization: production of cycloamylose', *Appl Environ Microbiol*, 65, 910-915.

Thompson J D, Higgins D G, and Gibson T J (1994), 'CLUSTAL W: improving the sensitivity of progressive multiple sequence alignment through sequence weighting, position-specific gap penalties and weight matrix choice', *Nucleic Acids Res*, 22, 4673-4680.

Toda H, Kondo K, and Narita K (1982), 'The complete amino acid sequence of Taka-amylase A', *Proc Japan Acad*, B58, 208-212.

Uitdehaag J C, Mosi R, Kalk K H, van der Veen B A, Dijkhuizen L, Withers S G, and Dijkstra B W (1999), 'X-ray structures along the reaction pathway of cyclodextrin glycosyltransferase elucidate catalysis in the α-amylase family', *Nature Struct Biol*, 6, 432-436.

van der Kaaij R M, Janecek S, van der Maarel M J, and Dijkhuizen L (2007), 'Phylogenetic and biochemical characterization of a novel cluster of intracellular fungal α-amylase enzymes', *Microbiology*, 153, 4003-4015.

Watanabe K, Kitamura K, Iha H, and Suzuki Y (1990), 'Primary structure of the oligo-1,6-glucosidase of *Bacillus cereus* ATCC7064 deduced from the nucleotide sequence of the cloned gene', *Eur J Biochem*, 192, 609-620.

Wind R D, Liebl W, Buitelaar R M, Penninga D, Spreinat A, Dijkhuizen L, and Bahl H (1995), 'Cyclodextrin formation by the thermostable α-amylase of *Thermoanaerobacterium thermosulfurigenes* EM1 and reclassification of the enzyme as a cyclodextrin glycosyltransferase', *Appl Environ Microbiol*, 61, 1257-1265.

Yuuki T, Nomura T, Tezuka H, Tsuboi A, Yamagata H, Tsukagoshi N, and Udaka S (1985), 'Complete nucleotide sequence of a gene coding for heat- and pH-stable α-amylase of *Bacillus licheniformis*: comparison of the amino acid sequences of three bacterial liquefying α-amylases deduced from the DNA sequences', *J Biochem*, 98, 1147-1156.

MOLECULAR MECHANISM OF α-GLUCOSIDASE

Masayuki Okuyama, Haruhide Mori, Hironori Hondoh, Hiroyuki Nakai, Wataru Saburi, Min-Sung Kang, Young-Min Kim, Mamoru Nishimoto, Jintanart Wongchawalit, Takeshi Yamamoto, Mee Son, Jin-Ha Lee, San San Mar, Kenji Fukuda, Seiya Chiba, and Atsuo Kimura

ABSTRACT

α-Glucosidase (EC 3.2.1.20), an exo-glycosylase to hydrolyze α-glucosidic linkage, is characterized by the variety of substrate specificity. Enzyme also catalyzes the transglucosylation, on which industrial interests focus due to the production of valuable glucooligosaccharides. α-Glucosidase is a physiologically important enzyme in most of organisms (microorganisms, insects, plants and animals including human). Therefore, there are many types of α-glucosidases to display unique functions, in which we are interested. This report describes the recently analyzed unique functions of α-glucosidases by mainly focusing on honeybee α-glucosidase isoenzymes, dextran glucosidase, multiple forms of rice α-glucosidases, and *Escherichia coli* α-xylosidase.

Key words: α-Glucosidase; structure; function; reaction mechanism; physiological role

INTRODUCTION

α-Glucosidase (EC 3.2.1.20) catalyzes a hydrolytic reaction of α-glucosidic linkage in the nonreducing end of substrate and releases α-glucose (Kimura, 2000; Nakai et al., 2005). Enzyme is characterized by the wide substrate specificity that exhibits the hydrolytic activity to α-glucobioses (e.c. maltose), oligosaccharides (e.c. maltotriose), heterogeneous substrate [e.c. sucrose and *p*-nitrophenyl α-glucoside (PNPG)], and polymer substrates (e.c. starch granules and glycogen). In the high concentration of substrate, the enzyme also displays a transferring reaction (transglucosylation) to form glucooligosaccharides. Commercially available oligosaccharides, comprising isomaltosyl structure or nigerosyl structure, are produced by the α-glucosidase-catalyzed transferring reactions (Nakakuki, 2005).

Deficiency of α-glucosidase activity in animal or plant tissue results in the serious metabolic problem in digestion of glycogen or starch granules (row starch), respectively. It is known that there are two types of glycogen degradation metabolisms in mammalian tissues. One is a combination of glycogen phosphorylase (EC 2.4.1.1) and glycogen debranching enzyme (EC 2.4.1.25; EC 3.2.1.33) to form glucose-1-phosphate from glycogen in liver and skeletal muscle. The other is the direct hydrolysis of lysosomal glycogen by acid α-glucosidase having an optimum pH at acidic range. Produced glucose moves to cytoplasm and metabolized. Less activity of acid α-glucosidase causes the glycogen accumulation in lysosome, which is the mechanism of serious glycogen storage disease (Pombe's disease). In plant tissue, a role of α-glucosidase is also important, since an α-glucosidase inhibitor strongly reduces the digestion of starch granules at germination of wheat seeds, resulting in the accumulation of maltose (Konishi et al., 1994).

α-Glucosidases are distributed in many organisms and show the diverse substrate specificities. On the basis of difference in substrate specificity, Chiba (Chiba, 1980, 1988, 1997) classified the enzymes into three groups: Type-I, -II, and -III. Type-Iα-glucosidase [enzymes from bacteria, yeast (*Saccharomyces cerevisiae*), and insects] recognizes the "α-glucosyl-structure" of substrate molecule, since this enzyme shows activity toward the heterogeneous substrates much more than homogeneous substrates (i.e. maltooligosaccharides) and polymer substrates. α-Glucosidase Type-II (enzymes from mold) hydrolyzes the homogeneous substrate more rapidly than heterogeneous substrate and polymer substrate by recognizing the "maltosyl-structure". Type-III α-glucosidase (enzymes from plant and animal), which has no activity to heterogeneous substrate, displays high hydrolytic activity to the homogeneous and polymer substrates, so that Type-III enzyme recognizes either "maltosyl-structure" or "polymer-structure". Structure-based classification indicates that Type-I enzyme belongs to glycoside hydrolase (GH) family 13 and both of Type-II and Type-IIIenzymes belong to of GH family 31 (Henrissat, 1991).

Above-described diversity of α-glucosidase suggests that there are many enzymes displaying unique functions, important in each organism. This review summarizes the recent studies of α-glucosidases, thereby allowing us to understand their interesting functions from the viewpoint of molecular level and physiological level.

GH FAMILY 13 α-GLUCOSIDASES

α-Glucosidase isoenzyms from honey bees

There are three kinds of α-glucosidases (HBGase I, II, and III) in the imagoes of European honeybees (*Apis mellifera*), which have the different substrate specificities (Takewaki et al., 1980, 1993; Kimura et al., 1990; Nishimoto et al., 2001). HBGase I, II, and III are monomeric enzymes, while HBGase I and II exhibit allosteric properties. HBGase III follows the typical Michaelis-Menten type hydrolytic reaction, and shows no cooperativity (Nishimoto et al., 2001). HBGase I is an interesting enzyme to display a negative allosteric cooperativity in reactions of the maltose, sucrose, and PNPG. In the

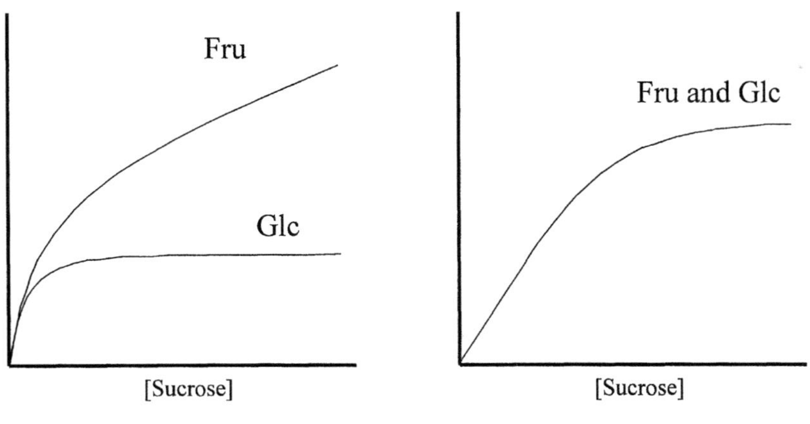

Figure 1 *s*-*v* curves for reaction of HBGase I (left) and HBGase III (right) with sucrose

Horizontal axis indicates the sucrose concentration. Vertical axis indicates the velocities of glucose and fructose released from sucrose. In HBGase I (left panel), the difference of velocities between glucose and fructose is the transglucosylation rate, so that the oligosaccharide production is activated at high concentration of sucrose. HBGase III (right panel) displays a typical Michaelis-Menten-type reaction.

Table 1 Kinetic parameters for hydrolysis of sucrose by HBGases and honey α-glucosidase

Enzyme	K_m (mM)	k_0 (s^{-1})	k_0/K_m (s^{-1} mM^{-1})
HBGase I	4.2	43.0	10.2
HBGase II	6.7	22.0	3.28
HBGase III	30	195	6.5
Honey α-glucosidase	30	160	5.3

high concentration of those substrates, the reaction velocity of HBGase I is activated and their *s*-*v* curves (*s*, substrate concentration; *v*, reaction velocity) do not reach plateau (Kimura et al., 1990) (Fig. 1). Enzyme catalyzes transglycosylation, so that the oligosaccharide formation is activated. A similar enzyme is also found in the eastern honeybees (*Apis cerana*) (Wongchawalit et al., 2006). Recently, it was found that HBGase I produced isomaltose from glucose through a combination of condensation and transglucosylation (Son et al., 2003).

Honeybees gather nectar and storage in their nest to make up honey. A main carbohydrate component in nectar is sucrose, which is converted to glucose and fructose (main components of honey) through the α-glucosidase-catalyzed hydrolysis of sucrose. Concentration of sucrose in nectar is 1.5 M to 2 M, so that honey enzyme should have the ability to hydrolyze sucrose of extremely high concentration, since an ordinary α-glucosidase decreases the activity in such a high substrate concentration. The first candidate was considered to be an HBGase I exhibiting allosteric activation at high sucrose concentration (Fig. 1). But our study clarifies that HBGase I does not involve in the honey formation. We purify an α-glucosidase from honey and find that its property is identical to that of HBGase III (Nishimoto et al., 2001; Kubota et al., 2004). The antiserum prepared from honey enzyme only reacts with HBGase III. Further immunological approach using specific antisera against each α-glucosidase elucidates the localization of three α-glucosidases in honeybee: HBGase I is in the ventriculus (corresponding to human stomach and intestine); HBGase II is in the ventriculus and the haemolymph (insect blood); and HBGase III is in the hypopharyngeal gland (a kind of salivary gland) (Kubota et al., 2004). Therefore, HBGase III is responsible for honey formation by secretion from hypopharyngeal gland to nectar. As shown in Table 1, both K_m and k_0 values of HBGase III for sucrose hydrolysis are 4 to 9-fold larger than those of HBGase I and II, allowing HBGase III to attack sucrose of high concentration efficiently.

Three HBGases are characterized by differences in their substrate specificities and immunological response, indicating that HBGase I, II, and III are isoenzymes with different amino-acid sequences. Three cDNAs encoding HBGase I, II, and III are cloned and their genes in chromosome are also analyzed (Nishimoto et al., 2007). Polypeptides are comprised of 588, 580, and 567 amino acid residues for HBGase I, II, and III, respectively, with high homologies of 38–44% identity. The chromosomal gene of HBGase I is composed of 8 exons and 7 introns; that of HBGase II includes no intron; and that of HBGase III is composed of 9 exons and 8 introns with extremely long intron 1 of 34,514 bp. HBGase II and III cDNAs are expressed in *Pichia pastoris*

(Nishimoto et al., 2007), enabling us to attempt the molecular-based analysis of those unique HBGases.

Dextran glucosidase from *Streptococcus mutans*

Dextran glucosidase from *Streptococcus mutans* (DGase), a GH family 13 enzyme, specifically hydrolyzes a non-reducing terminal α-1,6-glucosidic linkage of isomaltooligosaccharides and dextran to release α-glucose (Saburi et al., 2006). DGase has been originally classified to EC 3.2.1.70 enzymes together with the inverting exo-glycosylase of glucodextranase, which forms β-glucose from isomaltotriose (Takayanagi et al., 1987). DGase is a retaining enzyme to release α-glucose from substrate, so that the enzyme should belong to the α-glucosidase of EC 3.2.1.20 (Saburi et al., 2006). Similar α-glucosidase specific for α-1,6-glucosidic linkage is found in yeast (Matsusaka et al., 1977).

There is the high amino-acid sequence similarity between DGase and oligo-1,6-glucosidase (O16Gase; EC 3.2.1.10) more than other GH family 13 enzymes. But DGase and O16Gase differently recognize the substrate chain length. DGase efficiently hydrolyzes the substrates having short-chain and long-chain, although O16Gase hydrolyzes the short-chain substrate more than the long-chain substrate. Structural elements influencing the difference in substrate chain length specificity are analyzed by comparing the three-dimensional (3D) structures of DGase (Saburi et al., 2007; Hondoh et al., 2008) and O16Gase from *Bacillus cereus* (Watanabe et al., 1997). DGase and O16Gase consist of three domains, A, B, and C, which are commonly found in other GH family 13 enzymes. Two structural elements of DGase, i) Trp238 spanning subsites +1 and +2 and ii) short β→α loop 4, are predicted to be responsible for the high specificity to long-chain substrate. As shown in Table 2, both of Trp238 mutants (Trp238 is replaced by smaller amino acids) and β→α loop 4 mutant (short β→α loop 4 is replaced by the corresponding long loop of O16Gase) show a lower preference (k_0/K_m value) for long-chain substrates, indicating that these structural elements are important for the high activity toward long-chain substrates.

The crystal structure of DGase was determined very recently (Saburi et al., 2007; Hondoh et al., 2008). The enzyme composes of 536 amino acid residues with a molecular size of 62 kDa. The complex structure with the isomaltotriose and the catalytic mutant E236Q is also determined. Both native and complex structures are refined at 2.2 Å resolutions, resulting in a final *R*-factor of 18.5% and 19.2%, respectively [for the whole structures, see a reference (Hondoh et al., 2008)]. This is the first model of natural substrate (isomaltotriose) complex of α-1,6-glucosidic linkage hydrolyzing enzyme. As mentioned above, DGase has a conformation of three domains A, B, and C. Domain A comprises of a $(\beta/\alpha)_8$-barrel containing three catalytic residues: D194, E236, and D313. This $(\beta/\alpha)_8$-barrel are conserved among α-amylase family enzymes. Domain B has two helices and three strands. Domain C consists of a pseudo-antiparallel eight-stranded β-sheet with seven strands and one helix. Domains A and B form an active site pocket, on the bottom of which three catalytic amino acids are located. Isomaltotriose is also found on the bottom of active site pocket and occupies three subsites, -1 to +2. At subsite -1, the arrangement of amino-acid residues is common in the α-amylase family enzymes, so that the subsite +1 may contribute to the specificity to α-1,6-glucosidic linkage. Hydrogen bonds, which are formed by the side-chain atoms of E371 and K275 with O2 and O3 atoms of glucose unit at the subsite +1, appear to stretch the scissile bond at the cleavage site. Non-reducing terminal of substrate is recognized by hydrogen bonds from D60 and R398 to O4 atom of glucose

Table 2. Kinetic parameters of wild type, Trp238-mutated, or short β→α loop 4-having DGase

Substrate [b]	k_0 (s^{-1})	K_m (mM)	k_0/K_m (s^{-1} mM^{-1})	relative k_0/K_m [a] (%)
Wild type				
IG2	483±2	9.35±0.10	51.7	100
IG3	541±5	3.17±0.08	171	331
IG4	499±1	5.58±0.07	89.4	173
IG5	419±4	5.10±0.02	82.2	159
IG6	401±12	7.88±0.47	50.9	98.5
IG7	449±5	12.0±0.2	37.4	72.3
W238A				
IG2	34.5±1.6	15.8±1.3	2.18	100
IG3	33.1±0.6	11.4±0.4	2.90	133
IG4	22.3±0.5	20.9±0.5	1.07	49.1
IG5	22.6±0.4	38.3±0.6	0.590	27.1
IG6	36.0±1.6	76.9±4.8	0.468	21.5
IG7	33.2±2.4	81.7±7.3	0.406	18.6
W238P				
IG2	79.8±2.1	20.5±1.2	3.89	100
IG3	77.5±0.6	10.2±0.3	7.60	195
IG4	52.5±0.5	16.1±0.4	3.26	83.8
IG5	42.4±0.4	24.6±0.5	1.72	44.2
IG6	39.6±0.6	27.1±0.5	1.46	37.5
IG7	56.5±3.1	43.6±3.7	1.30	33.4
W238N				
IG2	8.88±0.21	8.91±0.06	0.997	100
IG3	12.2±0.1	4.96±0.09	2.46	247
IG4	8.96±0.11	7.77±0.21	1.15	115
IG5	9.21±0.10	15.1±0.3	0.610	61.2
IG6	7.80±0.14	19.9±0.7	0.392	39.3
IG7	9.81±0.06	26.9±0.1	0.365	36.6
Short β→α loop 4				
IG2	1.89±0.02	17.8±0.6	0.106	100
IG3	3.92±0.04	4.25±0.07	0.922	870
IG4	3.02±0.06	11.5±0.3	0.263	248
IG5	1.99±0.02	10.1±0.2	0.197	186
IG6	1.24±0.02	17.8±0.6	0.0697	65.8
IG7	2.43±0.05	53.9±1.6	0.0451	42.5

a), Relative value when k_0/K_m of isomaltose was 100.

b), IG2, isomaltose; IG3, isomaltotriose; IG4, isomaltotetraose; IG5, isomaltopentaose; IG6, isomaltohexaose; IG7, isomaltoheptaose.

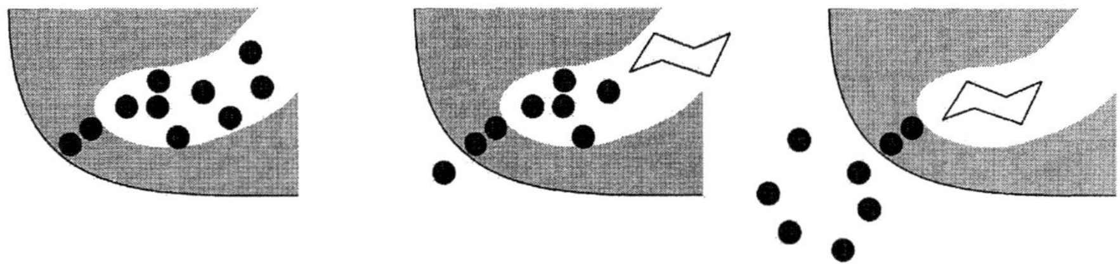

Figure 2 Schematic drawing of water drain

Bound water molecules (black circles) in the active pocket (A) are forced out by the incoming substrate (B). The water molecules are drained out from the active pocket (C).

unit at subsite -1. Similar manner is observed in other exo-type enzymes (Mirza et al., 2001, 2006). The comparison with α-amylase suggest that V195 of DGase prevents the binding of α-1,4 linked substrate by a possible steric hindrance, so that V195 may be essential for the accommodation of α-1,6 linked substrate to the active site. O16Gase, isoamylase and pullulanase conserve the bulky hydrophobic residues in the corresponding position of V195, also supporting the important role of V195. Interestingly, the row of water molecules that connects the inside and the outside of the enzyme molecule is found at the bottom of the active pocket (Fig. 2; Hondoh et al., 2008), meaning the presence of "very small pore" to accept only water molecule (water path). The function of this pore is obscure. But, the same water path is observed in O16Gase, amylose sucrase and sucrose phosphorylase, all of which have a "tight" active pocket, allowing the following prediction (Fig. 2). Bound water molecules stay in the active pocket, prior to entry of substrate. When a substrate molecule enters to the active pocket, the water molecules are drained out from water path. Possible function of water path is the water drain.

GH FAMILY 31 α-GLUCOSIDASES

Novel pathway of starch granule degradation in rice seeds by multiple forms of α-glucosidases.

Plant seeds store the insoluble starch granules, which are the energy source for a germinating stage. At germination, insoluble starch granules are degraded effectively by the combination of such amylolytic enzymes as α-amylase, debranching enzyme, β-amylase, and α-glucosidase. It is long believed that the initial attack by *de novo* synthesized α-amylase is a key event for the degradation, due to its starch-binding ability. Recently, we have shown that rice (*Oryza sativa* L., var Nipponbare) α-glucosidase isozymes (ONGase1, ONGase2, and ONGase3) are also capable of binding to and degrading starch granules directly, indicating the direct liberation of glucose from starch granules by α-glucosidase (Nakai et al., 2007a, 2007b, 2008).

ONGase2 is an important enzyme at germination, since a large amount of ONGase2 activity (0.14 unit per one seed) is observed before germination (i.e. dry seeds) (Nakai et al., 2007a). ONGase1 and ONGase3 are only observed at ripening stage and disappear in dry seeds (very small ONGase1 remains in dry seeds). In dry seeds, there are two isoforms of ONGase2 (ONGase2-I and ONGase2-II), that are generated by post-translational proteolysis at their N-terminal regions, resulting in the formation of two polypeptides of 6 kDa and 88 kDa (Fig. 3). Interestingly, two polypeptides of each isoform tightly bind each other and do not separate in a native enzyme. Two isoforms

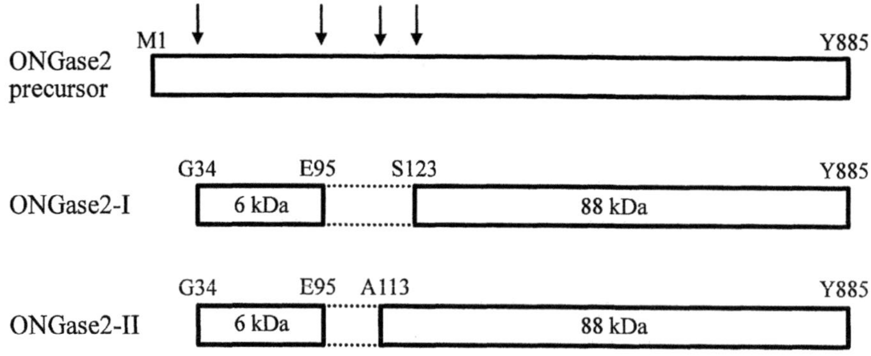

Figure 3 Post-translational proteolysis of ONGase2

ONGase2 precursor protein (M1-Y885) is digested by signal peptidase (a left arrow indicates position) and by post-translational proteolysis (other three arrows indicate positions), generating 6 kDa-polypeptide (G34-E95 for ONGase2-I and ONGase2-II) and 88 kDa-polypeptides (S123-Y885 for ONGase2-I and A113-Y885 for ONGase2-II).

share the same 6 kDa-polypeptides, while ONGase2-II has a 10 amino acids longer 88 kDa-polypeptide at its N-terminus than ONGase2-I does (Fig. 3). Both isoforms show the identical specificity for such small substrates as maltooligosaccharides and α-glucobiose. But the starch binding ability of ONGase2-II is about 2-fold higher than that of ONGase2-I, so that ONGase2-II digests starch granules more efficiently. N-terminal 10 amino acids of 88 kDa-polypeptide of ONGase2-II may contribute to starch binding (Nakai et al., 2007a).

mRNAs of three rice α-glucosidase isoenzymes (ONGase1, ONGase2, and ONGase3) are expressed in ripening seeds and ONGase1 mRNA is only observed in germination seeds at 48 h after imbibition (Nakai et al., 2007a). ONGase2 and ONGase3 mRNAs are generated by alternative splicing of exon 1. To know the function of α-glucosidases coded, the recombinant ONGase1 (rONGase1) and ONGase3 (rONGase3) are overproduced in *Pichia pastoris* (Nakai et al., 2007b). For small saccharides, two recombinant α-glucosidase exhibit the similar substrate specificities to ONGase2-I and ONGase2-II. Obvious difference is the binding and degrading abilities to starch granules: both of rONGase1 and rONGase3 display less abilities than ONGase2 isoforms, in the order of ONGase2-II > ONGase2-I > ONGase1 > ONGase3.

Starch granules digestion by ONGase2-II is the largest among rice α-glucosidase isozymes and isoforms. A main α-glucosidase in dry seeds is ONGase2-II. Therefore, ONGase2-II significantly contributes to the direct hydrolysis of starch granules at germination. This is a novel pathway of storage starch degradation by α-glucosidase in that stage before expression of α-amylase (during about 1.5 day after imbibition).

There is one more rice α-glucosidase homolog mRNA, which is expressed in the ripening and germinating seeds (Nakai et al., 2007a). Generation of this homolog mRNA is not synchronized with α-glucosidase activity, but with α-xylosidase activity at both seeds. Recombinant protein catalyzes the hydrolytic reaction to α-xylosides, in particular to xyloglucan oligosaccharides, and hardly hydrolyzes α-glucosides (Nakai et al., 2007b). Homolog is an α-xylosidase, which may play an important role in cell wall degradation during ripening and germinating stages. A gene encoding α-xylosidase is found for the first time.

α-Glucosidase converted from *Escherichia coli* α-xylosidase by mutation work

Mutation of the structural elements in target enzyme itself leads us to understand their functions directly. This approach was done using GH family 31 α-glucosidase from *Schizosaccharomyces pombe* (Okuyama et al., 2001, 2002, 2005). α-Xylosidase and α-glucosidase share the same 3D structures, so that we can learn more precise substrate recognition of GH family 31 enzymes, if we succeed in the conversion from α-xylosidase to α-glucosidase by mutation work. Recently such a challenge was attempted using *Escherichia coli* α-xylosidase (Okuyama et al., 2006).

E. coli has two GH family 31 α-glucosidase homologs (YihQ and YicI). YihQ is an α-glucosidase with quite small activity (Okuyama et al., 2004). YicI is a strict α-xylosidase, which scarcely splits α-glucosidic linkage (Okuyama et al., 2004; Kitamura et al., 2005), while its amino-acid primary sequence is similar to GH family 31 α-glucosidases. The first mutation study is focused on the 4 amino acids different between YicI and α-glucosidases at subsite -1 (Okuyama et al., 2006). Replacement of those 4 residues (F277Y/W, C307I, W345I/M, and K414W) results in decrease of original hydrolytic velocity and in no gain of α-glucosidase activity. Key factor is a $6CH_2$-OH group of glucose which is not seen in xylose, so that the favorable mutation is an introduction of $6CH_2$-OH-recognizing structure(s) to YicI. YicI is successfully converted to α-glucosidase by two mutations: i) replacement of C307/F308 by I307/D308 and ii) replacement of long loop 1 of $(\beta/\alpha)_8$-barrel by short loop (Table 3) (Okuyama et al., 2006). Possible role of C307/F308 is the fixation of O6 of non-reducing terminal glucose unit. Short loop 1 probably accepts $6CH_2$-OH moiety, due to the reduction of steric hindrance formed by original long loop.

In transxylosylation of YicI, an α-xylosyl unit is transferred to C-6 OH of either glucose, mannose, or allose, indicative of the interesting acceptor specificity recognizing equatorial 4-OH of aldohexopyranose (Kang, 2007, 2008). Among disaccharide transfer products formed, α-D-Xyl*p*-(1→6)-D-Man*p*, α-D-Xyl*p*-(1→6)-D-Fru*f*, and α-D-Xyl*p*-(1→3)-D-Fru*p* are novel saccharides. Both of α-D-Xyl*p*-(1→6)-D-Man*p* and α-D-Xyl*p*-(1→6)-D-Fru*f* display the moderate inhibitory activity to rat intestinal α-glucosidases.

Table 3 Hydrolytic velocities [a] of substrates by wild type and mutated YicI

Substrate [b]	Wild type	Short loop 1	C307I/F308D
Isoprimeverose	20,500	5.45	0
PNPX	18.7	1.97	0
Maltose	0.208	2.0	25.0
Isomaltose	2.77	3.27	21.4
Nigerose	0	0	40.2
Kojibiose	0	0	30.9
PNPG	0.213	29.5	28.2

a), Hydrolytic velocity is expressed as nmol/min/mg of protein.

b), Isoprimeverose (α-D-Xyl*p*-(1→6)-D-Glc*p*) and PNPX (*p*-nitrophenyl α-xyloside) are substrates for α-xylosidase. Others are for α-glucosidase.

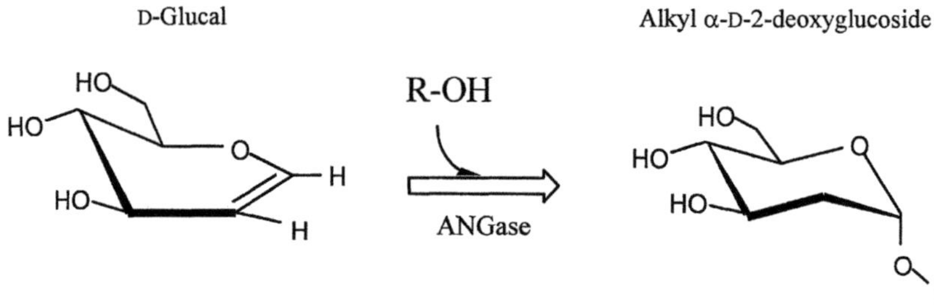

Figure 4 ANGase-catalyzed alkoxy-hydro-addition of D-glucal

Unique oligosaccharide formation catalyzed by GH family 31 α-glucosidases

α-Glucosidase from *Acremonium implicatum* (AIGase) is characterized by the high regioselectivity for the α-1,3-glucosidic linkages at hydrolysis and transglucosylation (Yamamoto et al., 2004). AIGase hydrolyzes nigerose (α-1,3-linked glucobiose) and maltose (α-1,4-linked) rapidly, but slowly kojibiose (α-1,2-linked) and slightly isomaltose (α-1,6-linked). The best substrate is a nigerose. The enzyme has the α-1,3- and α-1,4-glucosyl transfer activities to form oligosaccharides, and no ability to form α-1,2- and α-1,6-glucosidic linkages. From maltose, eight kinds of the disaccharide to tetrasacchaides, comprising of α-1,3- and/or α-1,4-glucosidic bond(s), are synthesized, in which 3^{II}-*O*-α-nigerosyl-maltose and 3^{II}-*O*-α-maltosyl-maltose are novel saccharides. It is known that nigerose and nigerosyl-maltooligosaccharides having α-1,3-glucosidic linkage show the immunopotentiating activity for helper T cell (T_H1)-like immune response (Murosaki et al., 1999, 2002) as well as anticariogenic activity (Imai et al., 1984). AIGase is a remarkable enzyme to produce the useful α-1,3-glucosidic glucooligosaccharides.

Aspergillus niger α-glucosidase (ANGase) (Kita et al., 1991; Kimura et al., 1992) shows a high stability in the high concentration of alkyl alcohols: for instance, the enzyme maintains 80% activity, even after incubation with 70% methanol at 35 °C for 4 h (Kim et al., 2005). Alkyl alcohol-resistant property is applied to the enzymatic alkoxy-hydro-addition of D-glucal (Fig. 4). ANGase displays the efficient regio-selective synthesis of alkyl α-D-2-deoxyglucosides (A2DG) from D-glucal in the presence of alkyl alcohols. About 90% of 500 mM D-glucal is converted to methyl α-D-2-deoxyglucoside in 70% methanol. Fourteen kinds of novel A2DG are produced by ANGase-catalyzed alkoxy-hydro-addition (Kim et al., 2005).

It was mentioned that a starch-hydrolyzing enzyme from *Schwanniomyces occidentalis* was a novel glucoamylase, but is devoid of the conclusive proof that it is glucoamylase. We purify an enzyme (SOGase) having the hydrolytic activity toward soluble starch from *S. occidentalis* (Sato et al., 2005). SOGase produces α-glucose, differing from glucoamylase as β-glucose producing enzyme, indicative of α-glucosidase. The primary amino-acid structure deduced from cloned DNA is highly similar to GH family 31 α-glucosidase. Production of recombinant SOGase is high (Sato et al., 2005), so that molecular approach is currently attempted.

REFERENCES

Chiba S (1980), 'Action and classification of α-glucosidase', *J Jap Soc Starch Sci*, 27, 84-90.

Chiba S (1988), 'α-Glucosidase', in The Amylase Research Society of Japan, *Handbook of Amylase and Related Enzymes*, Pergamon Press, Oxford, 104-105.

Chiba S (1997), 'Molecular mechanism in α-glucosidase and glucoamylase', *Biosci Biotechnol Biochem*, 61, 1233-1239.

Henrissat B (1991), 'A classification of glycosyl hydrolases based on amino-acid sequence similarities', *Biochem J*, 280, 309–316.

Hondoh H, Saburi W, Mori H, Okuyama M, Nakada T, Matsuura Y, and Kimura A (2008), 'Substrate recognition of α-1,6-glucosidic linkage hydrolyzing enzyme, dextran glucosidase from *Streptococcus mutans*', *J Mol Biol*, 378, 911-920.

Imai S, Takeuchi K, Shibata K, Yoshikawa S, Kitahata S, Okada S, Araya S, and Nishizawa T (1984), 'Screening of sugars inhibitory against sucrose-dependent synthesis and adherence of insoluble glucan and acid production by *Streptococcus mutans*', *J Dent Res*, 63, 1293-1297.

Kang M S, Okuyama M, Yaoi K, Mitsuishi Y, Kim YM, Mori H, Kim D, and Kimura A (2007), 'Aglycone specificity of *Escherichia coli* α-xylosidase investigated by transxylosylation', *FEBS J*, 274, 6074-6084.

Kang M S, Okuyama M, Yaoi K, Mitsuishi Y, Kim YM, Mori H, and Kimura A (2008), 'Glycoside hydrolase family 31 *Escherichia coli* α-xylosidase', *Biocatal Biotransfor*, 26, 96-103.

Kim Y M, Okuyama M, Mori H, Nakai H, Saburi W, Chiba S, and Kimura A (2005), 'Enzymatic synthesis of alkyl α-2-deoxyglucosides by alkyl alcohol resistant α-glucosidase from *Aspergillus niger*', *Tetrahedron: Asymmetry*, 16, 403-409.

Kimura A, Takewaki S, Matsui H, Kubota M, and Chiba S (1990), 'Allosteric properties, substrate specificity, and subsite affinities of honeybee α-glucosidase I', *J Biochem*, 107, 762–768.

Kimura A, Takata M, Sakai O, Matsui H, Takai N, Takayanagi T, Nishimura I, Uozumi T, and Chiba S (1992), 'Complete amino acid sequence of crystalline α-glucosidase from *Aspergillus niger*', *Biosci Biotechnol Biochem*, 56, 1368-1370.

Kimura A (2000), 'Molecular anatomy of α-glucosidase', *Trends Glycosci Glycotechnol*, 12, 373-380.

Kita A, Matsui H, Somoto A, Kimura A, Takata M, and Chiba S (1991), 'Substrate specificity and subsite affinities of crystalline α-glucosidase from *Aspergillus niger*', *Agric Biol Chem*, 55, 2327-2335.

Kitamura M, Ose T, Okuyama M, Watanabe H, Yao M, Mori H, Kimura A, and Tanaka I (2005), 'Crystallization and preliminary X-ray analysis of α-xylosidase from *Escherichia coli*', *Acta Cryst F*, 61, 178-179.

Konishi Y, Okamoto A, Takahashi J, Aitani M, and Nakatani N (1994), 'Effects of Bay m 1099, an α-glucosidase inhibitor, on Starch metabolism in Germinating wheat seeds', *Biosci Biotechnol Biochem*, 58, 135-139.

Kubota M, Tsuji M, Nishimoto M, Wongchawalit J, Okuyama M, Mori H, Masui H, Surarit R, Svasti J, Kimura A, and Chiba S (2004), 'Localization of α-glucosidases I, II, and III in organs of European honeybee, *Apis mellifera* L., and origin of α-glucosidase in honey', *Biosci Biotechnol Biochem*, 68, 2346-2352.

Matsusaka K, Chiba S, and Shimomura T (1977), 'Purification and substrate specificity of Brewer's yeast α-glucosidase', *Agric Biol Chem*, 41, 1917-1923.

Mirza O, Skov L K, Remaud-Simeon M, Potocki de Montalk G, Albenne C, Monsan P, and Gajhede M (2001), 'Crystal structures of amylosucrase from *Neisseria polysaccharea* in complex with D-glucose and the active site mutant Glu328Gln in complex with the natural substrate sucrose', *Biochemistry*, 40, 9032 -9039.

Mirza O, Skov L K, Sprogøe D, van den Broek L A M, Beldman G, Kastrup J S, and Gajhede M (2006), Structural rearrangements of sucrose phosphorylase from *Bifidobacterium adolescentis* during sucrose conversion', *J Biol Chem*, 281, 35576–35584.

Murosaki S, Muroyama K, Yamamoto Y, Kusaka H, Liu T, and Yoshikai Y (1999), 'Immunopotentiating activity of nigerooligosaccharides for the T helper 1-like immune response in mice', *Biosci Biotechnol Biochem*, 63, 373-378.

Murosaki S, Muroyama K, Yamamoto Y, Liu T, and Yoshikai Y (2002), 'Nigerooligosaccharides augments natural killer activity of hepatic mononuclear cells in mice', *Int Immunopharmacol*, 2, 151-159.

Nakai H, Okuyama M, Kim Y-M, Saburi W, Wongchawalit J, Mori H, Chiba S, and Kimura A (2005), 'Molecular analysis of α-glucosidase belonging to GH-family 31', *Biologia*, 60(S16), 131-135.

Nakai H, Ito T, Hayashi M, Kamiya K, Yamamoto T, Matsubara K, Kim Y-M, Wongchawalit J, Okuyama M, Mori H, Chiba S, Sano Y, and Kimura A (2007a), 'Multiple forms of α-glucosidase in rice seeds (*Oryza sativa* L., var Nipponbare)', *Biochimie*, 89, 49-62.

Nakai H, Tanizawa S, Ito T, Kamiya K, Yamamoto T, Matsubara K, Kim Y-M, Sakai M, Sato H, Imbe T, Okuyama M, Mori H, Sano Y, Chiba S, and Kimura A (2007b), 'Function-unknown glycoside hydrolase family 31 proteins expressed in rice ripening and germinating stages are α-glucosidase and α-xylosidase', *J Biochem*, 142, 491-500.

Nakai H, Tanizawa S, Ito T, Kamiya K, Yamamoto T, Matsubara K, Kim YM, Sakai M, Sato H, Imbe T, Okuyama M, Mori H, Sano Y, Chiba S, and Kimura A (2008), 'Rice α-glucosidase isozymes and isoforms showing different starch granules-binding and -degrading ability', *Biocatal Biotransfor*, 26, 104-110.

Nakakuki T (2005), 'Present status and future prospects of functional oligosaccharide development in Japan', *J Appl Glycosci*, 52, 267-271.

Nishimoto M, Kubota M, Tsuji M, Mori H, Kimura A, Matsui H, and Chiba S (2001), 'Purification and substrate specificity of honeybee, *Apis mellifera L.*, α-glucosidase III', *Biosci Biotechnol Biochem*, 65, 1610-1616.

Nishimoto M, Mori H, Moteki T, Takamura Y, Iwai G, Wongchawalit J, Surarit R, Svasti J, Kimura A, and Chiba S (2007), 'Molecular cloning of cDNAs for three α-glucosidases from European honeybee, *Apis mellifera* L.', *Biosci Biotechnol Biochem*, 71, 1703-1716.

Okuyama M, Okuno A, Shimizu N, Mori H, Kimura A, and Chiba S (2001), 'Carboxyl group of Asp-647 residues as possible proton donor in catalytic reaction of α-glucosidase from *Schizosaccharomyces pombe*', *Eur J Biochem*, 268, 2270-2280.

Okuyama M, Mori H, Watanabe K, Kimura A, and Chiba S (2002), ' α-Glucosidase mutant catalyzes "α-glucosynthase"-type reaction', *Biosci Biotechnol Biochem*, 66, 928-933.

Okuyama M, Mori H, Chiba S, and Kimura A (2004), 'Overexpression and characterization of two unknown proteins, YicI and YihQ, originated from *Escherichia coli*', *Pro Expres Purif*, 37, 170-179.

Okuyama M, Tanimoto Y, Ito T, Anzai A, Mori H, Matsui H, Chiba S, and Kimura A (2005), 'Purification and characterization of hyper-glycosylated α-glucosidase from *Schizosaccharomyces pombe* culture medium', *Enzyme Microb Technol*, 37, 472-480.

Okuyama M, Kaneko A, Mori H, Chiba S, and Kimura A (2006), 'Structural elements to convert α-xylosidase into α-glucosidase', *FEBS Lett*, 580, 2707-2711.

Saburi W, Mori H, Saito S, Okuyama M, and Kimura A (2006), 'Structural elements in dextran glucosidase responsible for high specificity to long chain substrate', *Biochim Biophys Acta*, 1764, 688-698.

Saburi W, Hondoh H, Unno H, Okuyama M, Mori H, Nakada T, Matsuura Y, and Kimura A (2007), 'Crystallization and preliminary X-ray analysis of *Streptococcus mutans* dextran glucosidase', *Acta Cryst F*, 63, 774-776.

Sato F, Okuyama M, Nakai H, Mori H, Kimura A, and Chiba S (2005), 'Glucoamylase originating from *Schwanniomyces occidentalis* is a typical α-glucosidase', *Biosci Biotechnol Biochem*, 69, 1905-1913.

Son M, Mori H, Okuyama M, Kimura A, and Chiba S (2003), 'Evidence of intramolecular transglucosylation catalyzed by an α-glucosidase', *J Appl Glycosci*, 50, 41-44.

Takayanagi T, Okada G, and Chiba S (1987), 'Quantitative study of anomeric forms of isomaltose and glucose produced by isomalto-dextranase and glucodextranase', *Agric Biol Chem*, 51, 2337-2341.

Takewaki S, Chiba S, Kimura A, Matsui H, and Koike Y (1980), 'Purification and properties of α-glucosidases of the honey bee *Apis mellifera L.*', *Agric Biol Chem*, 44, 731–740.

Takewaki S, Kimura A, Kubota M, and Chiba S (1993), 'Substrate specificity and subsite affinities of honeybee α-glucosidase II', *Biosci Biotechnol Biochem*, 57, 1508–1513.

Watanabe K, Hata Y, Kizaki H, Katsube Y, and Suzuki Y (1997), 'The refined crystal structure of *Bacillus cereus* oligo-1,6-glucosidase at 2.0 Å resolution: structural characterization of proline-substitution sites for protein thermostabilization', *J Mol Biol*, 269, 142-153.

Wongchawalit J, Yamamoto T, Nakai H, Kim Y-M, Sato N, Nishimoto M, Okuyama M, Mori H, Saji O, Chanchao C, Wongsiri S, Surarit R, Svasti J, Chiba S, and Kimura A (2006), 'Purification and characterization of α-glucosidase I from Japanese honeybee (*Apis cerana japonica*), and molecular cloning of its cDNA', *Biosci Biotechnol Biochem*, 70, 2889-2898.

Yamamoto T, Unno T, Watanabe Y, Yamamoto M, Okuyama M, Mori H, Chiba S, and Kimura A (2004), 'Purification and characterization of *Acremonium implicatum* α-glucosidase having high regioselectivety for α-1,3-glucosidic linkage', *Biochim Biophys Acta*, 1700,189-198.

STRUCTURE, FUNCTION AND APPLICATIONS OF MICROBIAL β-GALACTOSIDASE (LACTASE)

Byong Hoon Lee

ABSTRACT

A β-galactosidase (EC 3.2.1.23, β-D-galactoside galactohydrolase), commonly known as lactase is known to catalyze not only hydrolyze β-D-galactoside linkage of lactose or other β-galactosides into monosaccharides, glucose and galactose but also has transgalactosylation activity to synthesize galacto-oligosaccharides. Both reaction activities are well characterized and applied in many food industries. Although β-galactosidases are widely distributed in nature, the most thoroughly studied β-galactosidases are obtained from *E. coli,* and commercially used β-galactosidases are from mainly fungi and yeasts. Galacto-oligosaccharides, so-called prebiotics, have been shown to employ this growth-stimulating effect on probiotic bacteria, including bifidobacteria. Biochemical and molecular aspects of the β-galactosidase genes from different microorganisms have been studied, but the known structure and function of different β-galactosidases are limited. β-Galactosidases which belongs to the 4/7 superfamily of the glycoside hydrolase families (GHs) are currently divided into GH-1, GH-2, GH-35 and GH-42, and yet the four families are so distantly related to each other, and the hydrolytic and transgalactosylation activities of the isoenzymes appears to be different. The known 3D structures and functions of β-galactosidases from prokaryotic *E. coli* (mesophilic), three extremophiles like *Thermus thermophilus* (thermophilic), *Sulfolobus solfataricus* (thermophilic) , *Arthrobacter* (psychrotrophic), and eukaryote *Penicillium* were compared. The sequence, homology and multiple isozymes of lactic bacterial β-galactosidases were supplemented for their phylogenetic analysis. The applications of this enzyme on the oligosaccharides and prebiotics were discussed in details.

Key words: β-galactosidase (lactase); microorganisms; structure; function; applications

INTRODUCTION

The Carbohydrate-Active Enzymes database (CAZy) provides a continuously updated list of the glycoside hydrolase families, GHs (http://www.CAZY.org/). This group of enzymes was classified based on functional similarity, but today they are classified into 108 GHs on the basis of amino acid sequence similarity. Despite their similarities to enzymes with known functions, their primary functions are still unclear. Based on these criteria, β-galactosidase activities are now divided into four different families: GH-1, GH-2, GH-35 and GH-42, among which the better studied GH-2 includes β-galactosidases from *Escherichia coli*, *Aspergillus, Bacillus megatherium*, and *Sulfolobus solfataricus*, while those from thermophilic (Hirata et al., 1986; Moracci et al., 1992; Hidaka et al., 2002), psychrophilic (Gutshall et al, 1995; Skálová et al;., 2005) , and halophilic (Holmes et al., 1997) organisms belong to GH-42. Lactase is often confused as an alternate name for β-galactosidase, but it is actually simply a sub-class (small subunit) of β-galactosidase. β-Galactosidase (β-D-galactoside galactohydrolase, EC 3.2.1.23) catalyses hydrolysis of the galactosyl moiety from lactose, non-reducing termini of oligosaccharides or from glycisides, as well as those of related chromogens, *o*-nitrophenyl-β-D-galactopyranoside (ONPG), *p*-nitrophenyl-β-D-galactopyranoside (PNPG) and 6-bromo-2-naphthyl-galactopyranoside (BNG).

Most genes encoding GH-42 enzymes are from prokaryotes that are unable to grow on lactose as a sole carbon source (Daniel et al., 1997; Holmes et al., 1997), and at least two GH-42 β-galactosidases do not cleave lactose in vitro (Holmes et al., 1997; Van Laere et al., 2000). The determination of growth on lactose can be complicated by the multiple β-galactosidases because not all of the β-galactosidases are acting as lactases in vitro (Shipkowski and Brenchley, 2006). This enzyme has been purified and characterized from various sources, including plants (Ogasawara et al. 2007), animals (Gaur et al., 2000; Taniguchi and Takano, 2004), and many microorganisms (Hung et al., 2001; Lauro et al., 2008). In humans, lactase is present predominantly along the brush border membrane of the differentiated enterocytes lining the villi of the small intestine. Lactase is essential for digestive hydrolysis of lactose in milk. Deficiency of the enzyme causes lactose intolerance. Several 3D structures are available for the GH-1 and GH-2 β-galactosidases for which the catalytic residues have experimentally been determined (Matthiews, 2005), while three 3D structures of β-galactosidases from *E. coli* (Jacobson et al., 1994), *Penicillium* sp.(Roja et al., 2004) and *Thermus thermophilus* A4 (Hidaka et al., 2002) are available for either the GH-35 and GH-42 families. The 3D structure of cold temperature active β-galactosidase from *Arthrobacter* has been reported (Skálová et. al., 2005) and the 3D structure of related 6-phospho-β-galactosidase from *Lactococcus lactis* was also published (Wiesmann et al., 1997). Lactic acid bacteria including *Streptococcus thermophilus* (Lee et al., 1990), *Lactobacillus reuteri* (Halbmayr et al., 2008), *L. acidophilus* (Nguyen et al., 2007) and several *Bifidobacterium* strains contained several β-galactosidases, all of which have different properties (Van Laere et al., 2000; Hung et al., 2001; Møller et al., 2001, Hung and Lee, 2002, Hinz et al., 2004).

The enzyme lactase (β-galactosidase) has three main biotechnological uses in the dairy industry, for (1) the removal of lactose from milk for lactose-intolerant people to meet the needs of the large percentage of the world population afflicted with lactose intolerance in the field of medicine, (2) the production of prebiotic galacto-oligosaccharides for use in probiotic foodstuffs (Splechtna et al., 2007), and (3) the prevention of lactose crystallization and production of whey syrup sweeteners (Gul-Guven et al., 2007). This enzyme catalyzes both the hydrolysis and transgalactosylation; the hydrolysis reaction are widely used to improve digestibility, solubility, and sweetness of lactose, the main milk sugar, and the transgalactosylation synthetic reaction from lactose or structurally related galactosides produced galacto-olgosacchardies, which are prebiotics (growth factor) for *Bifidobacterium* (Hung et al., 2001; Splechtna et al., 2007; Baek and Lee, 2008).

Lactase commercial preparations are mainly from the dairy yeast, *Kluyveromyces lactis*, but to apply lactase to dairy and whey products, the development of highly thermostable enzymes is desirable from the viewpoints of preventing microbial contamination and of elevating lactose solubility during the processing. Many studies have been carried out on the extremophilic β-galactosidases from microbes including *Thermus aquaticus* (Berger et al., 1996), *Thermus thermophilus* (Hidaka et al., 2002),

Sulfolobus solfataricus (Moracci et al., 1995), *Bacillus coagulans* (Batra et al., 2002), *Bacillus cirvulans* (Vetere and Paoletti, 1998), *Bacillus stearothemophilus* (Chen et al., 2008), and thermoacidophile, *Alicyclobacillus acidocaldarius* (Lauro et al., 2008). Cold temperature-active lactases have been studied to produce low lactose milk from refrigerated milk to save energy, protect thermosensitive substances and reduce the risk of contamination by mesophilic microbes (Abdelrahim and Lee, 1991; Cieslinski et al., 2005, Makowski et al., 2006). Besides, enzymes that are efficient catalysists at about $0^{o}C$ are characterized by a lower conformational stability of molecules and a higher flexibility of catalytic domains in comparison with their mesophilic counterparts (Georlette et al., 2004). This manuscript is the comprehensive overview of the structure, function and applications of microbial β-galactosidase (lactase) including the author's own data.

STRUCTURE OF GH-2 ß-GALACTOSIDASE OF *ESCHERICHIA COLI* (EC-β-GAL)

The enzyme from *Escherichia coli* (GH-2) is routinely used as a marker enzyme, for a complementation-based blue-white screening and in other applications of molecular biology, and therefore its structure and function were intensively studied. It is able to hydrolyse a variety of natural and synthetic substrates, and its catalytic cycle was tracked by a crystallographic study (Juers et al, 2001). This enzyme is known to be active only in its tetrameric form due to an allosteric effect of the assembly, and this property was exploited in design of complementation-based blue-white screening, enzymatic sensors of antibodies, and in protein–protein interaction studies (Skálová et al., 2005). The best studied structure and function of *E. coli* β-galactosidase, which hydrolyzes lactose and other β-galactosides into monosaccharides, are the product of the *lacZ* operon, the central model by Jacob and Monad (1961). *E. coli* β-galactosidase belongs to the *4/7 superfamily* of the glycosyl hydrolases and is a 464-kDa tetramer of four identical 1,023 amino acids chains (Fig. 1). Each monomer (1023 amino acid residues) of β-galactosidase consists of five domains (A-D), the third of which is an eight-stranded a/β barrel that comprises much of the active sites at the C-terminal end of the central core of the TIM barrel, where Domain 1-One jelly-roll type barrel comprising of 170 residues, Domain 2, 4- two fibronectin type III-like barrels comprising of 110 residues each, Domain 3- central TIM barrel comprising of 300 residues, and domain 5-large b stranded sandwich with unique topology having 300 residues. The active sites also include portions of loops from the first, second, and fifth domains of the monomers (Juers et al., 2000, 2001). This enzyme can be split in two peptides, LacZα and LacZΩ, none of which is active by itself but both spontaneously reassemble into a functional enzyme. This characteristic is used in many cloning vectors to achieve α-complementation in specific laboratory strains of *E. coli*, where the small LacZα peptide is encoded by the plasmid while the large LacZΩ is encoded in *trans* by the bacterial chromosome. When DNA fragments are inserted in the vector and production of LacZα is disrupted, the cells exhibit no β-galactosidase activity: this allows the blue/white screening of recombinant clones. In the tetramer, the four monomers are grouped around three mutually-perpendicular two-fold axes of symmetry. These axes can be considered to form three distinct interfaces between different pairs of monomers. First, the "long" interface that formed by the horizontal two-fold axis relates the A and B monomers as well as C with D. Second, the "activating" interface involving surface area of about 4000 Å^2 that formed by the vertical two-fold axis relates A with D, and B with C. This interface involves a surface area of about 4600 Å^2. A third, much smaller interface, involving about 230 Å^2, relates A with C and B with D. Matthews (2005) reviewed the structure of *E. coli* β-galactosidase in details.

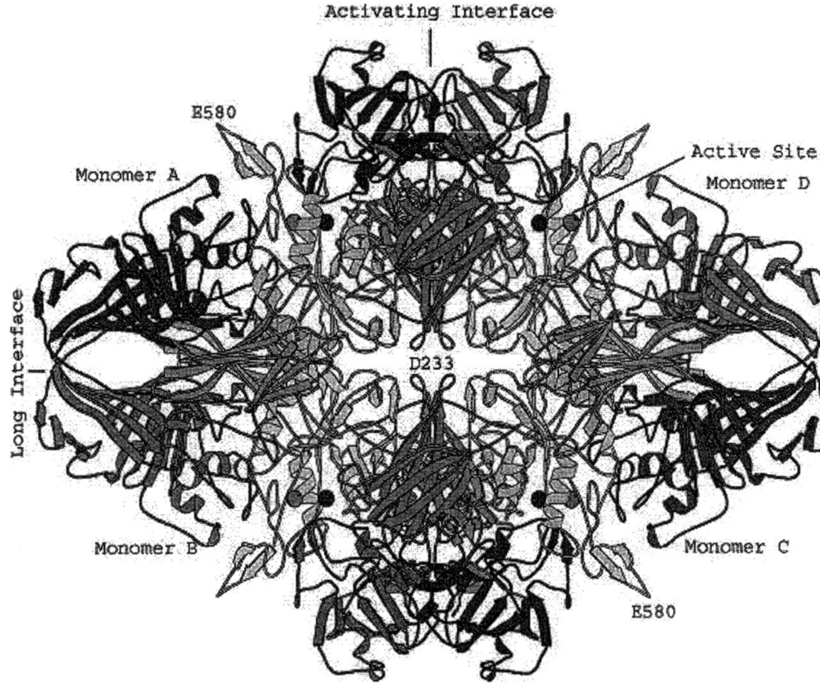

Figure 1 The *E. coli* β-galactosidase tetramer comprised of four polypeptide chains (labeled A–D) and one of the two-fold axes

Coloring is by domain: complementation peptide, orange; Domain 1, blue; Domain 2, green; Domain 3, yellow; Domain 4, cyan; Domain 5, red. Lighter and darker shades of a given color are used to distinguish the same domain in different subunits. The metal cations in each of the four active sites are shown as spheres: Na^+, green; Mg^{++}, blue (with permission of B. W. Matthews, Comptes Rendus Biologies 328: 549-556, 2005).

The active site and metal binding sites

The residues that form the active site are contributed by different segments of the polypeptide chain. Much of the active site was formed by a deep pit that intrudes well into the C terminal end of the TIM barrel in Domain 3. Besides there are loops that come from the first and fifth domain of the same monomer and there is a loop that comes from Domain 2 of a different monomer and extends into the neighboring active site across the "activating interface". While monomer A donates its Domain 2 loop to complete the active site of monomer D, similarly a loop from monomer D completes that active site of monomer A. Reciprocal donation and acceptance also occurs between monomer B and C to yield a total of four functional active sites. These active sites are well separated (Fig. 1) and presumably act independently, but because it takes two monomers to complete an active site, individual monomers of the enzyme are likely inactive. Glu461 acts as the acid carrying out protonation and G537 is considered as the nucleophile that has a role in stabilizing the intermediate by covalent catalysis (Juers et al., 2001). A magnesium ion was identified in the active site of both crystal forms of the native enzyme and both Mg^{2+} and Na^{+} were required for maximal activity of β-galactosidase. Mg^{2+} activates 5-100 fold, depending on the substrate, and can be substituted with Mn^{2+} with full activity. Studies of crystals soaked with potassium and rubidium identified five putative sodium-binding sites, among which one of these was in the active site and the remaining ones were on the surface of the protein (Juers et at., 2000).

Action mechanism

The *E. coli* β-alactosidase had two catalytic activities. Firstly, it hydrolyzes the disaccharide lactose to galactose and glucose by "shallow" and "deep" binding, and secondly, it converts lactose to another disaccharide, allolactose, which is the natural inducer for the *lac* operon. Although this enzyme has fairly strict specificity for the sugar in the galactosyl position, it can hydrolyze β-D-galactopyranosides with different aglycones. Allolactose was finally converted by the enzyme to galactose and glucose, and thus carrying out either hydrolysis or transgalactosylation (Fig. 2a). In catalytic reaction (hydrolysis or glycosidic bond cleavage), β-galactosidase is a retaining glycosidase, where the product retains the same stereochemistry as the starting state. The two step (double displacement) nature of the catalytic mechanism was first proposed by Koshland (1953) and experimentally proven later by methanol competition (Stokes and Wilson, 1972). A generalized scheme for the mechanism of action, based on a variety of evidence using the crystallographic analyses to visualize the presumed intermediates during catalysis (Juers et al., 2001) is shown in Fig. 2b. A series of crystallographic analyses were tried to visualize the presumed intermediates during catalysis (Juers et al., 2001). In *E. coli*, Glu-461 was thought to be the nucleophile in the substitution reaction, but it is now known that Glu-461 is an acid catalyst. Instead, Glu-537 is the actual nucleophile, binding to a galactosyl intermediate. In humans, the nucleophile of the hydrolysis reaction is Glu-268. The nucleophile, Glu537 had a role in stabilizing the intermediate by covalent catalysis, and the ligand stacks up on the indole of Trp999 (Juers et al., 2001) in Fig. 2a. Iodination of Tyr253 rendered the enzyme inactive, and thus Tyr253 was the most reactive tyrosine in the enzyme. Binding of the non-hydrolyzable substrate analogues, isopropyl thiogalactoside (IPTG) and 2-F-lactose (Huber and Brockbank, 1987) in this region brings about a conformational change and make the Tyrosine inaccessible to study the inactive complex. One such mutation is substitution of Glu537 with Gln537, the altered member, E537Q, is catalytically incompetent and helps in studying the analog enzyme complex. Thus, Glu461 is in a good position to donate a proton to the glycosidic oxygen, supporting its role as an acid catalyst for galactosylation (Juers et al., 2001). As the magnesium ion does interact directly with Glu461, it appeared to contribute to the role of this residue. To mimic the presumed oxocarbenium ion-like transition state, galactotetrazole and galactonolactone were used. These compounds are trigonal at the C1 carbon and bind tightly (Heightman

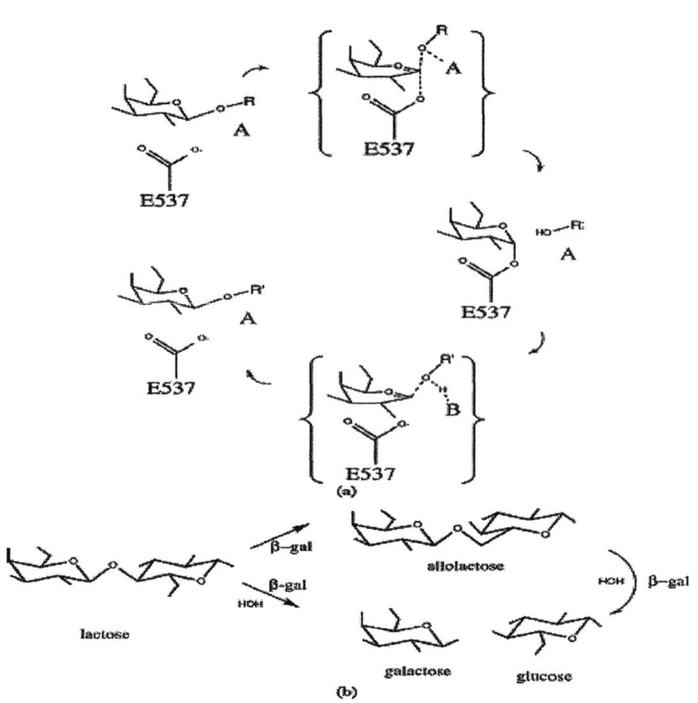

Figure 2 Mechanism of β-galactosidase

(a) Generalized outline for a double-displacement reaction catalyzed by β-galactosidase. In the first step (top), the substrate, a β-D-galactopyranoside with OR as the aglycon, forms a covalent α-D-galactosyl enzyme intermediate with the nucleophile Glu537 and with assistance from an acid, A (either Glu461 or a magnesium ion). Galactosyl transfer to the nucleophile is shown here as concerted with glycosidic bond cleavage, although this is controversial and may depend on the nature of the leaving group. In the second step (bottom), release of the intermediate is facilitated by a base, B (probably Glu461), which abstracts a proton from the acceptor molecule, R'OH. Galactosyl transfer from the enzyme is shown as stepwise, because there is substantial evidence that all substrates have a trigonal anomeric center in the transition state for this step. (b) General scheme for the action of β-galactosidase on the natural substrate, lactose. The enzyme can either perform hydrolysis (lower path) or transglycosylation (upper path) (with permission of B. W. Matthews et al., 2005)

et al., 1995). Both of these transition state analogues bind in a "deep" mode, which Glu461 contacting the atom that corresponds to the glycosidic oxygen. The product of the normal catalytic cycle, galactose, was also observed crystallographically to bind in the "deep" mode (Juers et al., 2001). Thus, in Step 1, the first half of the reaction cycle involves cleavage of the glycosidic bond, formation of the galactosyl enzyme intermediate, and release of the aglycon (glucose). Electrophilic attack of the glycosidic oxygen either by the Mg ion or by Glu461 results in cleavage of the glycosidic bond and departure of the aglycon (glucose). The transition state analogues suggest that Glu461 may donate to the glycosidic oxygen, thus explaining suggesting its role as an acid catalyst for galactosylation. Interaction between Mg^{2+} and Glu461 suggests that Mg^{2+} might have a role in helping Gly461 allowing it to act as a potent acid catalysist (Juers et al., 2001). In Step 2 on transgalactosylation reaction, which is the transfer of a glycan from one sugar molecule to another by using a retaining glycosidase mechanism, the glycan is transferred to any acceptor molecule other than water. In transgalactosylation of β-galactosidase, the galactosyl residue was transferred to the same glucose molecule that was cleaved especially at low concentration. Lactose can bind in the shallow mode. Once galactosylation has taken place with the galactosyl group moving deeper into the active site, reattaching the glucose at the 1-, 2- ,3-, or 4-positions appears to be sterically difficult (Juers et al., 2000, 2001). However, the 6-hydroxyl, which has an extra atom linking it to the galactose ring, can reach further into the active site pocket and more easily attack the enzyme-bound intermediate. This seems that if the glucose binding at the site was both specific and tight, then transglactosylation would be predominant, while the binding is very weak, then hydrolysis would dominate.

Structural basis of α-complementation and purpose of large enzyme size

The deletion of residues 23-31 or 11-41 from the N-terminus of β-galactosidase results in inactive dimers (so-called α-acceptors). By supplying the missing eptides (α-doners), the catalytic activity could be reversed. Two common α-donors (the residues 3-41 or 3-92) are the α-complementation phenomenon (Muller-Hill and Kania, 1974) based in large part on work of Zabin's group prior to the 3-D structure (Weinstock et al., 1982). The intact active β-gal is tetrameric and deletions at the N-terminus result in inactive dimers, thus indicating that the N-terminal residues must mediate dimer-dimer binding to form the active tetrameric enzyme. Fig. 3 based on knowledge of the structure showed that the interface disrupted by deletion of residues from the N-terminus is the so-called activation interface. This vertical interface includes a region where four-α helices come together (labeled 4α), and also a region of contact between residues 13–23 of the respective monomers. In addition, the residues 272–288 extend across the interface to complete the active site of the opposed monomer. Thus, the role of residues 11–41 in stabilizing the tetrameric structure seems straightforward. The segment 13–23 contributes directly to the dimer–dimer interface (Fig. 3). Also the segment 29–33 passes through a 'tunnel', which presumably the 'anchors' residues 13–23 to the rest of the monomer. The deletion of residues 23–31 in the alternative α-acceptor also disrupts both the dimer–dimer interface and the 'tunnel' interaction. The two classical α-donors (residues 3–41 and 3–92) also include the residues 13–23 plus 29–33. This suggests that a successful α-donor needs to both reconstitute the dimer–dimer interface and, in addition, provide residues 29–33 so that the α-donor can be anchored to the remainder of monomer. Its normal substrate is a simple disaccharide, but why then is such large size of the β-gal (464 kD tetramer). As the β-galactosidase belongs to the so-called ' superfamily' of GHs (Henrissat and Bairoch, 1993), the active site is built around the α /β barrel in Domain 3, that is common to many other GHs. β-Galactose arose from a prototypical, monomeric, single domain α/β barrel with an active site that could accommodate extended substrates (Juers et al., 1999). The subsequent addition and incorporation of elements from other domains could then have reduced the size of the active site to better hydrolyze the disaccharide lactose and, at the same time, to facilitate the production of inducer, allolactose. While such change may account for some increase in size of the enzyme, it is not certain why it needs to be tetrameric. Some β-galactosidases were dimers from *Lactobacillus delbrueckii* (homodimer of 110kDa; Adams et al., 1994), *Bifidobacterium breve* (homodimer of 76kDa; Yi, 2005), *Bifidobacterium infantis* HL96 (heterodimer of 113 and 76Kda; Hung et al., 2001) or *Bifidobacterium infantis* DSM20088 (homodimer of 140kDa, Møller et al., 2001).

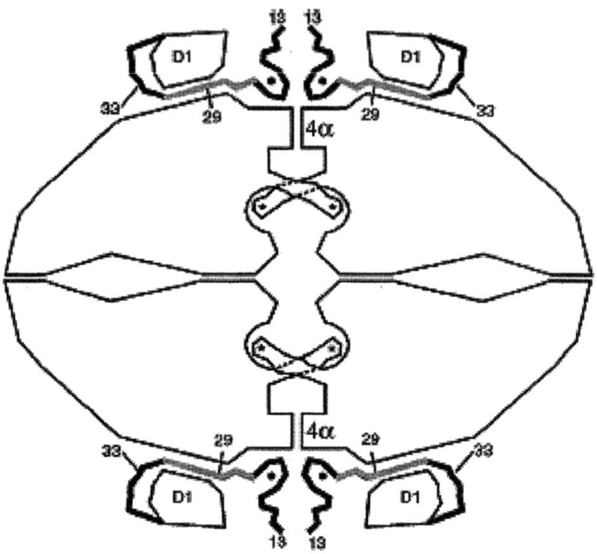

Figure 3 Scheme showing the key features of the β-galactosidase tetramer
At the N-terminus, residues 1–12 are not seen in the electron density map due to presumed disorder. Residues 13–50 (shown as thick lines) pass through a tunnel between the first domain (labeled D1) and the rest of the protein. The region shaded gray (residues 23–31) is deleted in one of the -donors. A magnesium ion (shown as a small solid circle) bridges between the complementation peptide and the rest of the protein. The four active sites are labeled with asterisks. The activation interface runs vertically through the middle of the figure. A part of this interface is a bundle of four α-helices in the region labeled 4α. When the activation interface is formed the four equivalent loops that include residues 272–288 extend across the interface to complete the active sites within the four recipient subunits (with permission of B.W. Matthews, Comptes Rendus Biologie 328: 549-556, 2005).

STRUCTURES OF GH-42 β-GALACTOSIDASE FROM *THERMUS THEMOPHILUS* A4 (TT-β-GAL)

Thermus thermophilus A4 β-galactosidase (Tt-β-gal) retained the full activity at 70 °C for 20 h (Ohtsu et al., 1998). This enzyme consists of 645 amino acid residues and belongs to GH-42. Today, 29 β-galactosidases are placed in GH-42, and some of the GH-42 enzymes can survive in various extreme conditions, such as psychrotrophic (Gutshall et al., 1995; Coombs et al., 1999; Cieslinski et al., 2005), thermophilic (Moore et al., 1994; Lauro et al., 2008), and halophilic (Holmes et al., 1997). These extreme enzymes appear to consist of 600–700 amino acid residues, and exhibit no sequence similarity with *E. coli*-β-gal of GH-2. These extremophilic β-galactosidases could be useful as new reporter enzymes in situations in which the Ec-β-gal can not function (Schrogel and Allmansberger, 1997).

Trimeric and monomeric structures

GH-42 belongs in Clan GH-A and the catalytic residues of Tt-β-gal were inferred to be Glu141 and Glu312 (Ohtsu et al., 1998). Jenkins et al. (1995) grouped five families having a TIM barrel fold (GH-1, GH-2, GH-5, GH-10, and GH-17) into the 4/7 superfamily on the basis of structural similarity, and all members of clan GH-A whose structures reported are classified into this superfamily. Recently, β-amylases (GH-14), which are inverting enzymes acting on axial glycosidic bonds, are considered to be members of the 4/7 superfamily (Juers et al., 1999; Nagano et al., 2001). The catalytic residues of the 4/7 superfamily are located at the C-termini of β-4 and β-7 of the TIM barrel fold, but there is no GH-42 enzyme whose three-dimensional structure is known to date. As the first known structures of a GH-42 enzyme, the crystal structures of free and galactose-bound enzyme at 1.6Å and 2.2Å resolution, respectively were revealed. The Tt-β-gal formed a homotrimeric structure resembling a flowerpot (Hidaka et al., 2002). Each monomer had an active site located inside a large central tunnel. The trimer was held together by numerous interactions at tight molecular interfaces. Fig. 4(a) shows a ribbon diagram of the structure of a Tt-β-gal monomer, which consists of three domains: (1) domain A, a TIM barrel fold domain (residues 1–389); (2) domain B, an α/β fold domain (390–589); and (3) domain C, α/β fold domain (590–644). The domain interfaces form a large cleft with bound MPD (2-methyl-2,4-pentanediol) molecules, an acetate group, and a galactose molecule (Fig. 4b). Domain A has a $(\beta/\alpha)_8$ (TIM) barrel supersecondary structure, consistent with the results of HCA (hydrophobic cluster analysis) regarding GH-42 enzymes.(Henrissat et al, 1995). Domain A appeared to possess a TIM barrel fold similar to that of a GH-14 enzyme, *Bacillus cereus* β-amylase (BCB) (Mikami et al., 1999) in Fig.4c. Both structures have an extra region (subdomain H) inserted between β-4 and α-4 of the TIM barrel, having a similar comma-like shape. Subdomain H consists of α-helices with a similar topology in Tt-β-gal and BCB. Tt-β-gal possessed domains B and C immediately after α-8, but BCB had a starch-binding domain after the TIM barrel domain. The domain has a fold similar to that of the TrpG subunit of *Sulfolobus solfataricus* anthranilate synthase (Knochel et al., 1999) that belongs to the "triad" glutamine amidotransferase family. However, all residues in the domain B, which are equivalent to the catalytic residues (Cys84, His175, and Glu177) of the TrpG subunit, are proline (Pro483, Pro571, and Pro573). Domain B was involved in the trimer formation. Domain C consisted of a β structure with no fold similar to any known structures, but its function is unknown.

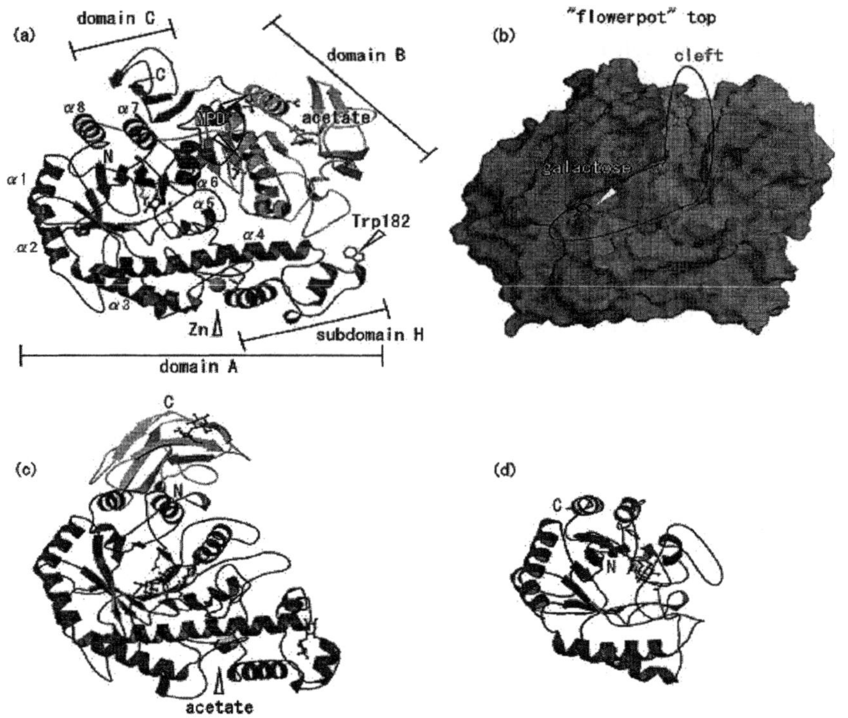

Figure 4 Structure of Thermus thermophilus β-galactosidase

(a) Overall structure of the Tt-β-gal monomer in a ribbon model (domains A, blue; B, yellow; C, red). Subdomain H inserted between β-4 and α-4 of the TIM barrel in domain A. The bound galactose (red), MPD molecules and acetate groups (green) as a ball-and-stick model, and the zinc atom as a sphere. The catalytic residues, Glu141 and Glu312, and Trp182, an active site through the subunit interaction, are shown as a wireframe model. (b) The molecular surface of Tt-β-Gal monomer (domains A, blue; B, yellow; C, red). The subunit interface is colored green. The domain interface forms a large cleft from the upper side of the "flowerpot" to the lower end of the active site, which is closed through the subunit interaction. The bound galactose, MPD molecules and an acetate group are located in the cleft: the view is from the same direction as in (a). (c) A ribbon diagram of *B. cereus* β-amylase (1B9Z; GH-14). The catalytic domain has a TIM barrel fold (blue) followed by a starch-binding domain (yellow). The catalytic residues, Glu172 and Glu367, are shown as a wireframe model. Bound maltose molecules are shown as a ball-and-stick model (red). (d) A ribbon diagram of the catalytic domain (domain 3) of Ec-β-Gal (1JZ7). The catalytic residues, Glu461 and Glu537 (a wireframe model) and the bound galactose molecule (a ball-and-stick model in red) are shown (with permission of Hidaka et al., 2002).

Metal-binding site

The domain A contained Cys106, Cys150, Cys152, and Cys155, which are highly conserved in GH-42 enzymes and formed a metal-binding cluster (Fig. 5). Inductive coupled plasma (ICP) measurement showed that while this enzyme contained 0.57 Zn atom and 0.16 Fe atom per monomer, Mg, Mo, Cu, Ni, and Co atoms were not detected. The native and selenomethionine-labeled datasets revealed strong electron density of the Zn atom, but the galactose-complex dataset revealed poor electron density around the metal cluster. Thus, the metal ion is likely bound loosely and the strength of the site is not related with the binding of galactose. As all four ligands for the zinc atom were cysteine residues and far from the galactose-binding site, this enzyme does not need any metal ions for its enzyme activity, this zinc ion appeared to be a structural feature (Auld, 2001). As shown in Figs. 4a and 5, loops and helices are combined together through coordination to the Zn atom at the root of subdomain H. In the BCB structure, the oxygen atoms of the acetate ion form hydrogen bonds with the side-chain of Asn131, and the methyl group of the acetate moiety is surrounded by hydrophobic residues, instead of the cysteine residues of the Tt-β-gal.

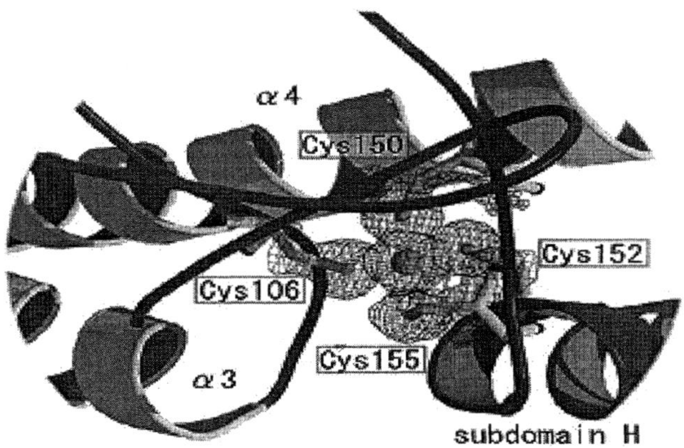

Figure 5 Structure around the metal-binding site of the native data

The $2|F_o|-|F_c|$ electron density around the metal-binding site is shown. Four cysteine residues and the zinc atom are shown as a wireframe model and a sphere, respectively. Loops and a helix containing the four cysteine residues are colored red (with permission of Hidaka et al., 2002).

Galactose-binding site and action mechanism

In the galactose-complex structure of this enzyme, one galactose molecule was bound to each domain A in the chair conformation and its electronic density O1 had the α-anomer configuration. Eleven direct H-bonds were involved between protein atoms and the OH groups of galactose, but in addition, several H-bond via water are also present. Among these, all five OH groups of galactose were H-bonded with two or more residues, indicating that the recognition of the galactose moiety appears to be very strict. Instead of the H-bonds, only one hydrophobic contact with Phe350 was found. Two putative catalytic residues, Glu141 and Glu312 close to C1 of galactose, are located at the C-terminal ends of β-4 and β-7 of the TIM barrel, respectively. The side chain of Try182 of the next subunit also recognized O3 of galactose via a water molecule. Trp182 and Phe181 residues are located at the top of subdomain H. The sequence alignment of the

Ec-β-gal and other GH-42 showed that the highly conserved 12 residues in domain A involved in catalysis and substrate binding. Domains B and C revealed low level of similarity among these β-galactosidases.

Structure comparison between the Tt-β-gal and other β-galactosidases

In the galactose-complex structure of the Tt-β-gal, galactose that was bound to domain A and Glu141 and Glu312 (catalytic residues) is close to the C1 atom and superimposed well with the catalytic residues of the Ec-β-gal. In the Ec-β-gal, catalytic nucleophile Glu537 formed hydrogen bonds with Tyr503 and Arg388 residue that are thought to control the pK_a of Glu537. While these two residues were necessary for the Ec-β-gal activity (Ring and Huber, 1990), the Tt-β-gal had corresponding residues (Tyr266 and Arg32). These structural similarities conclude that Glu312 of the Tt-β-gal is a nucleophile, and that Glu141 is an acid/base catalyst. When the two structures are superimposed with respect to all atoms of the catalytic residues (Glu141/Glu312 of the Tt-β-gal and Glu461/Glu537 of Ec-β-gal), galactose and other binding residues were overlapped. However, only two residues (Arg32/Asn140 of the Tt-enzyme and Arg388/Asn460 of the Ec-enzyme) other than the two catalytic residues were conserved in the active site. Among the two mechanisms of GH enzymes first proposed by Koshland (1953), both the Tt and Ec-β-galactosidases seem to be the retaining enzyme. For the recognition mechanisms, the overall folds of catalytic domains, domain A (1-389) in the Tt-β-gal and domain 3 (350-625) in the Ec-β-gal show considerable difference, though two have the same function. While the Tt-β-gal had an extended helix region (subdomain H) between β-4 and α-4, the Ec-β-gal had a compact TIM barrel structure. The major difference between two was that the Ec-β-gal contained two metal ions, Na^+ (recognizes 06 of galactose) and $Mg2^+$ (interacts directly Glu461 as acid/base catalyst) in its active site, but the Tc-β-gal contains no $Mg2^+$, and instead Glu360 of the Tt-enzyme played the role of Na^+ of the Ec- β-gal, which recognized 04 of galactose. Despite of very low homology in primary structure less than 10% identity, the Tt-β-gal showed all the 3D structural characteristics of the 4/7superfamily, such as two catalytic sites, distance of two carboxylate groups, and Asn-Glu motif. The GH families are known to classify into two types, cleft-type and pocket-type active sites (Davies and Henrissat, 1995), among which GH-1 and GH-2 like *E. coli* β-gal have pocket type, but GH-5, GH-10, and GH-17 seem to have cleft-type (Juers et al., 1999). Although the Tt-β-gal (GH-42) and Ec-β-gal (GH-2) even share a common ancestor of the 4/7 superfamily, GH-42 seems to be an unique member of the 4/7 superfamily that may provide a new insight on evolutionary relationships between "retaining enzyme" (retention of the configuration at the anomeric C of the substrate via a double-displacement mechanism) and "inverting enzyme" (inversion of the anomeric configuration via a single nucelophilic displacement).

In recent years, several thermostable β-galactosidases have been reported (Wanarska et al., 2005, Synowiecki et al., 2006, Chen et al., 2008). *Sulfolobus solfataricus* MT4, a hyperthermophilic archaeon first isolated from hot mud in the Solfatara crater north of Naples grows optimally at 87°C and expressed a GH activity, initially characterized as a β-galactosidase on the basis of hydrolysis of the chromogenic substrate, 5-bromo-4-chloro-3-indolyl-β-D-galactopyranoside (X-Gal; Moracci et al., 1992). Subsequent enzymatic analysis showed that this enzyme had several exo-β-glycosidase substrate specificity (Grogan, 1991). The amino acid sequence derived from the *lacS* gene put the enzyme in family-1 of the β-glycosylhydrolases, along with bacterial β-glycosidases, 6-phospho-β-galactosidases, cyanogenic β-glycosidases, plant myrosinases and mammalian gut lactases. The Ss-β-gal that had optimal activity with a half-life of 48h at 85C was not thermally denatured under 100°C (Moracci et al., 1995) and resistant to denaturation to organic solvents that can be very useful to synthesize a variety of glycosides by transglycosidation and condensation (Trincone et al 1994). Aguilar et al (1997) reported the structure of the native tetrameric enzyme, and site-directed mutagenesis as well as homology with other GH-1 glycosidases have allowed to identify the active site of the enzyme and define the substrate binding site. From analysis of the refined structure, the main feature that distinguishes this enzyme from mesophilic proteins was the presence of a large number of ion-pair networks which crosslink the surface of the protein. This feature, coupled with the observation of substantial numbers of buried water molecules, suggested that this enzyme (possibly other hyperthermophile proteins) may achieved its hyperthermostability by resilience rather than rigidity. Hyperthermostable β-galactosidases from *Pyrococcus furiosus* (EP 0687732, 2003; EP 0606008, 2004) and *Pyrococcus woesei* (Dabrowski et al., 1998, Dabrowski et al., 2000) have also cloned and sequenced, but the structures are not known. Their hydrolytic activities on lactose have not been studied, despite the high chromogenic activities on ONPG.

STRUCTURE OF GH-2 β-GALACTOSIDASE FROM PSYCHROTROPHIC *ARTHROBACTER* (AR-β-GAL)

Psychrotrophic bacterium *Arthrobacter* sp. C2-2 isolated in the Antarctic area (White et al., 2000) contained two genes coding β-galactosidase isozymes that were expressed in *E. coli* (Karasová-Lipovová et al., 2003). This enzyme hydrolysed oligosaccharides as well as synthetic galactosides and also synthesized galactosides by transglycosylation reactions. The enzyme required the bound ions for its activity and lost all its activity upon ion removal and desalting. The addition of dithiothreitol (DTT) or Mn^{2+} restored 90% of its activity; Co^{2+}, Mg^{2+}, $Ca,^{2+}$ and Zn^{2+} restored approximately 50% of the activity. Magnesium ions were present in the protein buffer in this structural study. The Ar-β-gal isozyme showed 81% sequence identity with β-galactosidase of *Arthrobacter psychrolactophilus* sp. B7 and 69% sequence identity with β-galactosidase of *Arthrobacter* sp. SB. Sequence homology data of β-galactosidases from GH-2 and a phylogenetic tree have been published (Karasová-Lipovová et al., 2003). The most similar protein with known 3D structure was β-galactosidase from *E. coli*, which has 33% sequence identity with the Ar-β-gal. Other β-galactosidases with known 3D structure of other GH families showed significantly lower levels of similarity. This is the first cold-active β-galactosidase with known 3D structure in the form of compact hexamers that can help to understanding low-temperature activity, reduced thermostability, transglycosylation capability and substrate specificity. The enzyme formed 660 kDa (hexamers; 6 x 110 kDa) consisting of the identical chains (A–F) in 1023 residues. The active sites opened to the central cavity of the hexamer and connected by eight channels with exterior solvent. The hexamer organization regulates access of substrates and ligands to six active sites. This enzyme belongs to GH-2, similar to *E. coli* β-galactosidase, forming tetramers necessary for its enzymatic function. However, significant differences were found between these two enzymes affecting the ability of tetramer/hexamer formation and complementation of the active site. This structure revealed a new insight into the cold-adaptation mechanism of enzymatic pathways of extremophiles.

Monomer structure and active site

The Ar-β-gal monomer was composed of five domains (Fig. 6). Domain 1 was identified as a GH-2 sugar binding domain with jelly-roll fold (residues 32–218). Domain 2 was a GH-2 immunoglobulin-like beta-sandwich domain (residues 219–312). Domain 3 was a GH-2 TIM barrel domain (eight-stranded a/β barrel, residues 313–609), which contained the active site. Domain 4 (residues 610–715) was similar to domain 2. Domain 5 was classified as a β-galactosidase small chain (residues 737–1023). In addition, two regions of the enzyme did not form classical domains: the N terminus (residues 1–31) and a small chain on the monomer surface (residues 716–736) connecting domain 5 with the rest of the protein. In all, 54 peptide bonds in the structure were found in the *cis* conformation, of which 11 were modeled alternatively in both *cis* and *trans* conformations (148–149 A–F, 59–60 A, 235–236 B, 862–863 C, D and F). In total, 103 residues contained atoms modeled in alternative conformations. Trp786 A–F was found in two conformations of its indole ring. The pairs of active sites, Glu442 and Glu521 was placed in a deep well of the TIM barrel in the centre of the monomer, and filled with a network of localized water molecules. Mg^{2+} present in the protein buffer was one of the ions critical to the activity.

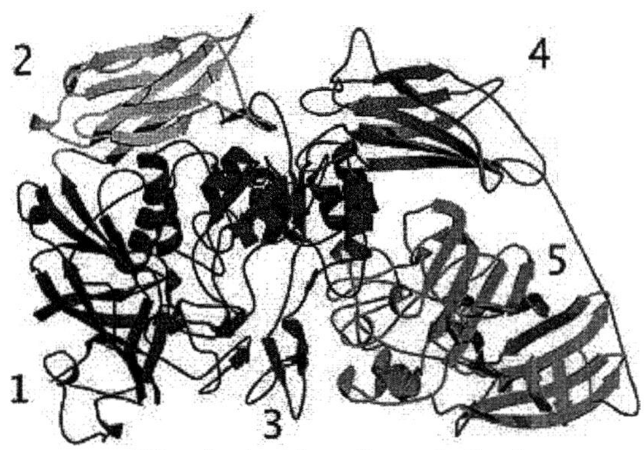

Figure 6 The monomer of β-galactosidase from *Arthrobacter* sp. C2-2 (Ar-β-gal) consisted of five domains

(1) GH-2 “sugar binding” domain with jelly-roll fold (blue, residues 32–218); (2) GH-2 immunoglobulin-like beta-sandwich domain (yellow, residues 219–312); (3) GH-2 TIM barrel domain (red, residues 313–609); (4) GH-2 immunoglobulin-like β-sandwich domain (magenta residues 610–715); (5) β-galactosidase small chain (green, residues 737–1023). The pair of catalytic residues Glu442 and Glu521 was placed in a deep well in the TIM barrel in the centre of the monomer, and the active site was filled with a network of localized water molecules. Magnesium, one of the ions critical for the activity was present in the protein buffer. A sodium ion in binding site I (residues NA 7501–7506 for chains A–F) was localized at a distance of ca 8 Å from carboxyl oxygen atoms of Glu442 and Glu521 (with permission of Skálová et al., J. Mol. Biol. 353:282-294, 2005).

Hexamer structure and channels

This Ar-β-gal formed hexamers (denoted as chains A–F) in the crystal structure with one hexamer per asymmetric unit. The hexamer was a compact sphere-like body composed of two trimers (A, B, C and D, E, F) with an altitude of the triangle of one trimer of ca 130 Å. The trimers are face-to-face oriented (A above F, B above E, C above D). The active sites of all monomers (one active site/monomer) are accessible from the interior cavity of the hexamer. Eight channels of three types (two channels of type I, three channels of type II and three channels of type III) were formed at the intermolecular interfaces and connected the inner space of the hexamer with surrounding solvent. Beside these eight intermolecular channels, two openings through each monomer can be found allowing passage of solvent molecules. The channel type I along the 3-fold axis of the trimer is visible from the top of the hexamer. The largest channel II is formed at the place of contact of two monomers each from a different trimer. The channel III was found along the non-crystallographic 2-fold axis in the place of contact of four monomers, two from the "upper" and two from the "lower" trimer. The outer side of the cylindrical opening was formed by side-chains of nine charged residues (Asp490, Arg494 and Lys542 of each monomer of the trimer). These nine charged side-chains are not involved in the trimer intermolecular contacts. The largest channel II of elliptical shape lied at the interface of two monomers. The surface of the narrowest part of the channel was formed by a mixture of hydrophobic and hydrophilic residues. While the channels I and II are suitable for passage of small ligands, e.g. lactose, its analogues and products, channel III provided a sufficient opening only for small solvent molecules. The contacts within the two trimers are mainly formed between domain 5 and the TIM barrel domain 3 of the neighbouring molecule. Conformations of individual monomers in the hexamer were similar but not identical.

Metal binding and solvent

One Mg^{2+} and two Na^{+} were localized in each monomer in which Mg^{2+} binds to Gly527 O, Gly529 O, Ala525 O and to three water molecules. The Na^{+} bound site I (residues NA 7501–7506 for chains A–F) in coordination of trigonal-bipyramid to Asp201 $O^{\delta 2}$, Phe585 O, Asp588 $O^{\delta 2}$ and to two water molecules. The Na^{+} in binding site II (residues NA 7511–7516 corresponding to chains A–F) bound to Asp100 O, Glu200 O, Asp199 $O^{\delta 1}$, Thr99 O, Thr99 $O^{\gamma 1}$ and to one water molecule. All three cations lied in the region of the TIM barrel entrance from the oxygen atoms of the pair of the catalytic glutamic acid residues. The structure contains 6471 localized water molecules, in which a total of 322 water molecules are completely buried within the six monomers of the enzyme. The entrance channels in the hexamer are partly filled with localized water molecules. Other solvent molecules were found to be bound to the enzyme: sulfate ions, a chloride ion and a dihydroxyethylether moiety, which is probably a part of a longer polyethylene glycol (PEG) chain used in crystallization.

Structural comparison with other β-galactosidases

To this date, besides this *Arthrobacter*-β-galactosidase (hexamers, Skálová et al., 2005), 3D structures of other four β-galactosidases are known: GH-1-β-galactosidase (β-glycosidase) from *Sulfolobus solfataricus* (tetramer, Aguilar et al., 1997), GH-2-β-galactosidase from *E. coli* (tetramer; Juers et al., 2001); GH-35-β-galactosidase from *Penicillium* sp. (monomer; Rojas et al., 2004); and GH-42-β-galactosidase from *Thermus thermophilus* A4 (trimer, Hidaka et al., 2002). Hexameric cage-like packing of this Ar-β-galactosidase was the first supramolecular arrangement of this type known in the galactosidase superfamily. The structure of one monomer in the Ar-β-Gal resembled that of the Ec-β-gal from the same GH-2 (sequence identity 33%). The other three β-galactosidases differed considerably and their similarity was limited mainly to the TIM barrel fold of the active site domain (Fig. 7).

Besides the Ec-β-gal, GH-2 included one more enzyme with its structure currently accessible in the PDB, β-glucuronidase from *Homo sapiens* (PDB code 1BHG). It consisted of three domains (similar to domains 1-3 of Ar-β-gal) and formed homo-tetramers. The total sequence identity between Ar-β-gal and β-glucuronidase from *H. sapiens* was 24% (Skálová et al., 2005). Some differences are associated with packing of monomers into tetramers (Ec-β-gal tetramer vs Ar-β-gal hexamer). Domain 1 of Ar-β-gal had an additional loop, 52–62, which does not exist in Ec-β-gal. Domain 2 is a part of the largest structural difference between both β-galactosidases. The Ec-β-gal had an outstanding loop (residues 276–287), which participates in contacts forming the tetramer. It was buried into the TIM barrel of the neighboring monomer that was one of the reasons why tetramerization was necessary for the Ec-β-gal activity. This loop did not exist in the Ar-β-gal immunoglobulin-like domain and an equivalent part of the domain (residues 270–273). The TIM barrel domains (domain 3) of this enzyme and Ec-β-gal differed in the residue range 484–514 (Ar-β-gal), which corresponds to residues 504–534 in Ec-β-gal. This is the part of Ec-β-gal, where the TIM barrel was completed by the outstanding loop of domain 2. In Ar-β-gal, this complementation did not occur. A loop of domain 5, residues 829–839, was placed in a different position near the TIM barrel in Ar-β-gal, but the contact was not very tight. The catalytic residues were placed at similar positions in both enzymes (Glu442 in Ar-β-gal *vs* Glu461 in Ec-β-gal and Glu521Ar-β-gal *vs* to Glu537 in Ec-β-gal) and the overall shape of the active site was mostly conserved. Trp999 in Ec-β-gal, which played an important role in ligand binding of shallow binding mode was placed by Cys999 in Ar-β-gal, that was a region of the largest differences in the active site between them. Trp568, in Ec-β-gal in close contact to ligands in the deep binding mode, was conserved as Trp552 in Ar-β-gal. Domain 4 of Ar-β-gal had two inserted loops, 617–623 and 659–666, in comparison with Ec-β-gal. Domain 5 differed in region 775–795 of Ar-β-gal in which the loop of Ec-β-gal formed no important interdomain or intermolecular contacts. In Ar-β-gal, the loop was oriented towards the TIM barrel of the same monomer and interacts with the TIM barrel by one hydrogen bond (Ser789 O–Ser485 O^{γ}) and one close van der Waals contact between Trp786 (in two alternative conformations) and Pro487. The Mg^{2+} in Ar-β-gal did not correspond to any magnesium binding site in Ec-β-gal. The active site Mg^{2+} of the Ec-β-gal structure did not resemble that of the Ar-β-gal. The sodium ion binding site I in Ar-β-gal was occupied by a sodium ion also in Ec-β-gal. The Ar-β-gal hexamer was a compact spherical body with active sites opened to the central cavity, which was connected by eight channels with exterior solvent. Unlike this, the shape of the Ec-β-gal tetramer resembled roughly a diamond, with individual molecules forming its sides. The active sites in Ec-β-gal tetramer are directly opened to solvent. The contacts between monomers in the hexamer and in the tetramer were achieved by completely different organization of the molecules. In general, although both enzymes showed a high level of sequence similarity, the temperature profiles of both enzymes differ significantly. Major differences between the two structures include oligomerization state, existence of internal cavity in the psychrotrophic case and active site sequence differences. Distinct oligomeric packing of molecules despite high levels of sequence similarity has to raise questions regarding its function. The structure reveals new insights into the cold-adaptation mechanism of enzymatic pathways of extremophiles.

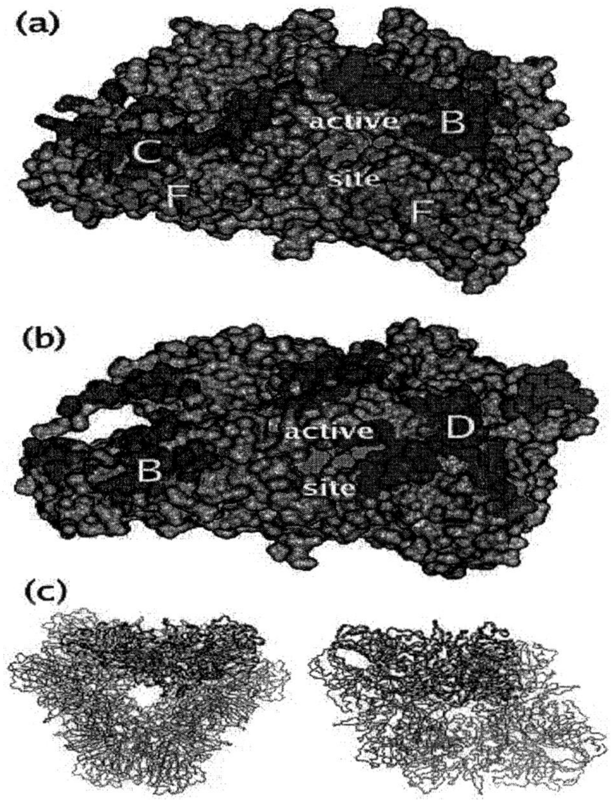

Figure 7 Surface contact regions of monomer A in (a) the hexamer of Ar-β-gal and (b) the tetramer of Ec-β-Gal, both in the same orientation of the monomer A

The active site residues are in green: Glu442, Glu521, Trp552 and Cys999 (C221-β-Gal); Glu461, Glu537, Trp568 and Trp999 (Ec-β-Gal). The capital letters mark the surface areas (distinguished by different colouring) involved in contacts with the corresponding monomers. The solvent-accessible surface was coloured on the basis of non-zero contact area of a given atom, (c) Comparison of monomer A placing in the hexamer of Ar-β-gal and the tetramer of Ec-β-gal (with permission of Skálová et al., 2005).

STRUCTURE OF GH-35 β-GALACTOSIDASE FROM *PENICILLIUM* SPECIES (PSP-β-GAL)

As all attempts to resolve the crystal structure of eukaryotic *Penicillium* β-galactosidase using the molecular replacement method of the Ec-β-gal and Tt-β-gal as search models were unsuccessful, the phase problem was solved by the single isomorphous replacement with anomalous scattering (SIRAS) method at 1.90Å and 2.10Å resolution, respectively (Rojas et al, 2004). The amino acid sequence of this 120 kDa protein was first assigned putatively by inspection of the experimental electron density maps and then determined by nucleotide sequence analysis. Primary structure alignments of *Penicillium sp.* β-gal belongs to GH-35, that is the first 3D structure for GH-35. Five distinct domains which comprise the structure are assembled in a way previously unobserved for other β-galactosidases. Superposition of this complex with other β-galactosidase complexed from several GH families allowed the identification of residue Glu200 as the proton donor and residue Glu299 as the nucleophile involved in catalysis. The Psp-β-gal consisted of a glycoprotein containing seven *N*-linked oligosaccharide chains. Based on the putative primary structure, different sets of degenerate primers were constructed and used to generate gene specific fragments from genomic *Penicillium* DNA through conventional PCR. The Psp-β-gal gene sequence containing a total of seven protein coding exons and six introns showed that the amino acid sequence translated from the Psp-β-gal gene had 1011 residues. When the primary structure of the Psp-β-gal was aligned with a number of homologous GH-35, the significant structural homology was found with other fungal β-galactosidases from *Aspergillus candidus*, *Aspergillus niger* and *Talaromyces emersonii*.

Monomeric structure

The refined Psp-β-gal model showed 971 amino acid residues, 16 mannoses (MAN) and 12 *N*-acetylglucosamines (NAG) in seven oligosaccharide chains, 1252 water molecules, three sodium ions, four phosphate ions and nine ethylene glycol molecules. These suggested a glycoprotein that is the only known glycosylated β-gal structure (Rojas et al., 2004). The extracellular Psp-β-gal was a 120 kDa monomer composed of five distinctive structural domains (Fig. 8). (1) The first domain containing the catalytic site was a distorted TIM barrel comprising 355 amino acid residues (Leu41-Gly395) that was different than the representative TIM barrel consisting eight β/α repeats, and the β/α barrel in the Psp-β-gal lacking the fifth helix. Furthermore, the hydrogen-bonding pattern around the barrel was irregular on one side, and the presence of distortions introduced by proline residues and a β-bulge in the seventh strand. A similar feature was observed in the crystal structure of the *E. coli*-β-gal (Jacobson et al., 1994). (2) The second domain comprising Tyr396-Tyr576 consisted of 16 antiparallel β-strands and an α -helix at its C-terminus. The fold of this domain appears to be unique by the DALI (Holm and Sander, 1996) structural similarity search. The last seven strands of the domain formed a subdomain with an immunoglobulin-like fold in which the first strand (conventionally labeled *a* strand) was divided between the two β-sheets. In the Psp-β-gal structure this strand was interrupted by a 12-residue insertion which forms an additional edge-strand to the second β-sheet of the sub-domain. (3) The third domain (Trp577-Tyr665) was much smaller than the second and consisted of an α helix at its N-terminus followed by eight antiparallel β-strands based on a Greek key β-sandwich. A DALI search against known structures revealed that despite possessing a commonly observed the β-sandwich, this domain had a previously unobserved topology. Upon leaving the third domain, a short stretch of the polypeptide chain (Thr666-Pro687) passed through domain 5, forming a short β-strand prior to entering domain 4. (4) The domain 4 comprised Glu688-Leu861 that was composed of eight β-strands in a β-sandwich. This is a best described class II right-handed jelly roll (Stirk et al., 1992), but in this work, the jelly roll sandwich was somewhat unusual in that it was composed of one five-stranded sheet and one three-stranded sheet rather than the more regular two four-stranded sheets. (5) The fifth domain was based on a class I jelly roll and consisted of a total of eight strands divided into five and three-stranded β-sheets. The first strand of a conventional jelly roll was missing explaining why one of the sheets possesses only three strands. The other sheet included an additional strand formed by part of the connecting peptide which runs between domains 3 and 4.

Comparison with two known bacterial β-gal structures

Both structures of the Psp-β-gal and Ec-β-gal have similar five domains, but with the exception of the catalytic domain, there is little or no structural similarity between them in terms of the individual domain folds (Fig. 8). Only the catalytic domain based on a TIM barrel was comparable, but its relative position in terms of sequence and domain orientation was very different in the two molecules. For example, in the Ec-β-gal, the catalytic domain was the third domain in the sequence, but in the Psp-β-gal, that domain was the first. One of the most notable differences between the two enzymes was the relative spatial disposition of those four domains with respect to the TIM barrel, which occupies a central position in both structures. Unlike the Psp-β-gal (monomer) and Ec-β-gal (tetramer), the Tt-β-gal structure was trimeric. Each monomer of the Tt-β-gal was composed of only three domains, a TIM barrel, an α /β fold and a β fold domain (Hidaka et al., 2002). The TIM barrel was more regular than that seen in either the Psp-β-gal or Ec-β-gal. Neither of the latter two domains showed any resemblance to the non-catalytic domains observed in the Psp-β-gal. As a result of the structural differences between the crystallographic structures of *Penicillium sp.*, *E. coli* and *T. thermophilus* β-galactosidases, a reasonable superposition of the monomers was not possible. Nevertheless, a superposition of TIM barrel domains of the three β-galactosidases was possible (Fig. 8). An rms deviation of 2.8Å and 3.3Å was calculated for 236 and 291 C atoms pairs for the Psp-β-gal and Ec-β-gal, and for the Psp-β-gal and Tt-β-gal, respectively.

Galactose-binding site and carbohydrate moieties

The structural analysis of the galactose-binding site was based on the comparison of the crystallographic models of the native enzyme and its complex with galactose. A single galactose molecule was bound to the TIM barrel domain of the Psp-β-gal in the chair conformation with its O1 in the β-anomer configuration. The electron density around the sugar molecule clearly indicated its presence in the catalytic site. Amino acid sequence comparison from β-galactosidases of fungi (Rojas et al., 2004) showed that all the nine residues involved in galactose binding are well conserved among the different

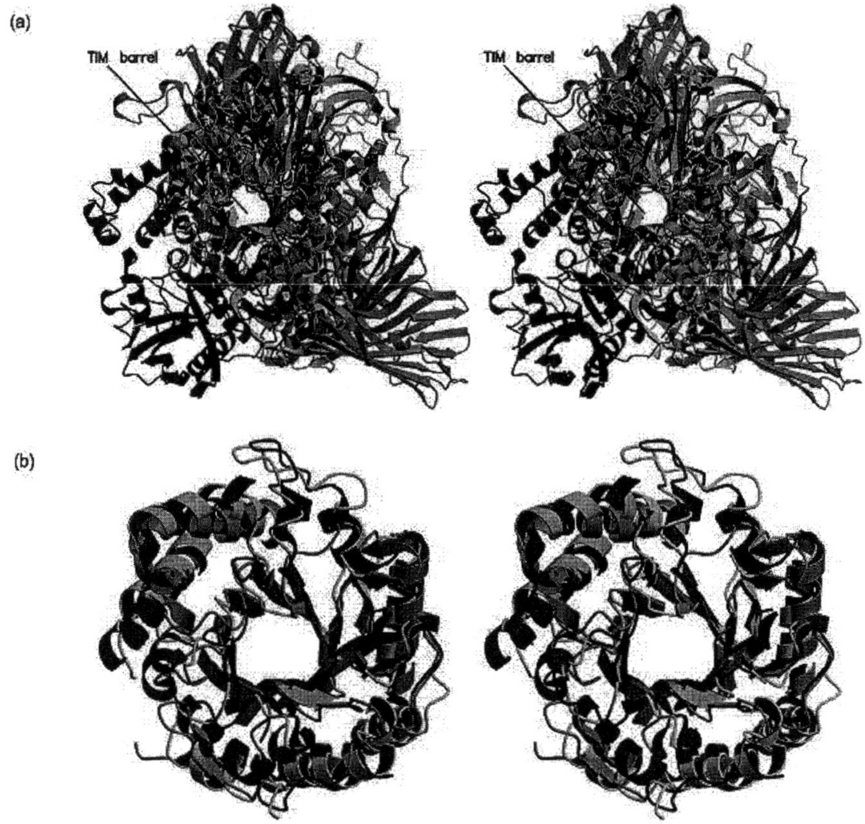

Figure 8 Comparison of structures of Psp-β-gal, Ec-β-gal and Tt-β-gal
(a) Stereo view of the structural superposition of the Psp-β-gal (in cyan), Ec-β-gal (in orange) and Tt-β-gal (in brown). (b) Stereo view of the superposition of their respective TIM barrel domains (with permission of Rojas et al., J. Mol. Biol. 343: 1281-1292, 2004).

species. Even the Tt-β-gal, whose TIM barrel catalytic domain was markedly different from that seen in the Psp-β-gal, presented a canonical heptapeptide, Asp-Ser-Tyr-Pro-Leu-Gly-Phe forming part of the sixth strand of the barrel. This was conserved in both sequence and conformation in the Psp-β-gal (residues 259–265). The Ec-β-gal did not have this heptapeptide sequence but had a tyrosine (Tyr503) in place of the serine whose orientation was inverted and pointed towards the galactose ligand substituting the tyrosine of the heptapeptide. Although the Psp-β-gal (GH-35) and Tt-β-gal (GH-42) belonged to distinct families within the GH-A clan, they conserved this heptapeptide sequence, despite showing a low sequence identify (only 17%) within the TIM barrel domains. A proton donor and a nucleophile/base, respectively, Glu200 and Glu299 were involved in the case of the Psp-β-gal. These residues resided on the fourth and seventh β-strand of the TIM barrel and the average distance between the four pairs of side-chain oxygen atoms was 4.9Å, within the appropriate range for retaining enzymes in a GH-35. This was confirmed by using ^{1}H NMR spectroscopy (Zinin et al., 2002). In the Psp-β-gal, the proton donor (Glu200) was orientated correctly with respect to the substrate by forming a hydrogen bond with the side-chain nitrogen of Trp809, which comes from the tip of the large finger-like loop between strands 6 and 7 in the fourth domain jelly roll. This large loop protruded from the jelly roll and entered the top of the TIM barrel domain from above its third β-strand. A loop in this position would be impossible in the structure of Ec-β-gal, due to the conformation of the connection between β-strand 3 and a-helix 3. However, in the Ec-β-gal this connection itself provides His418 which, by interacting with the proton donor Glu461, appears to play a role analogous to Trp809. This region of the TIM barrel was important for subunit–subunit contacts in Ec-β-gal tetramer. Specifically, a loop from domain 2 of a 2-fold related subunit in Ec-β-gal entered the top of the catalytic domain that may be essential for composing part of the active site (Jacobson et al., 1994). This may be the reason for quaternary structural integrity in Ec-β-gal for catalytic activity. In Psp-β-gal these interactions were prevented by a large insertion after the fourth strand of the barrel with respect to Ec-β-gal which would contain such subunit–subunit contacts. Thus, in the case of Psp-β-gal, the entire catalytic machinery would appear to be provided by a single subunit. Concerning carbohydrate moieties, the Protein Data Bank (PDB) showed about 70% of the deposited proteins as *N*-glycosylation sites (Asn-X-Ser/Thr, where X is not proline). Seven *N*-glycosylation sites have been localized in the electron density map of the Psp-β-gal, and three of them have 5, 7 and 9 monosaccharide residues each. Several oligosaccharides are wrapped around domains 3, 4 and 5 and a number of hydrogen bonds between amino acid residues at the protein surface and the carbohydrate moieties are observed. This is the first structural report of a glycosylated β-galactosidase with several long oligosaccharides attached to it.

β-GALACTOSIDASES FROM LACTIC ACID BACTERIA

The β-galactosidases (lactase) are found in most of food grade lactic acid bacteria (LAB) like *Streptococcus*, *Lactobacillus*, *Lactococcus* and *Bifidobacterium* species, though *Bifidobacterium* species currently belongs to *Actinomyces* with higher than 60% of G+C content (Biavati and Mattarelli 2001). LAB β-Galactosidases are known to catalyze not only β-D-galactoside linkage of lactose to produce glucose and galactose, but also have transgalactosylation activity to synthesize galacto-oligosaccharides. Both reaction activities are well characterized and applied in many food industries. Lactose hydrolyzed milk can reduce lactose intolerance problem, lactose hydrolyzed whey syrup and whey permeate can be utilized in frozen desserts, confectionary, bakery, fermentation products, and beverages (Crittenden and Playne, 1996). The galacto-oligosaccharides can also be employed as probiotic food ingredients (prebiotics), humectants, and emulsifiers, etc (Crittenden and Playne, 1996; Rastall and Martin, 2002). According to the classification of GHs, the sequence data from bifidobacteria and lactobacilli were found to be the GH-42 and GH-2, but none of them has been characterized extensively. Hung et al (2001) and Hung and Lee (2002) have characterized the GH-2 from *B. infantis* β-gal on the basis of substrate specificity, and yet only limited sequence data are available to conform which GHs they are classified (Hinz et al., 2004).

Comparison of amino acid sequence and phylogenetic analysis

Several inducible gene expression systems using LAB systems have also been developed for efficient and regulated overproduction of homologous and heterologous proteins (Halbmayr et al., 2008). β-Galactosidases from *Streptococcus thermphilus* (Lee et al., 1990; Schroeder et al., 1991), *Lactobacillus acidophilus* (Nguyen et al., 2007), *Lactobacillus reuteri* (Nguyen et al., 2007), *Lactobacillus sakei* (Obst et al., 1995), *Lactobacillus bulgaricus* (Schmidt et al., 1989),; *Bifidobacterium bifidum* (Møller et al., 2001), *B. infantis* (Hung et al., 2001, 2002), *Bifidobacterium longum* (Schell et al., 2002), *Bifidobacterium adolesscentis* (Hinz et al., 2004), and *B. breve* (Yi et al., 2005). Based on the role in carbohydrate metabolism, β-galactosidases can be classified in 4 groups like *LacA*, *LacZ*, *LacY*, and *LacG* family and their number of amino acids, source, and Genebank accession numbers are summarized in Table 1. Five conserved amino acid sequences, Try-57, Gly-83, Pro-104, Leu-107, and Tyr-161 were found in all *LacA* family protein. Highest similarity of *galA* with *LacA* family was found to *Bacillus halodurans* C-125 from *LacA* family β-galactosidase (35% identity). Eleven identical amino acid residues were found in β-galactosidases from bifidobacteria. Identical amino acid residues are Gly-23, 336, and 450, Glu-64 and 245, Try-98, Arg-23, Asp-236, Tyr-272, Met-372, and Prol-457.

In particular, β-gal III in *Bifidobacterium infantis* (Hung et al, 2001) and β-gal II in *Bifidobacterium adolescentis* (Hinz et al., 2004) differed from those known so far in that they are highly active towards Gal (β1-4) Gal-linkages, but showed very low transgalactosylation activity, resulting low galactooligosaccharides (GOS) from lactose or degrading prebiotic Gal-β-(1,4)-Gal-oligosaccharides. So far, the cloned β-galactosidases from bifidobacteria are classified into two families, GH-42 and GH-2. In our studies, β-gal I from *B. infantis* showing high activity towards lactose and synthesizing GOS appeared to be GH-2 family. Comparison of the deduced amino acid sequence of β-gal II in *Bifidobacterium adolescentis* (Hinz et al., 2004) with those of *Thermus thermophilus* (Hidaka et al., 2002) showed that the catalytic residues were all conserved and galactosde binding domain (W201) in both, suggesting a similar 3D structure as that of *T. thermophilus*. Another structure of 6-phospho-β-gal (PGALase) of *Lactococcus lactis* (Wiesmann et al., 1997) assigned to GH-1 showing lactose transport via phosphoenol-pyruvate dependent phosphotransferase (PTS) is known , but it is not related to β-galactosidase.

In the phylogenetic analysis of a β-galactosidase from *Bifidobacterium breve* B24 (Yi et al., 2005), *galA* was localized in unique branch indicating that a β-galactosidase from *Bifidobacterium breve* B24 was clearly distinguished from those of other *LacA* family β-galactosidases (Fig. 9). However, *galA* formed an unique subfamily in which *Bifidobacterium infantis* β-galactosidase III (Genebank accession number AAL02053), *Bifidobacterium longum* DJO10A (Genebank accession number ZP_00121008), *Bifidobacterium longum* NCC2705 β-galactosidase I (Genebank accession number NP_696337 691) were localized in the same branch (Yi, 2005).

Table 1 Classification of β-galactosidases on the basis of enzyme properties (Yi, 2005)

Family	Source	Genebank Accession #	Number of amino acids (identity; %)
Lac A	*Clostridium acetobutylicum* ATCC 824	NP_349128	982 (33)
	Bacillus halodurans C-125	NP_242888	689 (32)
	Bacillus halodurans C-125	NP_244568	672 (35)
	Bacillus subtilis subsp. *subtilis* 168	NP_391293	687 (31)
	Pyrococcus abyssi GE5	NP_127210	787 (21)
	Pyrococcus horikoshii OT3	NP_142480	778 (22)
	Sinorhizobium meliloti 1021	NP_437631	646 (29)
	Streptococcus pneumoniae TIGR4	NP_344609	595
	Thermotoga maritima MSB8	NP_228122	672 (34)
	Thermotoga maritima MSB8	NP_229000	649 (30)
	Xylella fastidiosa	NP_298130	612
	Yersinia pestis CO92	NP_404473	686 (33)
LacZ	*Streptococcus pyogenes* M1 GAS	NP_269647	1168
	Bacillus halodurans	Q9K9C6	1014
	Escherichia coli O157:H7	NP_308424	1024
	Escherichia coli O157:H7	NP_311985	1042
	Lactococcus lactis subsp. *lactis*	NP_268137	996
	Sinorhizobium meliloti 1021	NP_436544	755
	Sinorhizobium meliloti 1021	NP_386031	831
	Streptococcus pneumoniae TIGR4	NP_345155	2233
	Thermotoga maritima MSB8	NP_228998	1087
	Yersinia pestis CO92	NP_405234	1060
	Escherichia coli	P00722	1024
LacY	*Escherichia coli*	1PV6_A	417
	Escherichia coli	P16552	425
	Klebsiella oxytoca	P18817	416
	Escherichia coli	P30000	415
LacG	*Zea mays*	1HXJA	507
	Lactococcus lactis	1PBGA	468
	Sulfolobus acidocaldarius	P14288	491

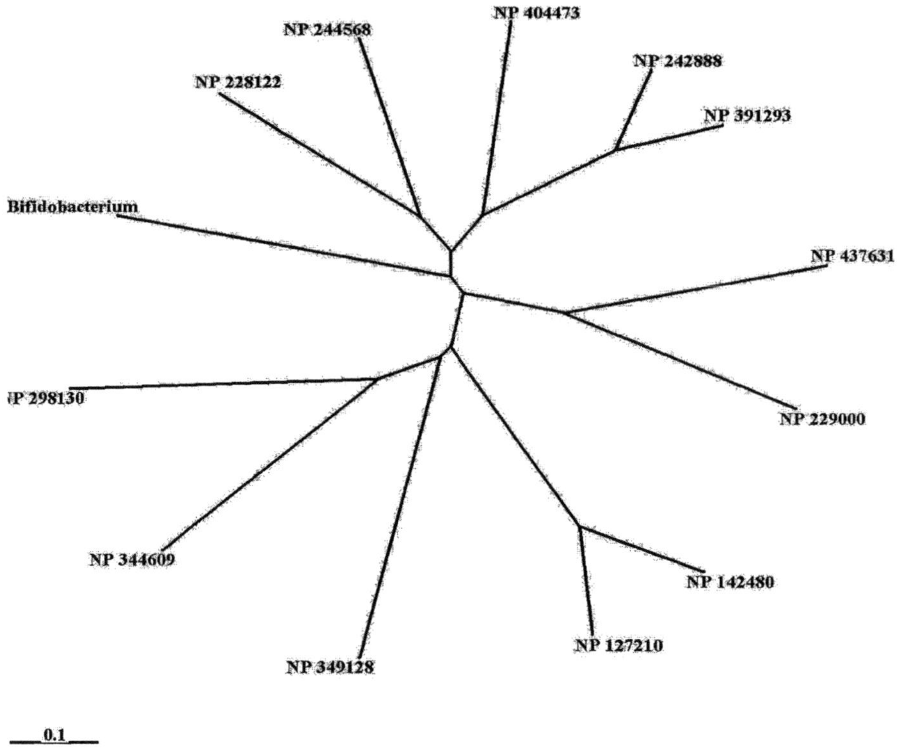

Figure 9 Phylogenetic tree of *galA* of *Bifidobacterium breve* B24 with other β-galactosidases from *LacA* family β-galactosidases from various bacteria
Genebank assession numbers for the published strains are follows: Bifidobacterium, *Bifidobacterium breve* B24; NP_298130, *Xylella fastidiosa* 9a5c; NP_344609, *Streptococcus pneumoniae* TIGR4; NP_349128, *Clostridium acetobutylicum* ATCC 824; NP_127210, *Pyrococcus abyss*i GE5; NP_142480, *Pyrococcus horikoshii* OT3; NP_229000, *Thermotoga maritima* MSB8; NP_437631, *Sinorhizobium meliloti* 1021, NP_391293, *Bacillus subtilis* subsp; NP_242888, *Bacillus halodurans* C-125; NP_404473, *Yersinia pestis* CO92; NP_244568, *Bacillus halodurans* C-125 NP_228122, *Thermotoga maritima* MSB8 (Yi, 2005).

ENZYME APPLICATIONS

The enzymatic hydrolysis can be accomplished by either free enzymes usually in a batch fermentation process, or by immobilized enzymes. An immobilized enzyme may be defined as the enzyme whose free movement has been restricted or somewhat confined to allow its use and reuse in a continuous catalytic process. The technology of enzyme immobilization has been applied successfully to the hydrolysis of lactose. Thus the inhibition of β-galactosidase by the accumulation of galactose formed during hydrolysis of lactose can be overcome by using these techniques. Many β-galactosidases have been immobilized on different types of matrix and their properties were studied (Li et al., 2007).

The properties of the enzyme and the final product specifications are the major factors that determine the exploitation of any particular immobilization technique. The enzymatic hydrolysis of lactose by β-galactosidase was found to be affected by the presence of some mineral ions naturally occurring in milk. The most important activators are magnesium and manganese, whereas sodium and calcium have a negative effect on the activity. Also the activity of β-galactosidase was found to be hampered by phytic acid present in soybean proteins, an important finding, since milk is incorporated with vegetable proteins in many food formulations. In the development of large-scale enzymatic manufacturing processes for lactose hydrolysis, the most important considerations are the purity, activity, non-toxicity, and the cost of the β-galactosidases. Several researches have been carried out to improve the microbial strains, among which one interest has been the increase of thermostability of strains and the β-galactosidase to allow lactose hydrolysis prior to and during pasteurization. These enzymes give higher conversion and are less prone to microbial contamination (Chen et al, 2008). Novel methods are disclosed for the enhanced expression and secretion of many lactases from filamentous fungi such as *Aspergillus* (Berka *et al.*, 1994).

Galacto-oligosaccharide formation by β-transgalactosyl-galactosidase

β-Galactosidase hydrolyzes terminal, non-reducing β-D-galactose residues in β-D-galactosides or lactose, but some of this enzyme catalyzes both hydrolytic and reverse transgalactosylation (EC 2.4.1.22: galactosyl transferase; GT) reaction. Apart from theoretical aspects, early research was prompted by nutritional concerns about the presence of these compounds as a flatulant factor in low-lactose milk (Burvall,1980), but more recently, interest in the reaction has been raised by observation that oligosaccharides may have beneficial effects as 'bifidus factors'-promoting the growth of desirable intestinal microflora. Also, the transferase reaction can be used to attach galactose to other chemicals and consequently has potential applications in the production of food ingredients, pharmaceuticals and other biologically active compounds (Raymond, 1998). Lactose hydrolysis catalysed by β-galactosidases has proven to be a very complex reaction. Apart from the actual hydrolysis product, glucose and galactose, many newly formed β-glycoside, mainly di-, tri, and tetrasaccharide, occur as kinetic intermediates, derived from so-called transgalactosylation reaction (Nakayama and Amachi, 1999). Because transgalactosylation products (galacto-oligosaccharides) are substrate of β-galactosidases-catalyzed hydrolysis, the composition of the product mixture changes quite significantly with progressing reaction time (Nakayama and Amachi, 1999).

The specific properties of oligosaccharides are very different depending on the formation of oligosaccharides, but some properties are common to almost all oligosaccharides. The sweetness of the oligosaccharide depends on structure and molecular mass of the oligosaccharides (Crittenden and Playne, 1996). Oligosaccharides are normally water soluble and mildly sweet, typically lower than sucrose and this low sweetness is useful in food production when reduced sweetness is desirable to enhance other food flavors. Compared with mono and disaccharides, the higher molecular weight of oligosaccharides provides increased viscosity, leading to improved body and mouthfeel. They can also be used to alter the freezing temperature of frozen foods, and

Table 2 Market volume of oligosaccharides

Class of oligosaccharide	Estimated production in 1995 (t)	Major manufacturers	Trade names
Galacto-oligosaccharides	15,000	Yakult Honsha (Jp)[a] Nissin Sugar Manufacturing Com.(Jp) Snow Brand Milk Products (Jp) Borculo Whey Products (NL)[b]	Oligomate Cup-Oligo P7L and others TOS-Syrup
Lactulose	20,000	Morinaga Milk Industry Co. (Jp) Solvay (Ger)[c] Milei GmbH (Ger) Canlac Corporation (Can)[d]	MLS/P/C
Lactosucrose	1,600	Ensuiko Sugar Refining Co. (Jp) Hayashibara Shoji Inc. (Jp)	Nyuka-Origo Newka-Oligo
Palatinose	5,000	Mitsui Sugar Co. (Jp)	ICP/O, IOS
Glucosyl sucrose	4,000	Hayashibara Shoji Inc. (Jp)	Coupling Sugar
Malto-oligosaccharides	10,000	Nihon Shokuhin Kako (Jp) Hayashibara Shoji Inc. (Jp)	Fuji-Oligo Tetrup
Isomalto-oligosaccharides	11,000	Showa Sangyo (Jp) Hayashibara Shoji Inc. (Jp) Nihon Shokuhin Kako (Jp)	Isomalto-900 Panorup Biotose & Panorich
Cyclodextrins	4,000	Nihon Shokuhin Kako (Jp) Ensuiko Sugar Refining Co. (Jp)	Celdex Dexy Pearl
Gentio-oligosaccharides	400	Nihon Shokuhin Kako (Jp)	Gentose
Soybean oligosaccharides	2,000	The Calpis Food Industry Co. (Jp)	Soya-oligo
Xylo-oligosaccharides	300	Suntory Ltd. (Jp)	Xylo-oligo
Chitosan oligosaccharides	200	Hubei Yufeng Bioeng. Co. Ltd. (China)	Chitosan-oligo

a) Japan; b) The Netherlands; c) Germany; d) Canada; e) France; f) Korea; g) Belgium
*Adapted from Crittenden and Playne, 1996; Baek and Lee, 2008.

to control the amount of browning due to Maillard reactions in heat-processed foods. Oligosaccharides provide a high moisture-retaining capacity, preventing excessive drying, and a low water activity, which is convenient in controlling microbial contamination. Although oligosaccharides possess these useful physicochemical characteristics, most of the interest in their use as food ingredients stems from their many beneficial physiological properties. Unlike starch and simple sugars, the currently available food-grade oligosaccharides are not utilized by mouth microflora to form acid or polyglucans. Hence, they are used as low-cariogenic sugar substitutes in confectionery, chewing gums, yogurts and drinks. Many oligosaccharides are not digested by humans and oligosaccharides have recently been described as one of several 'prebiotics', which can stimulate the growth of beneficial microflora (Gibson and Roberfroid, 1995).

Prebiotics are defined as "non-digestible food ingredients that beneficially affect the host by selectively stimulating the growth and/or activity of one or a limited number of bacteria in the colon" (Bomba et al., 2002). Some studies have shown that prebiotics target the activities of bifidobacteria and/or lactobacilli. These processes usually produce a range of oligosaccharides differing in their degree of polymerization and sometimes in the position of the glycosidic linkages. Residual substrates and monosaccharides are usually present after oligosaccharide formation, but such sugars can be removed by membrane or chromatographic procedures to form higher-grade products that contain pure oligosaccharides (Crittenden and Playne 1996).

Worldwide, there are 13 classes of food-grade oligosaccharides currently produced commercially (Table 2). Both the volume and diversity of oligosaccharide products are increasing very rapidly as their functional properties become further understood. Detailed production methods for various oligosaccharides have been reviewed by Playne (1994). The extensive research in this laboratory (Lee, unpublished) on the GOS production from 20-30 % lactose (w/v) using different lactases (native *vs* recombinant) showed that the recombinant lactases with high specific activity could produce more production than the native enzymes (Table 3). The solubility of high lactose appears to be a limiting factor to increase the yield of GOS in that the extremophilic enzymes could play an important role in this aspect.

Table 3. Production of galactooligosaccharides by native and recombinant lactases

Native extracts (%)	Expression level (fold)	Lactose (w/v, %)	GOS
Bacillus subtilis	NT	20	5
Thermus aquaticus	NT	20	30
Bifidobacterium	NT	20	25-34
Recombinant			
Lb. casei	200	20	35
Str. thermophilus	950	20	45
Bif. Infantis(β-galI)	500	20	55
Bif.breve	900	30	55

Author's compiled data (Lee, unpublished)

CONCLUSIONS

β-Galactosidase (lactase; EC 3.2.1.23), which hydrolyzes milk sugar lactose into glucose and galactose, is one of the well studied enzymes and is also the product of the lac oepron as well as the structuracl basis for the well knwn property of a-complementation. This enzyme, which belongs to the 4/7 superfamily of GHs is currently divided into GH-1, GH-2, GH-35 and GH-42, and yet the four families are so distantly related to each other. Their hydrolytic and transgalactosylation activities of multimeric β-galactosides appear to be different. Biochemical, molecular and phylogenetic aspects of the β-galactosidase genes from different microorganisms have been studied, but the known structure and function of different β-galactosidases are limited. The known 3D structures and functions of β-galactosidases from few microorganisms discussed will certainly reveal new insights into the structure-function relationships and protein stability under extreme conditions. However, further studies on β-galactosidases from other members of the group should not only address our understanding of structural properties of these proteins, but also have various industrial applications.

REFERENCES

Abdelrahim K A and Lee B H (1991) 'Production and characterization of ß-galactosidase from psychrotrophic *Bacillus subtilis* KL88', *Biotech Appl Biochem,* 13, 246-256.

Adams R M, Yoast S, Mainzer S E, Moon K, Palombella A L, Estell D A, Power S D, and Schmidt, B F. (1994), 'Characterization of two cold-sensitive mutants of the β-galactosidase from *Lactobacillus delbruckii* subsp. *Bulgaricus*', *J Biol Chem,* 269, 5666–5672.

Aguilar C F, Sanderson I, Moracci M, Ciaramella M, Nucci R, Rossi M, and Pearl L H (1997) 'Crystal structure of the beta-glycosidase from the hyperthermophilic archeaon *Sulfolobus solfataricus*: resilence as a key factor in thermostability', *J Mol Biol*, 271, 789-802.

Auld D S (2001), 'Zinc coordination sphere in biochemical zinc sites', *Biometals,* 14, 271–313.

Baek Y J and Lee B H, (2008) '*Chapter 13. Probiotics and Prebiotics as Bioactive Components in Dairy Products*' in Eds Guo M and Park Y W, Bioactive Components in milk and Dairy products, New York, Wiley-Blackwell Publishers, in Press.

Batra N, Singh J, Banerjee U C, Patnaik P R, and Sobt, R C (2002) 'Production of and characterization of a thermostable β-galactosidase from *Bacillus coagulans* RC553', *Biotech Appl Biochem,* 36, 1-6.

Berger J L, Lee B H, and Lacroix C (1995), 'Identification of new enzyme activities of several strains of *Thermus* species', *Appl Microbiol Biotech,* 44, 81-87.

Berka R M, Hucul J A, and Ward M (1998), 'Increased production of β-galactosidase *in Aspergillus oryzae*', United States Patent, 5736374.

Biavati B and Mattarelli P (2001) '*The family Bifidobacteriaceae*' Eds, Dworkin M, Falkow S, Rosenberg E, Schleifer K H, and Stackebrandt E, The Prokaryotes, New York, Springer, pp.1–70.

Bomba A, Nemcova R, Mudronova D, and Guba P, (2002) 'The possibilities of potentiating the efficacy of probiotics', *Trends Food Sci Technol*, 13,121-126.

Burvall A, Asp N, and Dahlqvist A (1980) 'Oligosaccharide formation during hydrolysis of lactose with "*Saccharomyces lactis*" lactase-part 3: digestibility by human intestinal enzymes in vitro', *Food Chem*, 5, 189–94.

Chen W, Chen H, Xia Y, Zhao J, Tian F, and Zhang H (2008), 'Production, purification, and characterization of a potential thermostable galactosidase for milk lactose hydrolysis from *Bacillus stearothemophilus*', *J Dairy Sci*, 91, 1751-1758.

Cieslinski H, Kur J, Bialkowska J, Baran I, Makowski K, and Turiewicz, M (2005), 'Cloning, expression, and purification of a recombinant cold-adapted β-galactosidase from Antarctic bacterium *Pseudoalteromonas* sp. 22b', *Protein Exp Purif*, 39, 27-34.

Coombs J M and Brenchley J E (1999), 'Biochemical and phylogenetic analyses of a cold-active β-galactosidase from the lactic acid bacterium *Carnobacterium piscicola* BA', *Appl Environ Microb,* 65, 5443–5450.

Crittenden R G and Playne M J (1996), 'Production, properties,and applications of food-grade oligosaccharides', *Trends Food Sci Technol*, 7, 353-361.

Dabrowski S, Maciunska J, and Synowiecki J (1998), 'Cloning and nucleotide sequence of the thermostable β-galactosidase gene from *Pyrococcus woesei* in *Escherichia coli* and some properties of the isolated enzyme', *Mol Biotechnol*, 10, 217-222.

Dabrowski S, Sobiewska G, Maciunska J, Synowiecki J, and Kur J (2000), 'Cloning, expression, and purification of the His_6-tagged thermostable β-galactosidase from *Pyrococcus woesei* in *Escherichia coli* and some properties of the isolated enzyme, *Protein Express Purif*, 19, 107-112.

Dainel R A, Haiech J, Denizot F, and Errington J (1997), 'Isolation and characterization of the *lacA* gene encoding beta-galactosidase in *Bacillus subtilis* and a regulatory gene, *lacR', J Bacteriol*, 179, 5636-5638.

EP0687732 (2004) *European patent on 'Hyperthermostable beta-galactosidase gene'.*

EP0606008 (2004) *European patent on 'Gene coding for hyperthermostable beta-galactosidase'.*

Gaur M N, Srivastava R, Prasad M, and Pereira B M J (2000), 'Evidence for the ionic modulation of β-galactosidase purified from goat epididymis', *J Anim Morphol Physiol*, 47, 25–32.

Georlette D, Blaise V, Collins T, D'Amico S, Gratia E, Hoyoux A, Marx J C, Sonan G, Feller G, and Gerday C (2004), 'Some like it cold: biocatalysis at low temperatures', *FEMS Microbiol Rev*, 28, 25-42.

Gibson C R and Roberfroid M B (1995), 'Dietary modulation of the human colonic microbiota: Introducing the concept of prebiotics', *J Nutr*, 125, 1401-1412.

Grogan D W (1991), ' Evidence that ß-galactosidase of *Sulfolobus solfataricus* is only one of several activities of a thermostable β-D-glycosidase', *Appl Environ Microbiol*, 57, 1644-1649.

Gul-Guven R, Guven K, Poli A, and Nicolaus B (2007), 'Purification and some properties of a β-galactosidase from the thermoacidophilic *Alicyclobacillus acidocaldarius* subsp. *rittmannii* isolated from Antarctica', *Enzyme Microb Tech*, 40, 1570-1577.

Gutshall K R, Trimbur D E, Kasmir J J and Brenchley J E (1995), 'Analysis of a novel gene and β-galactosidase isozyme from a psychrotrophic *Arthrobacter* isolate', *J Bacteriol*, 177, 1981–1988.

Halbmayr E, Mathiesen G, Nguyen T H, Maischberger T, Peterbauer C K, Eijsink V G H, and Haltrich D (2008), 'High level expression of recombinant β-galactosidase in *Lactobacillus plantarum* and *Lactobacillus sakei* using a sakasan P-based expression system', *J Agr Food Chem*, (Published Web 05/31/2008).

Heightman T D, Ermert P, Klein D, and Vasella A (1995), 'Synthesis of galactose- and *N*-acetylglucosamine-derived tetrazoles and their evaluation as β-glycosidase inhibitors', *Helv Chim Acta*, 78, 514–532.

Henrissat B, Callebaut I, Fabrega S, Lehn P, Mornon J P, and Davies G (1995), 'Conserved catalytic machinery and the prediction of a common fold for several families of glycosyl hydrolases', *Proc Natl Acad Sci US*, 92, 7090-7094.

Hidaka M, Fushinobu S, Ohtsu n, Motoshima H, Matsuzawa H, Shoun H, and Wakagi T (2002), 'Trimeric crystal structure of the glycoside hydrolase family 42 β-galactosidase from *Thermus thermophilus* A4 and the structure of its complex with galactose', *J Mol Biol*, 322, 79-91.

Hinz S W A, van den Broe L A M, Beldman K G, Vincken J P, and Voragen A G J (2004), 'β-Galactosidase from *Bifiodbacterium adolescentis* DSM20083 prefers β-(1,4)-galactosides over lactose', *Appl Microbiol Biotechnol*, 66, 276-284.

Hirata H, Fukazawa T, Negoro S, and Okada H (1986), 'Structure of a β-galactosidase gene of *Bacillus stearothermophilus*', *J Bacteriol*, 166, 722–727.

Holm L and Sander J, (1996), 'Mapping the protein universe. *Science* 273, 595-602.

Holmes M L, Scopes R K, Moritz R L, Simpson R J, Engler C, Pfeifer T F, and Dyall-Smith M L (1997), 'Purification and analysis of an extremely halophilic β-galactosidase from *Haloferax alicantei*', *Biochim. Biophys Acta*, 1337, 276–286.

Huber R E and Brockbank R L (1987), 'Strong inhibitory effect of furanoses and sugar lactones on β-galactosidase of *Escherichia coli*', *Biochemistry*, 26, 1526-1531.

Hung M N, Xia Z, Hu N T, and Lee B H (2001), 'Molecular and biochemical analysis of two-ß galactosidases from *Bifidobacterium infantis* HL96', *Appl Environ Microbiol*, 67, 4256-4263.

Hung M N and Lee B H (2002), 'Purification and characterization of recombinant β-galactosidases of *Bifidobacterium infantis* HL96', *Appl Microb Biochem*, 50, 257-264.

Jacob F and Monad J (1961), 'Genetic regulatory mechanisms in the synthesis of proteins', *J Mol Biol*, 3, 318–56.

Jacobson R H, Zhang X J, DuBose R F, and Matthews B W (1994), 'Three-dimensional structure of ß-galactosidase', *Nature*, 369, 761-766.

Jenkins J, Leggio L L, Harris G, and Pickersgill R (1995), 'β-Glucosidase, β-galactosidase, family-A cellulases, family-F xylanases and 2 barley glycanases form a superfamily of enzymes with 8-fold β/a-architecture and with 2 conserved glutamates near the carboxy terminal ends of β strand-4 and β strand-7', *FEBS Lett*, 362, 281–285.

Juers D H, Huber R E, and Matthews B W (1999), 'Structural comparisons of TIM barrel proteins suggest functional and evolutionary relationships between β-galactosidase and other glycohydrolases', *Protein Sci*, 8, 122–136.

Juers D H, Jacobson R H, Wigley D, Zhang X.J, Huber R E, Tronrud D E, and Matthews B W (2000), 'High resolution refinement of β-galactosidase in a new crystal form reveals multiple metal-binding sites and provides a structural basis for alpha-complementation', *Protein Sci*, 9, 1685–1699.

Juers D H, Heightman T D, Vasella A, McCarter J D, Mackenzie L, Withers S G, and Matthews B W (2001), 'A structural view of the action of *Escherichia coli* (*lacZ*) β-galactosidase', *Biochem*, 40, 14781–14794.

Karasová-Lipovová P, Strnad H, Spiwok V, Malá S, Králová B, and Russell N J (2003), 'The cloning, purification and characterisation of a cold-active β-galactosidase from the psychrotolerant Antarctic bacterium *Arthrobacter* sp. C2-2', *Enzyme Microb Technol*, 33, 836–844.

Knochel T, Iven A, Hester G, Gonzalez A, Bauerle R, Wilmanns M *et al* (1999), 'The crystal structure of anthranilate synthase from *Sulfolobus solfataricus*: functional implications', *Proc Natl Acad Sci US*, 96, 9479–9484.

Koshland Jr D E (1953), 'Stereochemistry and the mechanism of enzymatic reactions', *Biol Rev*, 28, 416-436.

Lauro B D, Strazzulli A, PeruginaG, Cara F La, Bedini E, Corsara M M, Rossi M, and Moracci M (2008), 'Isolation and characterization of a new family 42 β-galactosidase from the thermoacidophilic bacterium *Alicyclobacillus acidocaldarius*: Identification of the active site residues', *Biochem Biophy Acta – Prot Proteomics*, 1784: 292-301.

Lee B H, Robert N, Ricard N, and Jaques C (1990), 'Cloning and expression of β-galactosidase from *Streptococcus thermophilus* into *Saccharomyces cerevisiae*', *Biotechnol Lett*, 12, 499-504.

Li X, Zhou Q Z K, and Chen X D (2007), 'Pilot-scale lactose hydrolysis using ß-galactosidase immobilized on cotton fabric', *Chem Eng Process*, 46, 497-500.

Makowski, K, Białkowska A, Szczęsna-Antczak M, Kalinowska H, Kur J, Cieśliński H and Turkiewicz M (2006)', *FEMS Microbiol Ecol*, 170, 594-598.

Matthews B W (2005), 'The structure of *E. coli* ß-galactosidase', *Comptes Rendus Biol* 328, 549-556. Mikami B, Adachi M, Kage T, Sarikaya E, Nanmori T, Shinke, R and Utsumi S (1999), 'Structure of raw starch-digesting *Bacillus cereus* β-amylase complexed with maltose', *Biochemistry*, 38, 7050–7061.

Møller P L, Jørgensen F, Hansen O C, Madsen S M, and Stougaard P (2001), ' Intra- and extracellular beta-galactosidases from *Bifidobacterium bifidum* and *B. infantis*: molecular cloning, heterologous expression, and comparative characterization', *Appl Environ Microbiol*, 67, 2276-2283.

Moore J B, Markiewicz P, and Miller J H (1994), 'Identification and sequencing of the *Thermotoga maritima lacZ* gene, part of a divergently transcribed operon', *Gene*, 147, 101–106.

Moracci M, Volpe A L, Pulitzer J F, Rossi M, and Ciaramella M (1992), 'Expression of the thermostable β-galactosidase gene from the archaebacterium *Sulfolobus solfataricus* in *Saccharomyces cerevisiae* and characterization of a new inducible promoter for heterologous expression', *J Bacteriol*, 174, 873-882.

Müller-Hill B and Kania J (1974), 'Lac repressor can be fused to β-galactosidase', *Nature*, 249, 561–563.

Nagano N, Porter C T, and Thornton J M (2001), 'The $(\beta/a)_8$ glycosidases: sequence and structure analyses suggest distant evolutionary relationships', *Protein Eng*, 14, 845–855.

Nakayama T and Amachi T (1999) '*β-Galactosidases*' Eds, Encyclopedia of Bioprocess Technology: Fermentation Biocatalysis and Bioseparation, Vol. 3, John Wiley & Sons, New York, pp.1291.

Nguyen T H, Splechtna B, Krasteva S, Kneifel W, Kulbe K, Divine C, and Haltrich D (2007), 'Characterization and molecular cloning of a heterodimeric β-galactosidase from the probiotic strain *Lactobacillsu acidophilus* R22', *FEMS Microbio Lett*, 269, 136-144.

Ogasawara S, Abe K, and Nakajima T (2007), 'Pepper beta-galactosidase 1 (PBG1) plays a significant role in fruit ripening in bell pepper (*Capsium annuum*)', *Biosci Biotechnol Biochem*, 71, 309-322.

Obst M, Meding E R, Vogel R F, and Hammes W P (1995), 'Two genes encoding the β-galactosidase of *Lactobacillsu sakei*', *Microbiol*, 141, 3059-3066.

Ohtsu N, Motoshima H, Goto K, Tsukasaki F, and Matsuzawa H (1998), 'Thermostable β-galactosidase from an extreme thermophile, *Thermus* sp. A4: enzyme purification and characterization, and gene cloning and sequencing', *Biosci Biotechnol Biochem*, 62, 1539–1545.

Playne M J (1994), 'Production of carbohydrate-based functional foods using enzyme and fermentation technologies', *Int Chem Eng Symp Ser*, 137, 147-1 56.

Rastall R A and Maitin V (2002), 'Probiotics and synbiotics: towards the next generation', *Curr Opinion Biotech*, 13, 490-496.

Raymond R M (1998), 'Galactosyl-oligosaccharide formation during lactose hydrolysis: a review', *Food Chem*, 63:147-154.

Ring M and Huber R E (1990), 'Multiple replacements establish the importance of tyrosine-503 in β-galactosidase (*Escherichia coli*)', *Arch Biochem Biophys*, 283, 342–350.

Rojas A L, Nagem R A P, Neustroev K N, Arand M, Adamska E V, Eneyskaya A A, Kulmininskaya R C, Garratt A M, and Polikarpov I, (2004), 'Crystal structures of ß-galactosidase from *Penicillium* sp. and its complex with galactose', *J Mol Biol*, 343, 1281-1292.

Schell M A, Karmirantzou M, Snel B, Vilanova D, Berger B, Pessi G, Zwahlen M C, Desire F, Bork P, Delley M, Pridmore R D, and Arigoni F (2002), ' The genome sequence of Bifidobactero\ium longum reflects its adapatation to the human gastrointestinal tract', *Proc Natl Acad Sci*, 99, 14422-14427.

Schmidt B F, Adams R M, Requadt C, Power S, and Mainzer S E (1989), 'Expression and nucleotide sequence of the *Lactobacillus bulgaricus* beta-galactosidase gene cloned in *Escherichia coli'*, *J Bacteriol*, 171, 625-635.

Schroeder C J, Robert C, Lenzen G, McKay L L, and Mercenier A (1991), 'Analysis of the lacZ sequences from two *Streptococcus thermophilus* strains: comparison with the *Escherichia coli* and *Lactobacillus bulgaricus* β-galactosidase sequences', *J Gen Microbiol*, 137, 369-380.

Schrogel O and Allmansberger R (1997), 'Optimisation of the *BgaB* reporter system: determination of transcriptional regulation of stress responsive genes in *Bacillus subtilis*', *FEMS Microbiol Lett*, 153, 237–243.

Shipkowski S and Brenchley J E (2006), 'Bioinformatic, genetic, and biochemical evidence that some glycoside hydrolase family 42 ß-galactosidases are arabinogalactan type I oligomer hydrolases' *Appl Environ Microbiol*, 72, 7730-7738.

Skálová T, Dohnálek J, Spiwok V, Lipovová P, Vondráčková E, Petroková H, Dušková J, Strnad H, Králová B, and Hašek J (2005),' Cold-active β-galactosidase from *Arthrobacter* sp. C2-2 forms compact 660 kDa hexamers: Crystal structure at 1.9Å resolution', *J Mol Biol*, 353, 282-294.

Splechtna B, Nguyen T H, Zehetner R, Lettner H P, Lorenz W, and Haltrich D (2007)', Process development for the production of prebiotic glacto-oligosaccharides from lactose using ß-galactosidases from *Lactobacillus* sp', *Biotechnol J*, 2, 480-485.

Stirk H J, Woolfson D N, Hutchinson E G, and Thornton J M (1992), 'Depicting topology and handedness in jellyroll structures', *FEBS Lett*, 308, 1–3.

Synowiecki J, Grzybowska B, and Zdzieblo A (2006), 'Sources, properties and suitability of new thermostable enzymes in food processing', *Crit Rev Food Sci*, 46, 197-205.

Taniguchi A Y and Takano K (2004), 'Purification and properties of β-galactosidase from Tilapia intestine: digestive enzyme of Tilapia-X', *Fish Sci*, 70, 688–694.

Trincone A, Nicolaus B, Lama L, Nucci R, Rossi M, and Gambacorta A (1994), 'Enzymatic synthesis of carbohydrate derivatives using β-glycosidase of *Sulfolobus solfataricus*', *Biocatalysis*, 10, 195–210.

Van Laere K M J, Abee T, Schols H A, Beldman G, and Voragen A G J (2000), ' Characterization of a novel beta-galactosidase from *Bifidobacterium adolescentis* DSM 20083 active towards transgalactooligosaccharides', *Appl Environ Microbiol*, 66, 1379-1384.

Vetere A and Paoletti S (1998), 'Separation and characterization of three beta-galactosidases from *Bacillus circulans*. *Biochim Biophys Acta*, 1380, 223-231.

Wanarska M, Kur J, Pladzyk R, and Turkiewicz M (2005), 'Thermostable *Pyrococcus woesei* beta-galactosidase-high level expression, purification and biochemical properties', *Acta Biochem Pol*, 52, 781-787.

Weinstock G M, Berman M L, and Silhavy T J, (1982) '*Chimeric genetics with β-galactosidase*' Eds: T.S. Papas, M. Rosenberg and C. Chirikjian, Gene Amplification and Analysis vol. 3, New York, Elsevier, pp. 27–64.

Wiesmann C, Hengstenberg W, and Schulz G E (1997), 'Crystal structures and mechanism of 6-phospho- ß-galactosidases from *Lactococcus lactis*', *J Mol Biol*, 269, 851-860.

White P L, Wynn-Williams D D, and Russell N J (2000), 'Diversity of thermal responses of lipid composition in the membranes of the dominant culturable members of an Antarctic fellfield soil bacterial community', *Antarct Sci*, 12, 386–393.

Yi S H (2005), Biochemical *and molecular characterization of beta-galactosidase from Bifiodbacterium breve B24*, Ph.D thesis, McGill University, Montreal, Canada

Zinin A I, Eneyskaya E V, Shabalin K A, Kulminskaya A A, Shishlyannikov S M, and Neustroev K N (2002), '1-*O*-Acetyl-β-d-galactopyranose: a novel substrate for the transglycosylation reaction catalyzed by the β-galactosidase from *Penicillum sp.*' *Carbohydr Res*, 337, 635–642.

STRUCTURAL FEATURE OF THE ARCHEAL GLYCOGEN DEBRANCHING ENZYME FROM *SULFOLOBUS SOLFATARICUS*

Eui-Jeon Woo, Seungjae Lee, Hyunju Cha, Jong-Tae Park, Sei-Mee Yoon, Hyung-Nam Song, and Kwan-Hwa Park

ABSTRACT

TreX is an archaeal glycogen debranching enzyme which catalyses both the α-1,4-transferase and α-1,6-glucosidase activity, similar to GDEs in mammals and yeast. It exists in two oligomeric states in a solution, as a dimer and a tetramer, with its tetramer showing a four-fold higher catalytic efficiency compared to that of the dimer. TreX has a high specificity for hydrolysis of maltohexaosyl α-1,6-β-cyclodextrin, showing the high preference for side chains consisting of 6 glucose residues or more. The structure of TreX reveals the unique arrangement of the subunits with the substrate binding grooves connected each other in tetrameric form, adopting a suitable architecture for binding to the branched glycogen. The analysis of the active cleft shows that the helix at the substrate binding groove provides a platform for the stable binding to the longer substrate, explaining the substrate specificity of TreX. Based on the structural analysis and biochemical study, we suggest that the unique dual catalytic property of the archaeal debranching enzyme may be associated to the tetramer, giving rise to the modulation of the activities of TreX upon oligomerization of its subunits.

Key words: *Sulfolobus solfataricus*; crystal structure; glycogen debranching enzyme; oligomerization;

INTRODUCTION

Glycogen-debranching enzyme (GDEs) plays an important role in carbohydrate metabolism. A deficiency of this enzyme causes glycogen storage disease (Braun et al., 1996). GDEs in eukaryotes are known to be bifunctional, possessing both 4-α-glucanotransferase (EC 2.4.1.25) and amlyo-α-1,6-glucosidase (EC 3.2.1.33) activities within a single polypeptide chain (Bates et al., 1975; Nakayama et al., 2001). The enzyme transfers maltosyl groups out of the side chains of phosphorylase and limit glycogen to the non-reducing end glucose of the main chain to form an α-1,4-linkage. The enzyme then specifically hydrolyses the remaining glucosyl residue at the branch point to produce glucose and maltodextrins (Liu et al., 1991).

GDEs are distributed in some bacteria and archaea as well as in mammalian cells and yeast (Fig. 1). Although some genes encoding the glycogen debranching enzymes from bacteria have been cloned and sequenced, the properties of the corresponding enzymes and their roles in glycogen synthesis and degradation have not yet been fully elucidated (Maruta et al., 1996b; Maruta et al., 1996a; Abdullah and Whelan, 1963; Nelson et al., 1972; Walker and Whelan, 1960). The molecular mass of prokaryotic GDEs ranges from 75 to 90 kDa, which is approximately half the molecular mass of eukaryotic GDEs (Fig. 1). Although the size of the GDE from *S. solfataricus* is approximately half of the molecular mass of eukaryote GDE, some prokaryote GDE is of special interest as it carries out debranching on the amylopectin or glycogen substrate due to the cooperation of the two activities (Table 1).

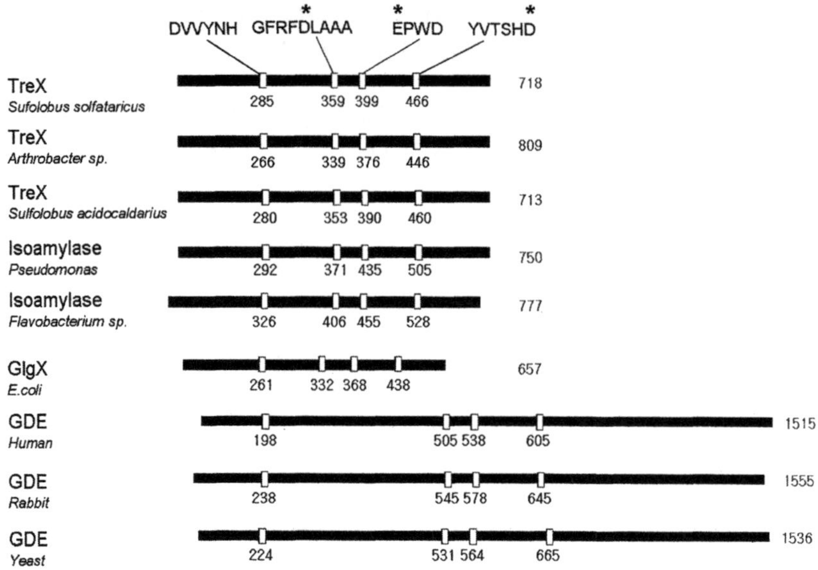

Figure 1 TreX and related enzymes with high sequence homology

White boxes indicate four conserved motifs in GH13 enzymes. Corresponding aa and total aa were given with number

Table 1 Catalytic properties of TreX and relative enzymes

	Isoamylase	GlgX	AmyX	TreX	Eukaryotic GDEs
Action mode					
α-1,6-hydrolyzing activity	+	+	+	+	+
α-glucanotransferase activity	-	-	-	+	+
Major products from					
Maltooligosaccharides	-	-	-	Series of maltooligo-saccharides (DP≥1)	Series of maltooligo-saccharides (DP≥2)

Isoamylase (Katsuya et al., 1998); GlgX, pullulanase from *E. coli* (unpublished data); AmyX, pullulanase from *Bacillus subtilis* (Hong et al., 2008).; Eukaryotic GDEs (Nakayama et al., 2001).

The *treX* gene is one of the three genes located in the trehalose operon consisting of *treX*, *treY* (malto-oligosyltrehalose synthase) and *treZ* (malto-oligosyltrehalose trehalohydrolase) (Maruta et al., 1996a; Maruta et al., 2000; Maruta et al.; 1996b). Glycogen is recognized as the major starting material for trehalose synthesis among archaea. TreX is known to debranch the side chain of glycogen into maltodextrin, which is further converted to trehalose by TreY and TreZ (Maruta et al., 1996a). Unlike other reported microbial glycogen debranching enzymes, TreX exhibits 4-α-glucanotransferase as well as the amlyo-α-1,6-glucosidase activity, catalyzing the transfer of α-1,4-glucan oligosaccharides from one molecule to another using various substrates such as glycogen and amylopectin (Park et al., 2007). It is the only glycogen debranching enzyme found in bacteria/archaea with 4-α-glucanotransferase activity to date.

A TYPICAL MOTIF FOR DEBRANCHING ENZYME IN THE ACTIVE SITE

The overall structure of TreX monomer is similar to its homologue debranching enzyme of *Pseudomonas* isoamylase which consists of three domains of the N-terminal domain, the central domain and the C-terminal domain (Fig. 2A; Woo et al., 2008). The N-terminal domain (aa 1-153), comprising six-β strands and forming a β-sandwich, was previously observed for several enzymes that act on a branched substrate (Jespersen et al., 1991; Katsuya et al., 1998). The central catalytic domain contains the characteristic $(\beta/\alpha)_8$-barrel motif found in a wide range of the α-amylase family, consisting of eight parallel β-strands surrounded by eight parallel α-helices. The C-terminal domain (aa 600-718) has a Greek-key motif showing local similarity to the C-terminal domain in CGTase. The superposition of TreX to the isoamylase of *Pseudomonas* shows similarity in most regions with r.m.s.d value of 1.13 Å over 562 Cα atoms while some part of the substrate binding groove, including the helix region of aa 231-237, an inserted region of aa 601-612, and deleted regions of aa 392-395 and aa 371-374 show variations (Fig. 2B).

The conserved calcium ion observed in isoamylase and other α-amylases is not found in the TreX structure, whereas it has two disulfide bridges between residue 505 and 519 and between residue 254 and 261.

The geometry of the active site of TreX is highly similar to those seen in isoamylases and related debranching enzymes with the three critical catalytic residues (Asp 363, Glu 399 and Asp 471) located at the bottom of the active-site cleft (Fig. 2C). Structural conservation observed in those debranching enzymes includes the deeply buried -1 subsite, such as Tyr 244, His 291, Arg 361, and His 470. The tyrosine residue Y244 in the motif NYWGYDP (residue 555-561) known to be essential for the van der Waals interactions with glucose rings at subsites -1 and -2 in most debranching enzymes was found to form a hydrogen bond to Asp 286 in TreX. The side chains of the three carboxylate residues Asp 363, Glu 399 and Asp 471 are located in close proximity to the subsite -1, possibly bridging sugar rings at -1 and +1. In the ligand complex structure, the side-chain of the catalytic nucleophile Asp 363 is bonded to the C1 atom of the -1 ring of the ligand. The Glu 471 is hydrogen bonded to both the O2 (3.0 Å) and O3 (3.8 Å) hydroxyl groups of the -1 ring, which may reduce the electronegativity of the O2 atom of the -1 glucosyl residue. The residues of Glu 399 and Asp 363 are located in the appropriate range for the retaining enzyme with a distance of 3.5 Å and 2.9 Å in the acarbose-free and acarbose complex structures, respectively (Henrissat and Davies, 1997). Additional Sugar-aromatic stacking interactions were found at site +2 with Trp 401 and at site + 3 with Tyr 408 in the complex structure. Residues of Trp 401 and Tyr

408 are located in a row next to Glu 399, suggesting that the branched substrate of TreX is likely to bind in a curved manner, making a sharp bend at the bond between the -1 and +1 rings, as observed in the pullulanase (Mikami et al., 2006). The fact that TreX recognizes the branched points in the common active site with isoamylase and pullulanase along with the structural conservation in the active site with other debranching enzymes shows that TreX follows the same catalytic mechanism for amylo-α-1,6-glucosidase activity (Fig. 2).

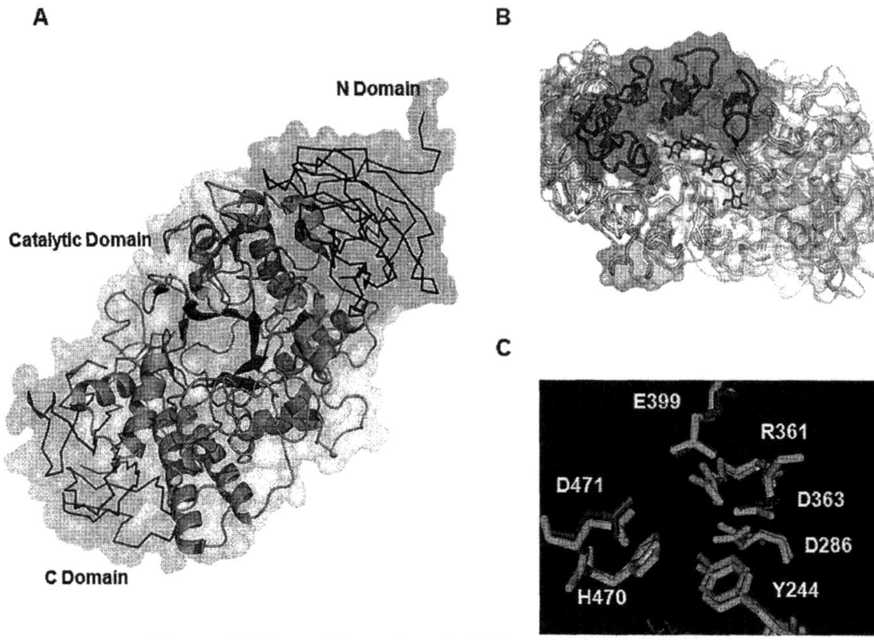

Figure 2 Three-dimensional (3D) structure of TreX

A, the monomer of TreX consist of 3 domains. Catalytic domain was drawn in cartoon model and the others drawn in ribbon; B, comparison of TreX monomer to that of isoamylase from *Pseudomonas amyloderamosa*. The difference in substrate cleft was highlighted in shaded surface. C, Superposition of essential catalytic residues of TreX to those of isoamylase

An analysis of the substrate binding region shows that a region of residues (aa 228-238) forms a helix α4 and protrudes at the bottom of the substrate groove, interacting with the bound acarbose ligand (Fig. 3A). The residue Phe 232, adjacent to Phe 557 and Phe 491, is in the position of subsite -4 and is involved in a stacking interaction with the longer substrate. The helix α4 is observed only in TreX but is not found in the isoamylase or pullulanase structure (Fig. 3B). Given the typical left-handed helical configuration of the substrate in a negative electrostatic pocket, this helix may provide a platform for the stable binding of the substrate. In comparison to isoamylase and pullulanase, TreX exhibits uniquely higher activity on the branched substrate with longer maltooligosaccharides (Fig. 3C; Park, 2008).

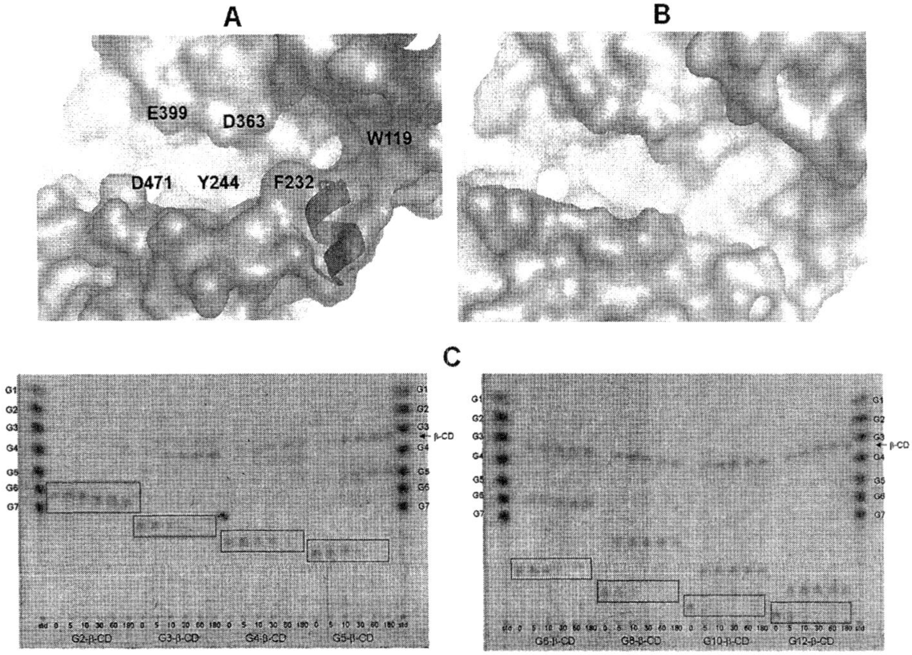

Figure 3 Substrate binding region of TreX (A) and isoamylase (B)
The helix uniquely observed in TreX substrate cleft was shown in cartoon. C, substrate preference of TreX determined with branched (DP2-DP12)-β-cyclodextrins

OLIGOMERIZATION OF TREX

TreX exists in two oligomeric forms in a solution, a dimer and a tetramer. The *Pseudomonas* isoamylase, the closest homologue of TreX, is a monomer with a single amylo α-1,6-glucosidase activity (Mikami et al., 2006, MacGregor et al., 2001). Gel permeation chromatography and sedimentation equilibrium analytical ultracentrifugation revealed that the enzyme exists as a dimer at pH 7.0, and as a mixture of dimers and tetramers at pH 5.5 (Fig. 4). TreX existed mostly as a tetramer in the presence of dimethyl sulfoxide (DMSO) at pH 5.5-6.5. Interestingly, the tetramer showed a 4-fold higher catalytic efficiency than that of the dimer. To date, there was no such a report of the activity modulation upon the oligomerization among the glycoside hydrolase family 13 (GH13) members.

Both the dimeric and the tetrameric structures have been determined and analysed. In the dimeric structure, each monomer is oriented side by side with a two-fold axis of rotation at the center in which the active site of each monomer faces the same side (Fig. 5A). In the tetrameric structure, two dimers form a tetramer in which two dimers face each other with a slight offset as a dimer of dimers with the two-fold rotation axis at the center between the dimmers (Fig. 5B, C).

This unique configuration of tetramer yields its substrate binding grooves of each subunit in such a way that they are connected inside the tetramer. The dimeric arrangement of TreX produces a buried interface corresponding to 5.6 % of the solvent accessible surface per monomer whereas the dimer of dimer arrangement produces the buried interface of additional 3.3 % suggesting the weaker binding interaction for tetramerization in compared to the dimerization.

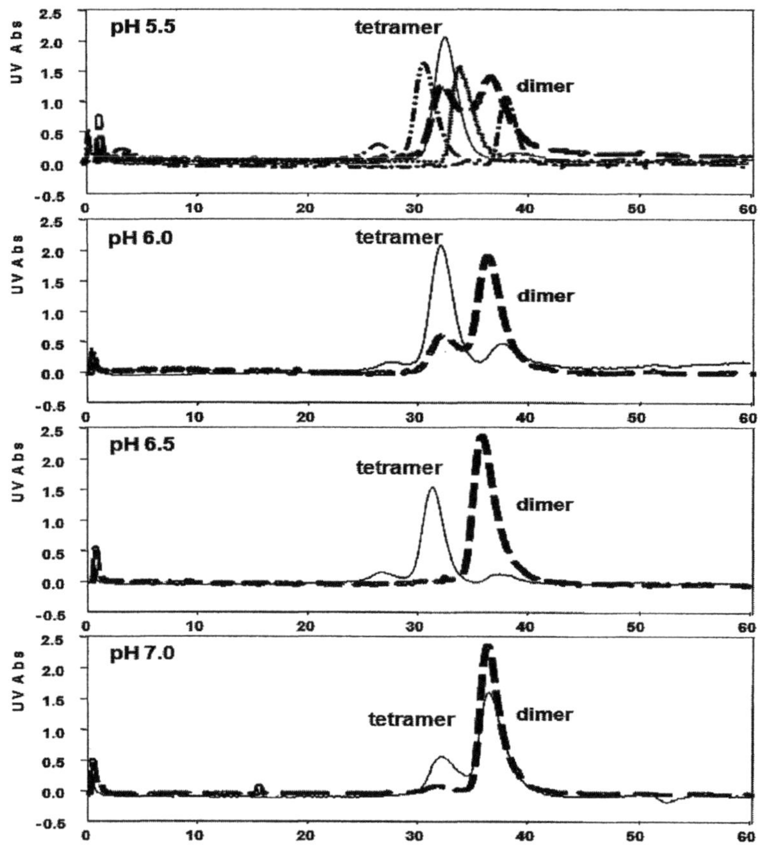

Figure 4 Oligomeric state of TreX in various pH conditions

Protein sample was analyzed by gel permeation chromatography. Solid, TreX (w/o DMSO); bald dash, TreX (10% DMSO); the other peaks in pH 5.0 were protein standards as order in appearance, 400kDa, 150kDa, and 66kDa respectively (adopted from Park et al., 2008).

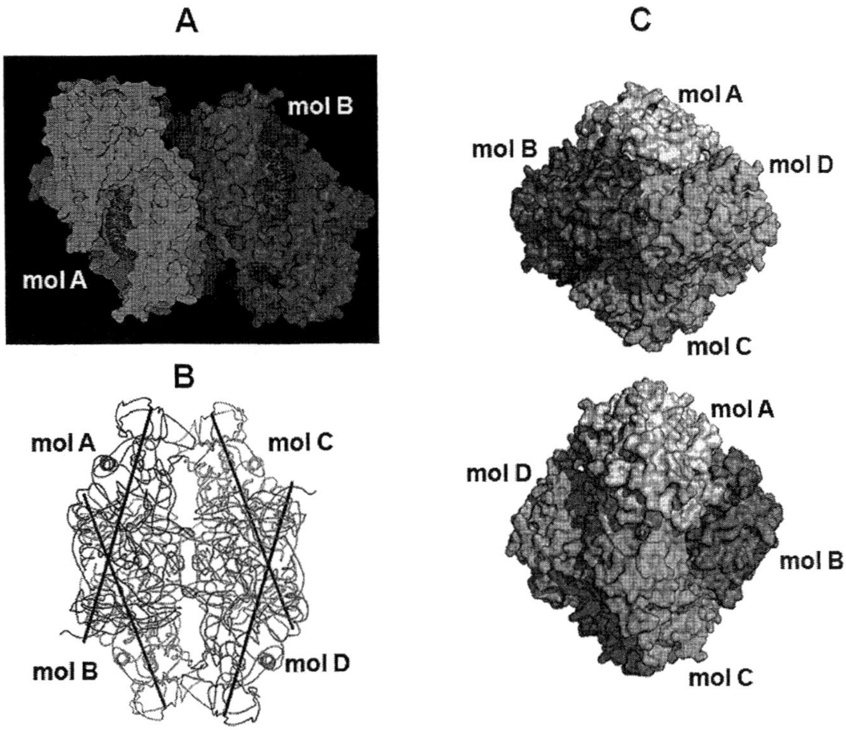

Figure 5 Oligomerization of TreX

A, dimerization of TreX: each active site faces the same side with the ligand in the active site drawn in mesh; B, tetramerization of TreX: two dimers (mol A-B and mol C-D) face each other orienting the active sites close; C, surface model of teterameric TreX.

In the tetrameric arrangement, two regions in a dimer subunit are placed at the active sites of the other dimer molecule, resulting in the change of the geometry to reshape the active site. Two regions, one from the loop region in the N-domain (aa 92-97) and the other from the loop region of the catalytic domain (aa 315-322), are placed right above the substrate binding cleft (Fig. 6A).

The structural feature of shaping the active site by oligomeric arrangement observed in TreX is reminiscent to other maltogenic amylases (Park et al., 2000;Yokota et al., 2001;Cheong et al., 2002). For example, the N-domain of one molecule is located near the active site of the other subunit in the pullulanase dimeric structure where several residues of N-domain interact with the bound sugar moieties at the other monomer. The formation of this dimer is thought to be realized only in the presence of longer sugar chains in a solution. ThMA also form a domain-swapped dimer, in which the N-domain of one subunit is involved in extensive interactions with the $(\beta/\alpha)_8$-barrel domain of the other subunit. This dimeric structure of ThMA provides monomer-dimer equilibrium to modulate substrate specificity of the enzyme toward starch, pullulan, and cyclodextrin (Park et al., 2005).

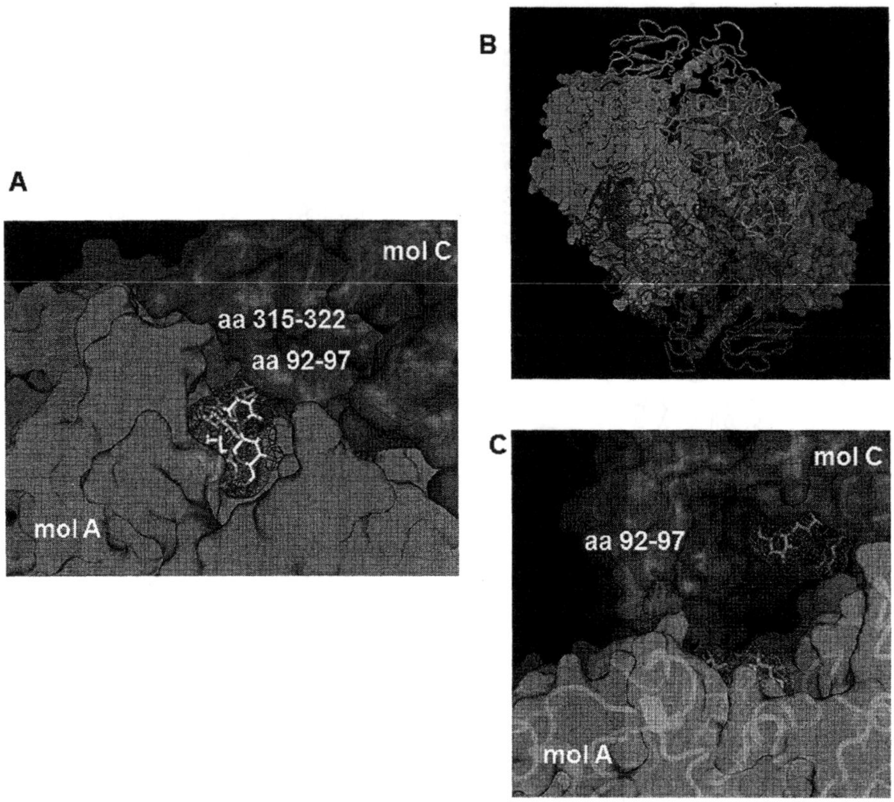

Figure 6 Reshaping the active site of TreX by tetramerization
A, interaction of two regions of one subunit (dimer C-D) to the active site of the other subunit (dimer A-B) in tetramer; B, Two dimers forms a tetramer with a slight offset. One dimer was drawn in cartoon with the other dimer drawn in surface; C, Connected substrate binding groove formed by tetramerization. The ligand was shown in sticks and mesh.

CHANGE OF THE SUBSTRATE BINDING CLEFT UPON OLIGOMERIZATION

By adopting a tetrameric arrangement, TreX not only reshapes the active site but also generates a connected hollow cavity inside the tetramer molecule. The tetrameric structure of TreX reveals a novel configuration that two dimers face each other such that the substrate binding cleft are close to each other with a slight offset (Fig. 6B). While the substrate binding cleft is wide and open at one side of the active site in the dimer, the association of this region to the active site in the tetramer results in a channel-like cavity with structural lids on the active site (Fig. 6C).

The formation of the channel-like cavity generated by the structural lids is accompanied by a conformational change of the loop (aa 399-416) in the active site (Fig. 7A). This loop shifts outward from the active center and moves away from the helix $\alpha7$ (aa 364-370) in the tetrameric form, causing the substrate binding groove to connect each other and extend. However, the corresponding loop is placed adjacent to the helix α7 in the dimeric form, closing the edge of the cleft (Fig. 7B). Due to the shift of the loop, the substrate binding groove is open at one side and extends to the nearby groove of the adjacent molecule to form connected channels with an approximate width of 10-15 Å (Fig. 7C). Given the strong electrostatic potential distribution of the negative charge along this groove and the geometrical length of the holes, the connected channel appears to be a suitable architecture for the binding of a branched substrate.

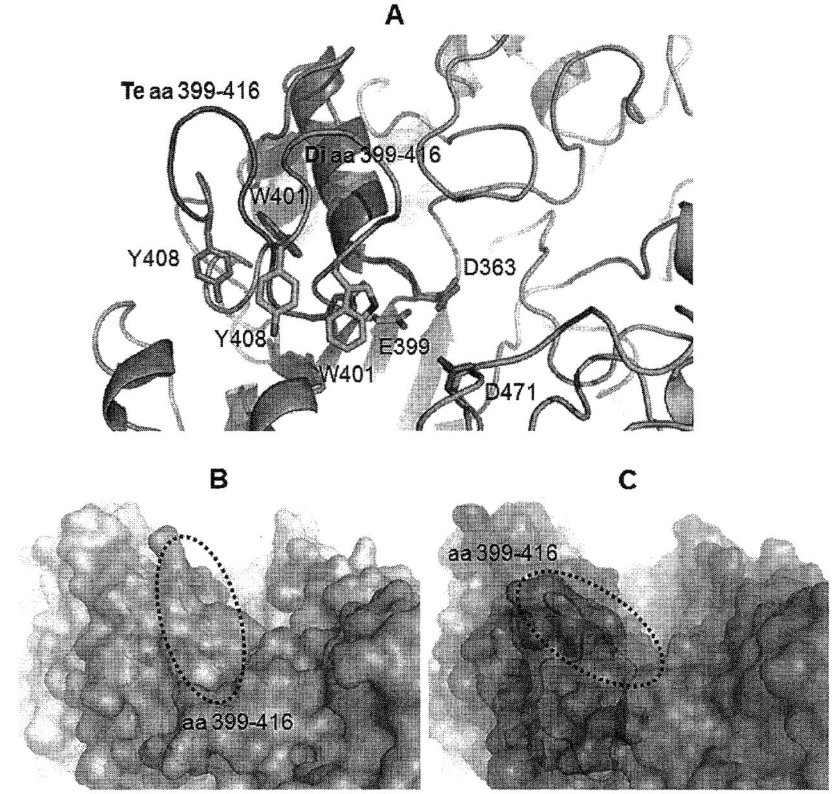

Figure 7 Conformational change at the active site of TreX by tetramerization
A, a flexible loop (aa 399-416) with W401 and Y408 were shown both in dimer (Di) and tetramer (Te). Three catalytic residues were additionally drawn in sticks. B and C; surface representation of the substrate binding groove. The loop in dimer (B) undergoes a significant conformational change generating an open and connected surface groove in tetramer (C).

The loop region (aa 399-416) was previously found to adopt a flexible conformation with an induced fit motion to accommodate the binding of different substrates in many enzymes of this family (Barends et al., 2007; Przylas et al., 2000; Mikami et al., 2006; Hondoh et al., 2003). However, the conformational change observed in the TreX tetramer takes a significantly different conformation with a large deviation of Cα. No conformational change has been reported upon oligomerization. This loop contains essential residues such as the catalytic nucleophile Glu 399 and substrate interacting residues Trp 401 and Tyr 408. In the dimeric form, the side-chain of Trp 401 and Tyr 408 rotates by approximately 90° to allow stacking interactions at subsites +3 and +2, whereas in the tetramer, the side chain of the Trp 401 folds back and the Glu 399 and the Tyr 408 move far away, at 4.5 Å and 8.3 Å respectively, resulting in Glu 399 being too far from the other catalytic residues to function as a nucleophile (Fig. 7A). It is assumed that TreX adopts this conformation of an extended groove to allow binding of branched substrates such as glycogen, but subsequent conformational change should occur for catalysis.

MODULATION OF THE ACTIVITY BY OLIGOMERIZATION

The existence of the dimer and the tetramer in solution and the existence of the structural lid in the active site generated by oligomerization whereas no similar structural component exists in the dimer suggest that the 4-α-glucanotransferase activity of this bi-functional enzyme may be related to the oligomeric arrangement. In fact, extensive structural analysis for enzymes with 4-α-glucanotransferase activity shows an interesting feature that they commonly contain a region of a structural lid above the substrate binding cleft, as observed in the TreX tetramer. For example, neopullulanase shows a narrow cleft at the active site with a structural lid that is formed through dimerization. This oligomerization is thought be essential for the enzyme to exhibit the unique substrate specificity (Hondoh et al., 2003; Kim et al., 1999; Lee et al.; 2002). In the structure of *Thermotoga maritima* 4-α-glucanotransferase, a loop in the N-terminal domain from the same molecule forms a clamp over the active site that captures the sugar rings of the substrate at the acceptor binding site (Roujeinikova et al., 2002). A lid (aa 627-630) protruding from the helix bundle significantly moves in the active site upon binding of the substrate in 4-α-glucanotransferase (Imamura et al., 2003), while the loop region (aa 246-258) is involved in acceptor binding in amylomaltases structure (Barends et al., 2007). In CGTase, tyrosine or phenylalanine protrudes in that region while various mutants at this position affect the cyclization (Schmidt et al., 1998).

The postulation for involvement of the lid generated by the oligomerization to the 4-α-glucanotransferase activity of TreX was supported by the fact that the mutational change in this region, Glu 93, showed a significant change in the 4-α-glucanotransferase activity without considerable alteration of the amylo-α-1,6-glucosidase activity. Since this region is located a considerable distance from the active site of its own subunit (45 Å), it is not possible for this region to participate into the catalytic mechanism of the same monomer.

In general, the active site of 4-α-glucanotransferase is composed of the glycon binding site, the aglycon binding site, and the acceptor binding site whereas the aglycon binding site may be identical to the acceptor binding site (Watanabe et al., 2006). The 4-α-glucanotransferase activity catalyses a disproportionation reaction in which a segment of 1,4-α-D-glucan is transferred to the 4-hydroxyl group in the acceptor sugar molecule. The location of the acceptor binding site to the region corresponding to the structural lid in TreX implies a possible role of this region for interacting with and stabilizing the acceptor molecule during the catalysis. In neopullulanase and CGTase structures, significant density of the putative acceptor has been reported in the region corresponding to the site under the structural lid (Hondoh et al., 2003; Strokopytov et al., 1996). Thus, through the effective use of these two activities, TreX may produce long maltooligosaccharides from a glycogen substrate in the process of trehalose biosynthesis. The modulation of the enzyme activities by oligomerization of its subunit is not new to amylases. For example, it was known that dimerization of ThMA monomer modifies the wide active site cleft by the N-terminal domain to form a narrow and deep groove-shaped substrate-binding site, making the access of slim h-CD or maltooligosaccharides preferable to bulky starch or pullulan.

CONCLUSIONS

TreX is of special interest as it carries out debranching on the amylopectin or glycogen substrate due to the cooperation of the two activities, the 4-α-glucanotransferase (EC 2.4.1.25) and amlyo-α-1,6-glucosidase (EC 3.2.1.33). Functional features observed in various amylases appear to be combined in this archaeal GDE to yield the efficient cleavage of the branched substrate (Fig. 9). An analysis of the structure reveals the unique tetrameric configuration of TreX reshaping the active site with an additional region from the N-terminal domain. This result provides the structural basis for the modulation of activity upon oligomerization and the bi-functional mechanism.

The fact that TreX catalyzes dual activities and the strict structural conservation of the motif for debranching enzyme in TreX indicates this enzyme is functionally close to the isoamylase and the neopullulanase. The dual specificity of the neopullulanase enzyme toward α-1,4 and α-1,6-glucosidic linkage takes place at the same catalytic site via the same mechanism (Hondoh et al., 2003; Takata et al., 1992). Thus, both 4-α-glucanotransferase and amylo-α-1,6-glucosidase activities of TreX may occur at the same catalytic cleft while the 4-α-glucanotransferase activity is closely associated to the structural lid generated by the tetramerization. Previously, the two activities of GDE in mammals and yeast were suggested to be independent and located at different sites on a single polypeptide chain. The N-terminal and C- terminal regions of yeast GDE are involved in the transferase and glucosidase activities, respectively (Liu et al., 1991; Nakayama et al., 2001). The structural analysis yielded the modulation of the dual activity of TreX may occur upon the oligomerization. In this respect, it would be interesting to see whether glycogen debranching enzymes in higher organisms have a similar configuration at the active site to that of TreX from *Sulfolobus solfataricus*.

ACKNOWLEDGEMENT

This study was supported by a grant from KRIBB Research Initiative, by the Marine and Extreme Genome Research Center Program of the Ministry of Land, Transportation and Maritime Affairs, Republic of Korea, and in part by the Korea Research Foundation Grant (KRF-2006-005-J04703)

REFERENCES

Abdullah M and Whelan W J (1963), 'A new pathway in rabbit muscle for the enzymatic debranching of glycogen', *Nature,* 197979-197980.

Barends T R, Bultema J B, Kaper T, van der Maarel M J, Dijkhuizen L, and Dijkstra B W (2007), 'Three-way stabilization of the covalent intermediate in amylomaltase, an alpha-amylase-like transglycosylase', *J Biol Chem,* 282(23), 17242-17249.

Bates E J, Heaton G M, Taylor C, Kernohan J C, and Cohen P (1975), 'Debranching enzyme from rabbit skeletal muscle; evidence for the location of two active centres on a single polypeptide chain', *FEBS Lett,* 58(1), 181-185.

Braun C, Lindhorst T, Madsen N B, and Withers S G (1996), 'Identification of Asp 549 as the catalytic nucleophile of glycogen-debranching enzyme via trapping of the glycosyl-enzyme intermediate', *Biochemistry,* 35(17), 5458-5463.

Cheong K A, Kim T J, Yoon J W, Park C S, Lee T S, Kim Y B, Park K H, and Kim J W (2002), 'Catalytic activities of intracellular dimeric neopullulanase on cyclodextrin, acarbose and maltose', *Biotechnol Appl Biochem,* 35(Pt 1), 27-34.

Henrissat B and Davies G (1997), 'Structural and sequence-based classification of glycoside hydrolases', *Curr Opin Struct Biol,* 7(5), 637-644.

Hondoh H, Kuriki T, and Matsuura Y (2003), 'Three-dimensional structure and substrate binding of Bacillus stearothermophilus neopullulanase', *J Mol Biol,* 326(1), 177-188.

Hong J S (2008), 'Characterization and rule of AmyX in glycometabolism of Bacillus subtilis 168', MS Thesis, Seoul National University.

Imamura H, Fushinobu S, Yamamoto M, Kumasaka T, Jeon B S, Wakagi T, and Matsuzawa H (2003), 'Crystal structures of 4-alpha-glucanotransferase from Thermococcus litoralis and its complex with an inhibitor', *J Biol Chem,* 278(21), 19378-19386.

Jespersen H M, MacGregor E A, Sierks M R, and Svensson B (1991), 'Comparison of the domain-level organization of starch hydrolases and related enzymes', *Biochem J,* 280, 51-55.

Katsuya Y, Mezaki Y, Kubota M, and Matsuura Y (1998), 'Three-dimensional structure of *Pseudomonas* isoamylase at 2.2 A resolution', *J Mol Biol,* 281(5), 885-897.

Kim J S, Cha S S, Kim H J, Kim T J, Ha N C, Oh S T, Cho H S, Cho M J, Kim M J, Lee H S, Kim J W, Choi K Y, Park K H, and Oh B H (1999), 'Crystal structure of a maltogenic amylase provides insights into a catalytic versatility', *J Biol Chem,* 274(37), 26279-26286.

Lee H S, Kim M S, Cho H S, Kim J I, Kim T J, Choi J H, Park C, Lee H S, Oh B H, and Park K H (2002), 'Cyclomaltodextrinase, neopullulanase, and maltogenic amylase are nearly indistinguishable from each other', *J Biol Chem,* 277(24), 21891-21897.

Liu W, Madsen N B, Braun C, and Withers S G (1991), 'Reassessment of the catalytic mechanism of glycogen debranching enzyme', *Biochemistry,* 30(5), 1419-1424.

MacGregor E A, Janecek S, and Svensson B (2001), 'Relationship of sequence and structure to specificity in the alpha-amylase family of enzymes', *Biochim Biophys Acta,* 1546(1), 1-20.

Maruta K, Hattori K, Nakada T, Kubota M, Sugimoto T, and Kurimoto M (1996a), 'Cloning and sequencing of trehalose biosynthesis genes from *Arthrobacter* sp. Q36', *Biochim Biophys Acta,* 1289(1), 10-13.

Maruta K, Kubota M, Fukuda S, and Kurimoto M (2000), 'Cloning and nucleotide sequence of a gene encoding a glycogen debranching enzyme in the trehalose operon from *Arthrobacter* sp. Q36', *Biochim Biophys Acta,* 1476(2), 377-381.

Maruta K, Mitsuzumi H, Nakada T, Kubota M, Chaen H, Fukuda S, Sugimoto T, and Kurimoto M (1996b), 'Cloning and sequencing of a cluster of genes encoding novel enzymes of trehalose biosynthesis from thermophilic archaebacterium *Sulfolobus acidocaldarius*', *Biochim Biophys Acta,* 1291(3), 177-181.

Mikami B, Iwamoto H, Malle D, Yoon H J, Demirkan-Sarikaya E, Mezaki Y, and Katsuya Y (2006), 'Crystal structure of pullulanase: evidence for parallel binding of oligosaccharides in the active site', *J Mol Biol,* 359(3), 690-707.

Nakayama A, Yamamoto K, and Tabata S (2001), 'Identification of the catalytic residues of bifunctional glycogen debranching enzyme', *J Biol Chem,* 276(31), 28824-28828.

Nelson T E, White R C, and Watts T E (1972), 'The action of the glycogen debranching enzyme system in a muscle protein particle', *Biochem Biophys Res Commun,* 47(1), 254-259.

Park H S, Park J T, Kang H K, Cha H, Kim D S, Kim J W, and Park K H (2007), 'TreX from *Sulfolobus solfataricus* ATCC 35092 displays isoamylase and 4-alpha-glucanotransferase activities', *Biosci Biotechnol Biochem,* 71(5), 1348-1352.

Park J T, Park, H. S., Kang, H. K., Hong, J. S., Cha, H., Woo, E. J., Kim, J. W., Kim, M. J., Boos, W., Lee, S., and Park, K. H. (2008), 'Oligomeric and functional properties of a debranching enzyme (TreX) from the archaeon *Sulfolobus solfataricus* P2', *Biocatalysis and Biotransformation,* 26(1), 76-85.

Park K H, Kim T J, Cheong T K, Kim J W, Oh B H, and Svensson B (2000), 'Structure, specificity and function of cyclomaltodextrinase, a multispecific enzyme of the alpha-amylase family', *Biochim Biophys Acta,* 1478(2), 165-185.

Park S H, Cha H, Kang H K, Shim J H, Woo E J, Kim J W, and Park K H (2005), 'Mutagenesis of Ala290, which modulates substrate subsite affinity at the catalytic interface of dimeric ThMA', *Biochim Biophys Acta,* 1751(2), 170-177.

Przylas I, Terada Y, Fujii K, Takaha T, Saenger W, and Strater N (2000), 'X-ray structure of acarbose bound to amylomaltase from Thermus aquaticus. Implications for the synthesis of large cyclic glucans', *Eur J Biochem,* 267(23), 6903-6913.

Roujeinikova A, Raasch C, Sedelnikova S, Liebl W, and Rice D W (2002), 'Crystal structure of *Thermotoga maritima* 4-alpha-glucanotransferase and its acarbose complex: implications for substrate specificity and catalysis', *J Mol Biol,* 321(1), 149-162.

Schmidt A K, Cottaz S, Driguez H, and Schulz G E (1998), 'Structure of cyclodextrin glycosyltransferase complexed with a derivative of its main product beta-cyclodextrin', *Biochemistry,* 37(17), 5909-5915.

Strokopytov B, Knegtel R M, Penninga D, Rozeboom H J, Kalk K H, Dijkhuizen L, and Dijkstra B W (1996), 'Structure of cyclodextrin glycosyltransferase complexed with a maltononaose inhibitor at 2.6 angstrom resolution. Implications for product specificity', *Biochemistry,* 35(13), 4241-4249.

Takata H, Kuriki T, Okada S, Takesada Y, Iizuka M, Minamiura N, and Imanaka T (1992), 'Action of neopullulanase. Neopullulanase catalyzes both hydrolysis and transglycosylation at alpha-(1,4)- and alpha-(1,6)-glucosidic linkages', *J Biol Chem,* 267(26), 18447-18452.

Walker G J and Whelan W J (1960), 'The mechanism of carbohydrase action. 8. Structures of the muscle-phosphorylase limit dextrins of glycogen and amylopectin', *Biochem J,* 76264-76268.

Watanabe Y, Makino Y, and Omichi K (2006), 'Activation of 4-alpha-glucanotransferase activity of porcine liver glycogen debranching enzyme with cyclodextrins', *J Biochem (Tokyo),* 140(1), 135-140.

Woo E J, Lee S, Cha H, Park J T, Yoon S M, Song H N, and Park K H (2008), ' Structural insight into the bifunctional mechanism of the glycogen debranching enzyme TreX from archaeon *Sulfolobus solfataricus*', *J Biol Chem*, In press

Yokota T, Tonozuka T, Kamitori S, and Sakano Y (2001), 'The deletion of amino-terminal domain in *Thermoactinomyces vulgaris* R-47 alpha-amylases: effects of domain N on activity, specificity, stability and dimerization', *Biosci Biotechnol Biochem,* 65(2), 401-408.

MOLECULAR CLONING OF THE AMYLOSUCRASE GENE FROM A MODERATE THERMOPHILIC BACTERIUM *DEINOCOCCUS GEOTHERMALIS* AND ANALYSIS OF ITS DUAL ENZYME ACTIVITY

Dong-Ho Seo, Jong-Hyun Jung, Suk-Jin Ha, Sang-Ho Yoo, Tae-Jip Kim, Jaeho Cha, and Cheon-Seok Park

ABSTRACT

Amylosucrase (AS; EC 2.4.1.4) is an interesting enzyme to form an insoluble amylose-type polymer from sucrose. A putative AS gene was cloned from *Deinococcus geothermalis* (*dgas*) and efficiently expressed in *E. coli* using the glutathione S-transferase fusion system. The optimal reaction temperature and pH for the sucrose-hydrolysis activity of DGAS were determined to be 45°C and pH 8, respectively. When compared with other ASs, DGAS has exceptionally thermostable characteristics, as demonstrated by a half-life of 6.8 h at 55°C, indicating this enzyme is the most active and thermostable AS known to date. DGAS also showed transglucosylation activity using sucrose as a sole substrate. However, the optimal temperature for trans-glucosylation was observed to be lower (25°C) than that for hydrolysis. These results suggest that two different catalytic activities in one enzyme demonstrate an opposite trend in its optimal reaction conditions. The transglucosylation reaction of DGAS with sucrose and maltooligosaccharides (G1-G7) as donor and acceptor molecules reveals that maltooligosaccharides need to be longer than maltotriose in order to be efficient acceptors for the glucan polymerization by DGAS. It was also found that some monosaccharides other than sucrose (galactose, xylose, methyl-α-D-gluco-pyranoside, and methyl-β-D-glucopyranoside) can be used as acceptors for the DGAS trans-glucosylation reaction.

Key words: amylosucrase; *Deinococcus geothermalis*; molecular cloning; insoluble glucan; transglucosylation

INTRODUCTION

Amylosucrase (AS, EC 2.4.1.4) is a transglucosidase member of family 13 of the glycoside hydrolases (GH13) that transfers the glucose moiety of sucrose to the non-reducing end of an α-1,4 glucan chain of the acceptor molecule, which leads to the formation of an insoluble amylose-type polymer from sucrose (Potocki de Montalk et al., 2000b; Albenne et al., 2004; Potocki-Veronese et al., 2005). The use of this enzyme as an industrial tool for the synthesis or modification of various polysaccharides originates from the catalytic property of requiring sucrose, a readily available inexpensive substrate, instead of an expensive activated sugar, such as ADP- or UDP-glucose, to transfer the glucose unit to the acceptor molecule (van der Veen et al., 2006; Pizzut-Serin et al., 2005).

Many studies focused on the AS from *Neisseria polysaccharea* (NPAS) whereas AS was first discovered in cultures of *N. perflava* which formed a polysaccharide very similar to glycogen (MacKenzie et al., 1977; Tao et al., 1988; Potocki de Montalk et al., 1999; Skov et al., 2001). A detailed study of the catalytic properties of NPAS demonstrates that several reactions are catalyzed by a single enzyme in the presence of sucrose alone. The versatile reactions catalyzed by NPAS were sucrose hydrolysis, the synthesis of maltooligosaccharides such as maltose and maltotriose by consecutive transfers of the glucosyl moiety of sucrose to the released glucose, the synthesis of sucrose isomers, and the production of an insoluble amylose-type polymer (Potocki de Montalk et al., 2000a). The gene for the AS from *N. polysaccharea* (ATCC 43768) was also cloned and its involvement in the production of linear α-1,4-glucans from sucrose was proven by the functional expression in *Escherichia coli* (Buttcher et al., 1997, 1999). The formation of insoluble α-1,4 amylose-type glucan is initialized by sucrose hydrolysis. The glucose molecule that is formed by sucrose hydrolysis is consequently used as the first acceptor molecule. The successive transfer of glucose at the non-reducing end of the acceptor molecule produces maltooligosaccharides and an insoluble amylose-type polymer by reaching a critical size (Potocki-Veronese et al., 2005).

Recent genomic search tools have shown that there are some gene sequences of putative ASs in various microorganisms including *N. meningitidis*, *Deinococcus radiodurans*, *Rhodopirellula baltica* SH 1, *Xanthomonas campestris*, and *Caulobacter crescentus* (Lin et al., 1999; White et al., 1999; Parkhill et al., 2000; Nierman et al., 2001; Thieme et al., 2005). First, the gene encoded as a putative AS in the *D. radiodurans* genome was expressed, and the catalytic properties of the recombinant protein (DRAS) were investigated (Pizzut-Serin et al., 2005). DRAS performed typical AS catalytic activities such as sucrose hydrolysis and the formation of an insoluble amylose polymer, sucrose isomers, and soluble maltooligosaccharides. However, the results showed that DRAS, as well as NPAS, were extremely heat-labile enzymes (Pizzut-Serin et al., 2005; van der Veen et al., 2006). Since low thermostability limits the industrial use of AS, there are many efforts to improve the thermostability of AS using directed evolution (van der Veen et al., 2006). Recently, a general procedure for the high-throughput isolation of AS variants displaying a higher thermostability or increased resistance to organic solvents was developed (Emond et al., 2007). By using this optimized and automated protocol, a mutant displaying a 25-fold increased-stability at 50°C was selected.

D. geothermalis is a Gram-positive, remarkably radiation resistant, and moderately thermophilic bacterium closely related to the mesophilic *D. radiodurans*. It can grow at temperatures as high as 55°C (Ferreira et al., 1997). There are many examples of enzymes originating from thermostable microorganisms that exhibit highly thermostable characteristics, including a thermostable single-stranded DNA-binding protein (SSB) from *D. geothermalis* (Filipkowski et al., 2006). This study describes the isolation and expression of a gene corresponding to an AS of *D. geothermalis* (DGAS), and the enzymatic characterization of the recombinant DGAS. The results showed that DGAS has extremely thermostable characteristics compared with previously identified NPAS and DRAS. Furthermore, most interestingly, the optimal temperatures for the two separate catalytic activities of sucrose hydrolysis and polymer synthesis were distinct from each other. These properties are important factors for the industrial use of this enzyme.

MATERIALS AND METHODS

Enzymes and chemicals

Restriction endonucleases and other modifying enzymes, such as T4 DNA ligase and *Pfu* DNA polymerase, were purchased from New England Biolabs (Beverly, MA, USA) or Solgent (Seoul, Korea). The GeneAll™ total DNA purification kit (GeneAll Biotechnology, Seoul, Korea) was used for the purification of the genomic DNA of *D.*

geothermalis. The PCR products or DNA restriction fragments were purified by the QIAquick gel extraction kit from Qiagen (Valencia, CA, USA). All other chemicals used were of reagent-grade quality obtained from Sigma Chemical Co (St. Louis, MO, USA).

Bacterial strains, media and plasmids

D. geothermalis DSM 11300 was obtained from Dr. Michael Daly at the Uniformed Services University of the Health Sciences. The cells were grown in TGY broth (0.5% tryptone, 0.3% yeast extract, and 0.1% glucose) or on TGY agar at 50°C for 24 h and used as the source of chromosomal DNA. *Escherichia coli* DH5a [F$^-$Φ80*lac*Z M15 (*lac*ZYA-*arg*F)U169 *rec*A1*end*A1 *hsd*R17(r_k^+, m_k^+) *pho*A *sup*E44 *thi*-1 *gyr*A96 *rel*A1λ^-] was employed as a host for DNA manipulation and transformation. *E. coli* BL21 [F$^-$ *ompT hsdS*$_B$ (r_B^-, m_B^-) *gal dcm* λ(DE3)pLysS T1R] was used as a host for the expression plasmid. The selection of recombinant clones was carried out on an LB agar plate with100 μg/mL of ampicillin, 0.5 mM of isopropyl-β-D-thiogalactopyranoside (IPTG) and 40 μg/mL of 5-bromo-4-chloro-3-indolyl-β-D-galactopyranoside (Xgal). The plasmid pGEM-T Easy (Promega Co., Madison, WI, USA) and pGEX-4T-1 vector (Amersham Biosciences, Buckinghamshire, UK) were utilized for the cloning of PCR products and construction of the expression vector, respectively.

Cloning of the AS-encoding gene from *D. geothermalis*

ORF Dgeo0572 (Genbank accession number, ABF44874), which is homologous to the *npas* gene, was isolated by PCR using *D. geothermalis* genomic DNA for a template. The genomic DNA of *D. geothermalis* was isolated using the GeneAll™ total DNA purification kit. The gene corresponding to DGAS was obtained by PCR using two primers designed based on the known putative α-amylase nucleotide sequence (Genbank accession number, ABF44874) of *D. geothermalis*. Two oligonucleotide primers containing *Eco*RI and *Bam*HI recognition sites (underlined) were Dgasl (5'-<u>GGA TCC</u> ATG CTG AAA GAC GTG CTC ACT-3') and Dgas2 (5'-<u>GAA TTC</u> TTA TGC TGG AGC CTC CCC GGC-3'). DNA fragments were amplified by *Pfu* DNA polymerase using the following protocol: initial denaturation at 94°C for 5 min, followed by 40 cycles of denaturation at 94°C for 1 min, annealing at 60°C for 1 min, and an extension at 72°C for 1 min with an additional extension at 72°C for 5 min during the final cycle. The amplified PCR product was purified using the QIAEX II Gel Extraction Kit (Qiagen Inc.) and cloned into the T-easy vector. The nucleotide sequence of the PCR-generated gene was determined using the BigDye Terminator Cycle Sequencing Kit for an ABI377 PRISM (Perkin-Elmer Inc., Boston, MA, USA). The DNA fragment corresponding to the *dgas* gene was inserted into an *Eco*RI- and *Bam*HI-cut pGEX-4T-1 vector (Amersham Biosciences) to create pGEX-DGAS.

DGAS production and purification

Recombinant *E.coli* BL21 harboring pGEX-DGAS was grown in 1 L of LB medium supplemented with 0.1 mg/mL ampicillin at 37°C with agitation at 250 rpm. The cells were harvested by centrifugation at 4,000 rpm for 20 min at 4°C after a 3 hr induction of *dgas* gene expression by adding IPTG to a final concentration of 1 mM when the optical density at 600 nm (OD_{600}) reached 0.5 to 0.6. The pellet was resuspended in 5 mL of phosphate-buffered saline (PBS) buffer [140 mM NaCl, 2.7 mM KCl, 10 mM

Na_2HPO_4, and 1.8 mM KH_2PO_4, pH7.3] per gram (wet weight) of cells. The resuspended cells were incubated on ice for 30 min after adding lysozyme (1 mg/ml) and disrupted by sonication (Sonifier 450, Branson, Danbury, CT, USA; output 4, 6 times for 10s, constant duty) in an ice bath. The cell lysate was centrifuged at 12,000 rpm for 10 min at 4°C. The clarified supernatant was applied directly onto a Glutathione-Sepharose™ High Performance affinity column (Amersham Biosciences) pre-equilibrated with PBS buffer. The column was washed with two volumes of PBS buffer. The bound recombinant DGAS was eluted with a 50-mM Tris-Cl buffer (pH 8.0) containing 30 mM of reduced glutathione. Finally, DGAS was purified to apparent homogeneity, and the purity was confirmed by SDS–PAGE analysis. Protein content was determined by the Bradford reagents kit (Bio-Rad, Hercules, CA, USA) with bovine serum albumin as a standard. The elimination of fused glutathione S-transferase protein from the purified recombinant DGAS was carried out using thrombin treatment.

Enzymatic assay of DGAS

The activity assay of DGAS was performed using 100 mM sucrose at 45°C in 50 mM Tris-HCl buffer, pH 8.0. One unit of DGAS was defined as the amount of enzyme that catalyses the production of 1 μmole of fructose per minute in the assay conditions. Fructose concentration was determined by the dinitrosalicylic acid method using fructose as a standard (Kikuchi et al., 1999; Oh et al., 2005). The effect of pH on the activity was investigated within the range of pH 4.0 to pH 10.0 (0.1 M sodium acetate buffer for pH 4 and 5, 0.1 M sodium citrate buffer for pH 5, 6, and 7, 0.1 M Tris-HCl buffer for pH 7, 8, and 9, and 0.1 M glycine-NaOH for pH 9 and 10) at 45°C. The effect of temperature on the activity was studied between 20°C and 70°C at pH 8.0. The thermostability of the purified DGAS was observed by preincubating the enzyme in the absence of substrate at 45°C, 50°C, 55°C, and 60°C. After various time durations, the residual activities were measured under standard assay conditions.

Thin layer chromatography and high performance anion exchange chromatography (HPAEC) analysis

The detection and identification of hydrolysis and transglycosylation products produced after enzyme reaction were done by TLC and HPAEC analyses. TLC analysis was performed as follows. The reaction products were spotted on Whatman K5F silica gel plates (Whatman) activated at 110°C for 30 min. Aliquot (5μL) of the reaction mixture was spotted onto a Silica gel K5F plate and developed with a solvent system of n-butanol/acetic aicd/water (3:1:1, v/v/v) in a TLC developing tank. Ascending development was repeated twice at room temperature. The plate was allowed to air-dry in a hood and then developed by being soaked rapidly in 0.3% (w/v) N-(1-naphthyl)-ethylenediamine and 5% (v/v) H_2SO_4 in MeOH. The plate was dried and placed in an 110°C oven for 10 min to visualize the reaction spots.

HPAEC analysis was carried out with an analytical column for carbohydrate detection (CarboPac PA100, Dionex Co., Sunnyvale, CA, USA) and an electrochemical detector (ED40, Dionex Co.). Filtered samples were eluted with a linear gradient from 100% buffer A (100 mM NaOH in water) to 60% buffer B (500 mM of sodium acetate in buffer A) over 70 min. The flow rate of the mobile phase was maintained at 1.0 ml/min. Detection was carried out by a Dionex ED50 module (Park et al., 2005). Soluble maltooligosaccharides were quantified using the linear relationship existing between the detector response per mole α-1,4-glucan chains and the degree of polymerization as described by Pizzut-Serin et al. (2005).

RESULTS AND DISCUSSION

Sequence analysis, expression, and purification of AS encoding gene from *D. geothermalis*

Recently, with the help of whole genome sequence databases, the presence of many putative ASs have been discovered in various genera of microorganisms including *D. geothermalis, D. radiodurans, N. meningitidis, Rhodopirellula baltica* SH 1, *Xanthomonas campestris*, and *Caulobacter crescentus* (Lin et al., 1999; White et al., 1999; Parkhill et al., 2000; Nierman et al., 2001; Thieme et al., 2005). Among those, an AS homologue designated as an α-amylase in the GenBank database was found in *D. geothermalis* genome. Sequence analysis revealed that its open reading frame encoded a protein of 650 amino acids that was only 41% identical to the previously characterized NPAS of *N. polysaccharea* (Buttcher et al., 1997; Potocki de Montalk et al., 1999). However, it showed 74% identity with the DRAS (Pizzut-Serin et al., 2005). Here, this AS homologue of *D. geothermalis* was designated as DGAS.

Further amino acid sequence analysis of DGAS reveals that it possesses a signature sequence of AS activity involves in sucrose specificity and polymerase activity. These conserved residues are Asp144, Asp394, Arg415, Arg446, and Arg509 in NPAS. Two residues (Asp144 and Arg509) confer to AS its sucrose specificity by forming a salt bridge at the bottom of the active site pocket, in contrast with typical α-amylases (Mirza et al., 2001; Albenne et al., 2002). Oligosaccharide and sucrose complexes of NPAS show that the interactions between NPAS and its substrates, sucrose and α-1,4-linked glucosyl chains, are intimately related to the B'-domain (residues 394–450 in NPAS) (Skov et al., 2002). Three residues, Asp394, Arg415, and Arg446, in the B'-domain ensure the elongation of the maltooligosaccharides initially produced, and preventing the hydrolysis reaction. Asp394 and Arg446 are involved in the transglucosylation reaction, whereas Arg415 is related to the anchoring and guidance of the chains to be elongated. Those amino acids corresponding to these conserved residues are all conserved in DGAS, demonstrating DGAS belongs to the group of AS.

In order to confirm the enzymatic activity of DGAS, the *dgas* gene was amplified by PCR from *D. geothermalis* genomic DNA and expressed in *E. coli* (Seo et al., 2007a, 2007b). When the *dgas* gene of *D. geothermalis* was fused to a six-histidine tag in the pRSET-B, the expression of the *dgas* gene in *E. coli* was not successful. However, it was efficiently expressed in *E. coli* with the glutathione S-transferase fusion system in the pGEX-4T-1 vector, as shown in Fig. 1. The recombinant *E. coli* strain showed the blue-staining halo around the colony when the cells were grown on sucrose-containing medium and stained with iodine vapor, indicating that an amylase-like polymer was produced by these cells (data not shown).

The *dgas gene* consists of 1,953 nucleotides, encoding 651 amino acids with a theoretical molecular mass of 73,296 Da. The one-step affinity purification using a Glutathione-Sepharose^{TM} High Performance affinity column was enough to purify the recombinant DGAS to homogeneity (Fig. 1). The recombinant DGAS fused with glutathione S-transferase was shown as a single band with a molecular mass of 100 kDa after affinity chromatography. The fused glutathione S-transferase protein was effectively removed by treatment with thrombin (Fig. 1, lanes 4 and 5). The molecular mass of the cleaved DGAS was about 73 kDa, which is in good agreement with the

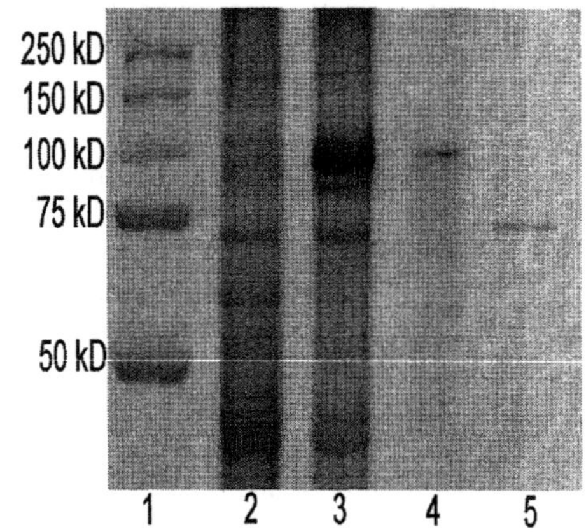

Figure 1 SDS-PAGE analysis of DGAS expressed in *E. coli*
Lane 1, molecular size marker; lane 2, crude cell extract of *E. coli* BL21 (control); lane 3, crude cell extract of *E. coli* BL21 harboring pGEX-DGAS; lane 4, GST-tagged DGAS purified by use of a Glutathion-Sepharose™ High Performance affinity column; lane 5, the purified DGAS, whose GST-tag was removed by thombin treatment.

predicted molecular mass of DGAS. The specific activity of GST-DGAS (28,710 U/g) and purified DGAS (69,444 U/g) was determined to be much higher than the specific activities of previously known ASs from *D. radiodurans* R1 (4,000 U/g) and *N. polysaccharea* (9,565 U/g) in terms of the hydrolysis activity (Pizzut-Serin et al., 2005). In the case of AS from *D. radiodurans* R1, the stability of the enzyme was found to be reduced after the last purification step. However, this instability was not observed in DGAS despite both enzymes having originated from *Deinococcus* strains and very similar the deduced amino acid sequences. The functional expression of DGAS in *E. coli* shown in this study confirms that *dgas* is an AS-encoded gene in *D. geothermalis*.

Biochemical properties of DGAS

The sucrose hydrolytic activity of the purified DGAS was detected with sucrose as a sole substrate. As the reaction proceeded, the opacity of the reaction mixture changed from clear to cloudy, indicating the formation of insoluble glucans. This result, together with the blue-staining halo around the colony (Buttcher et al., 1997), confirms that *dgas* encodes a functional enzyme with AS activity and that an amylose-like material is produced. The optimal reaction temperature and pH of DGAS are determined to be 45°C and pH 8, respectively, in terms of sucrose-hydrolytic activity (Fig. 2). *D. geothermalis* propagates in a temperature range from 45 to 50°C and a pH range from 4.5 to 8.5, experiencing optimal growth at 47°C and a pH of 6.5 (Ferreira et al., 1997). Since the enzymatic activity was maintained at more than 80% of its optimal activity in the range from 40 to 50°C, this result corresponds to the optimal growth condition of *D. geothermalis*. The sucrose-hydrolytic activity was at maintained about 55% of its maximum activity at 60°C. The most thermostable AS reported up until now was the E9 variant of NPAS obtained from a molecular evolution strategy (van der Veen et al., 2006). However, even the E9 variant showed almost no activity at 60°C. Furthermore, a heat stability experiment showed that DGAS has half-lifes of 28.1 hr and 6.8 hr at temperatures of 50°C and 55°C, respectively (Fig. 2). The activity was maintained for

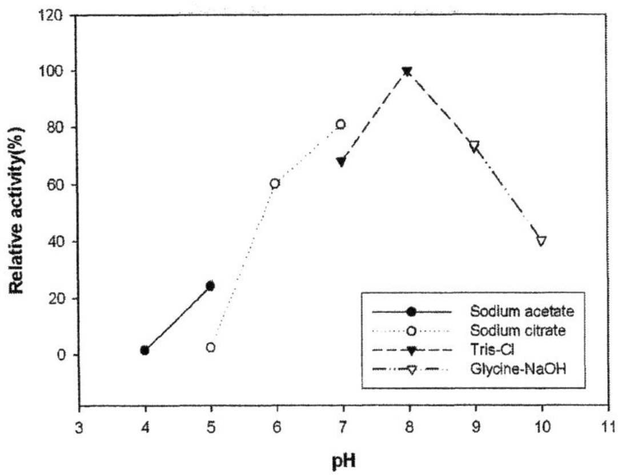

(A)

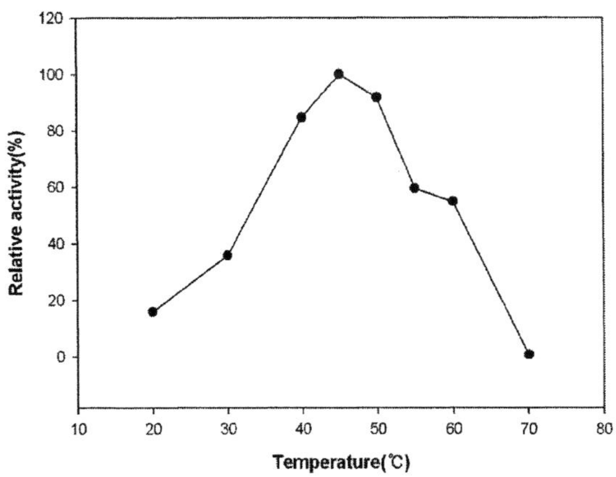

(B)

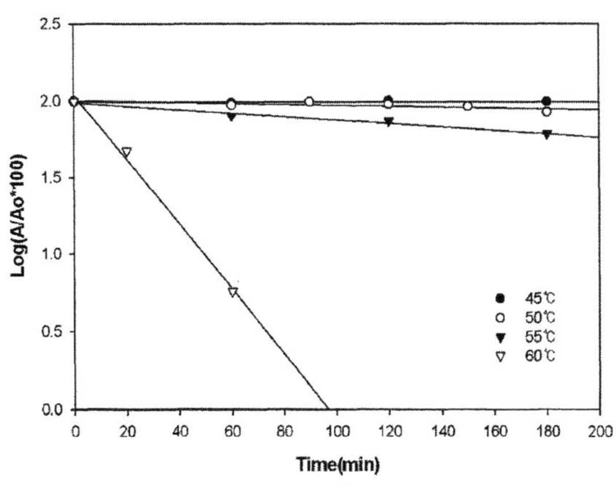

(C)

Figure 2 Optimum pH (A), temperature (B) and thermostability (C) of purified DGAS

Table 1 Yields of the reaction products synthesized by various amylosucrases such as NPAS, DRAS, and DGAS using sucrose (100 mM) as a substrate

Reaction products	NPAS (30°C)	DRAS (30°C)	DGAS (30°C)	DGAS (45°C)
Glucose	5.4	9.6	7 ± 1	33 ± 4
Fructose	-	-	28 ± 3	27 ± 3
Maltose	6.6	9.3	2 ± 1	6 ± 1
Maltotriose	16.8	11.0	2 ± 1	2 ± 1
Soluble maltooligosaccharides (4= DP ≤ 33)	12	28.8	29 ± 3	4 ± 1
Sucrose isomer	14.5	33.5	22 ± 3	25 ± 2
Insoluble glucan	44.7	7.8	6 ± 1	N.D
Unknown			4 ± 1	3 ± 1
Ratio between products (hydrolysis:isomerization:transglucosylation)	5.4:14.5:80.1	9.6:33.5:56.9	35:22:39*	60:25:12*
References	21	21	This work	This work

* Ratios were calculated from the mean values.

more than a month in a refrigerator. However, the activity was dramatically reduced above 60°C. Compared with the half-life times of NPAS and DRAS (21 hr and 15 hr at 30°C, respectively), DGAS has extremely thermostable characteristics, representing that DGAS is the most thermostable AS ever reported to date (Pizzut-Serin et al., 2005). The enzyme activity was measured in the presence of various metal ions (data not shown). The hydrolytic activity of DGAS was strongly inhibited by Zn^{2+} and Co^{2+}. Interestingly, Mg^{2+} was able to increase the activity by up to 40% at a concentration of 1 mM. The positive effect of Mg^{2+} on the DGAS activity was not concentration-dependent above 1 mM.

Hydrolysis, isomerization, and transglucosylation activities of DGAS

Similar to NPAS, sucrose is the only substrate for the hydrolysis activity of DGAS. Other disaccharides, such as maltose, lactose, and isomaltulose are not degraded by DGAS. AS catalyzes the transfer of a D-glucose moiety onto glycogen branches when glycogen exists as an acceptor molecule, although, in the absence of glycogen, the reaction is more complex (Potocki de Montalk et al., 2000a; Albenne et al., 2007). Typical AS reaction products in the absence of glycogen can be divided into three groups: 1) the hydrolysis products, glucose and fructose, 2) the transglucosylation products, maltose, maltotriose, soluble maltooligosaccharides (degree of polymerization, DP ≥ 4), and insoluble glucans, and 3) the isomerization products of sucrose, which are turanose and trehalulose (Potocki de Montalk et al., 2000b; Pizzut-Serin et al., 2005).

The reaction products resulting from the activity of DGAS on sucrose as the sole substrate were analyzed at 45°C (Table 1). The products were composed of glucose, fructose, maltose, maltotriose, sucrose isomers (turanose and trehalulose), and soluble maltooligosaccharides. When the yields of each product were analyzed, the ratio of the reaction products were 28% of glucose, 35% of fructose, 25% of sucrose isomers (turanose and trehalulose), and 4% of soluble maltooligosaccharides. Interestingly, the insoluble glucans, which were the representative reaction products of AS, were not observed in a significant amount in the DGAS reaction at 45°C, although the change of opacity from clear to cloudy was noticeable in the reaction mixture. Pizzut-Serin *et al* analyzed the reaction products resulting from DRAS and NPAS action on sucrose as the sole substrate (Pizzut-Serin et al., 2005). The reaction contained various products including glucose, maltose, maltotriose, turanose, trehalulose, soluble maltooligosaccharides and insoluble glucan, but not fructose. This is interesting since the hydrolysis action of AS on the sucrose releases glucose and fructose as reaction products. Then, the released glucose is transferred to other acceptor molecules and the fructose should be left (Potocki de Montalk et al., 2000a). Compared to the yields of the product synthesized by DRAS and NPAS, DGAS produces more glucose and fructose while soluble maltooligosaccharides and insoluble glucans are produced in relatively small amounts (Table 1). This result implies that DGAS possesses a relatively strong hydrolytic activity on sucrose at 45°C.

Interestingly, however, when the reaction is performed at 30°C, a significant amount of white precipitate corresponding to water-insoluble glucans is formed. The determination of the composition of the reaction mixture synthesized by DGAS at 30°C reveals that the ratio of reaction products is 7% glucose, 28% fructose, 21% sucrose isomers, 29% soluble maltooligosaccharides, and 6% insoluble glucans (Table 1). Two interesting features are the decrease of glucose and the increase of soluble maltooligosaccharides and insoluble glucans at 30°C compared to the reaction at 45°C. It is obvious that this observation is related to the transglucosylation activity of AS. Because the transglucosylation reaction is active, the amount of glucose, a donor molecule, is reduced while the amount of transglucosylation products (soluble maltooligosaccharides and insoluble glucans) increases, suggesting that glucose generated from sucrose by the hydrolysis activity of AS is immediately transferred to the acceptor molecules in the enzyme and, as a result, soluble maltooligosaccharides and insoluble glucans are formed.

When the ratio of the three groups of reaction products, hydrolysis products (glucose and fructose), isomerization products, and transglucosylation products (maltose, maltotriose, soluble maltooligosaccharides, and insoluble glucan) are investigated at various temperatures (25, 30, 35, 40 and 45°C), it is obvious that the transglucosylation activity is dominant over the hydrolysis activity at low temperature (25 and 30°C), whereas the hydrolysis reaction is favorable at a high temperature (45°C) (Fig. 3). The ratio among hydrolysis, isomerization, and transglucosylation products is 35:22:39 at 30°C, whereas it is 60:25:12 at 45°C. Interestingly, the ratio of the isomerization products of sucrose does not change significantly between those temperatures. Also, it is found that both ASs from *Deinococcus* strains (DRAS and DGAS) produced more soluble maltooligosaccharides than insoluble glucans compared with NPAS (Table 1).

The effect of the enzyme quantity on the two different activities of AS is examined. When 5 units (U) of enzyme are added into the reaction mixture, the decrease in transglucosylation products is more obvious compared with a reaction mixture containing 20 U of enzyme at 45°C (Fig. 3). The absolute yield of the transglucosylation products is higher when more enzymes are added to the reaction mixture. The

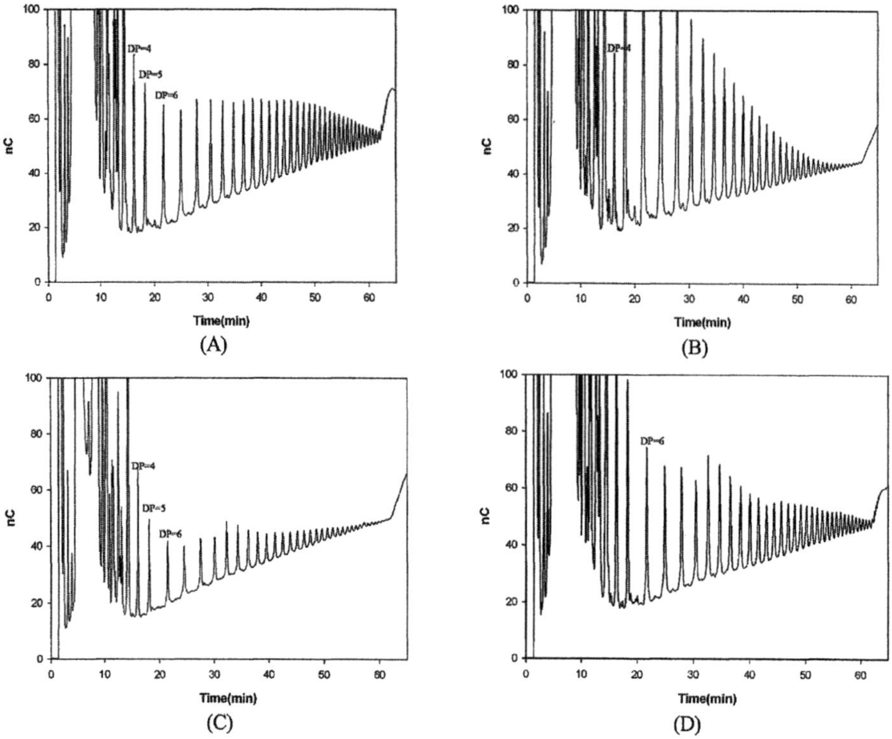

Figure 3 High performance anion exchange chromatography (HPAEC) analysis of soluble maltooligosaccharides on the reaction products formed with DGAS
Sucrose (100 mM) was used as a sole substrate. Reactions were performed with 5 U (A and C) and 20 U (B and D) of enzymes at 30 (A and B) and 45°C (C and D).

higher yields of transglucosylation products including insoluble glucans and soluble maltooligosaccharides are observed at lower temperatures and with a high quantity of enzyme. This result implies that the production of insoluble glucans or soluble malto-oligosaccharides by DGAS can be controlled by modulating the temperature and the amount of enzyme.

Unfortunately, this observation can not be compared with NPAS or DRAS since the temperature-based reaction profile data for NPAS or DRAS are not available due to their low thermostability (Pizzut-Serin et al., 2005). However, glycoside hydrolases are theoretically capable of forming as well as breaking glycosidic bonds in the substrate molecule as any chemical reaction can proceed in either direction, depending on the thermodynamic equilibrium. In general, a reaction breaking glycosidic bonds by glycoside hydrolases is more favorable at a high temperature until it reaches to an unfolding temperature (Cote et al., 1990). However, under suitable conditions, such as with an optimal water content, the addition of BSA, the presence of an organic solvent, and a high acceptor concentration, hydrolytic enzymes can be induced to form glycosides. The examples of differential effect of temperature on the hydrolytic and transferase activities were reported in other glycoside hydrolases (Lang et al., 2006; Turner et al., 2007). Murata et al., found that the production of poly-N-acetyllactosamines β-glycosides by *Escherichia freundii* endo-β-galactosidase was significantly enhanced by a low-temperature condition (Murata et al., 2005). This result was caused by the efficient transglucosylation activity of *E. freundii* endo-β-galactosidase and the decreased rate of hydrolysis of transglucosylation product at a low temperature. Similarly, the synthesis of alkyl glucosides by the transglucosylation activity of *Thermotoga neapolitana* β-glucosidase was optimal at a temperature of 60°C although the enzyme has an apparent optimum for hydrolysis at 90°C (Turner et al., 2007).

Transglucosylation activity towards various acceptor molecules

The transglucosylation activity of DGAS (20 U) is investigated in the presence of sucrose (100 mM) and various maltooligosaccharides (G1-G7, 50 mM) as donor and acceptor molecules, respectively, at 30°C (Fig. 4). When the reaction was performed without an acceptor molecule (only 100 mM of sucrose), small amounts of soluble maltooligosaccharides with a DP ≥ 4 and insoluble glucans started to emerge after 6 hrs, whereas maltotriose and maltose primarily appear earlier in the reaction. When glucose or maltose is present in the reaction as an acceptor, the result was similar to the result of a sucrose-only reaction. DGAS can not efficiently transfer the glucose molecules to these acceptor molecules, implying that glucose or maltose are not good acceptors for DGAS. Under the same reaction conditions, however, a variety of long maltooligosaccharides are produced by DGAS using maltotriose (G3) and longer maltooligosaccharides (G4-G7) as acceptor molecules (Fig. 4). The result indicates that maltooligosaccharides longer than maltotriose are effective acceptors for the transglucosylation of DGAS. The transglucosylation reaction is accelerated as the concentration of an acceptor is increased up to 200 mM (data not shown). It can be assumed that the production of maltotriose is the rate-limiting step in the polymerization reaction of DGAS in a sucrose only reaction.

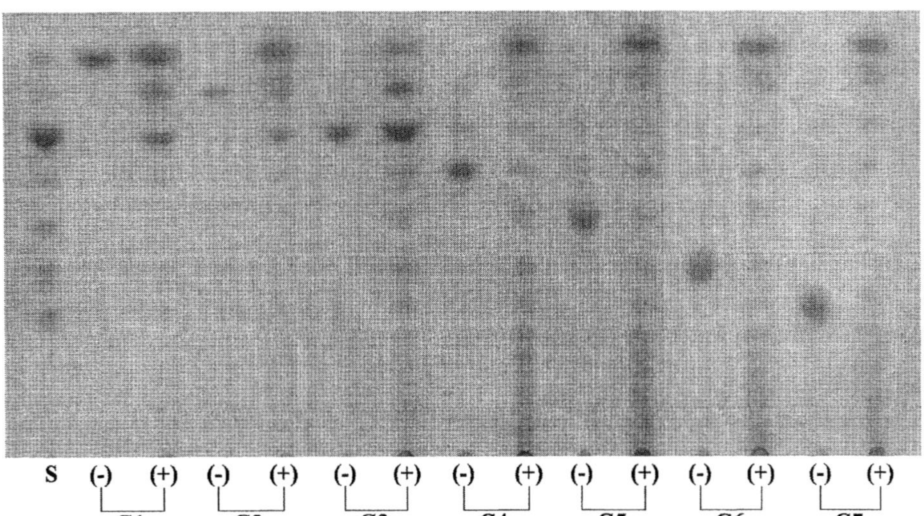

Figure 4 TLC analysis of transglucosylation activity of DGAS (20 U) on various maltooligosaccharides (G1-G7, 50 mM) in the presence of sucrose (100 mM)
The lanes designated with (-) are the reactions without the addition of enzyme, whereas those marked with (+) are the samples reacted with DGAS.

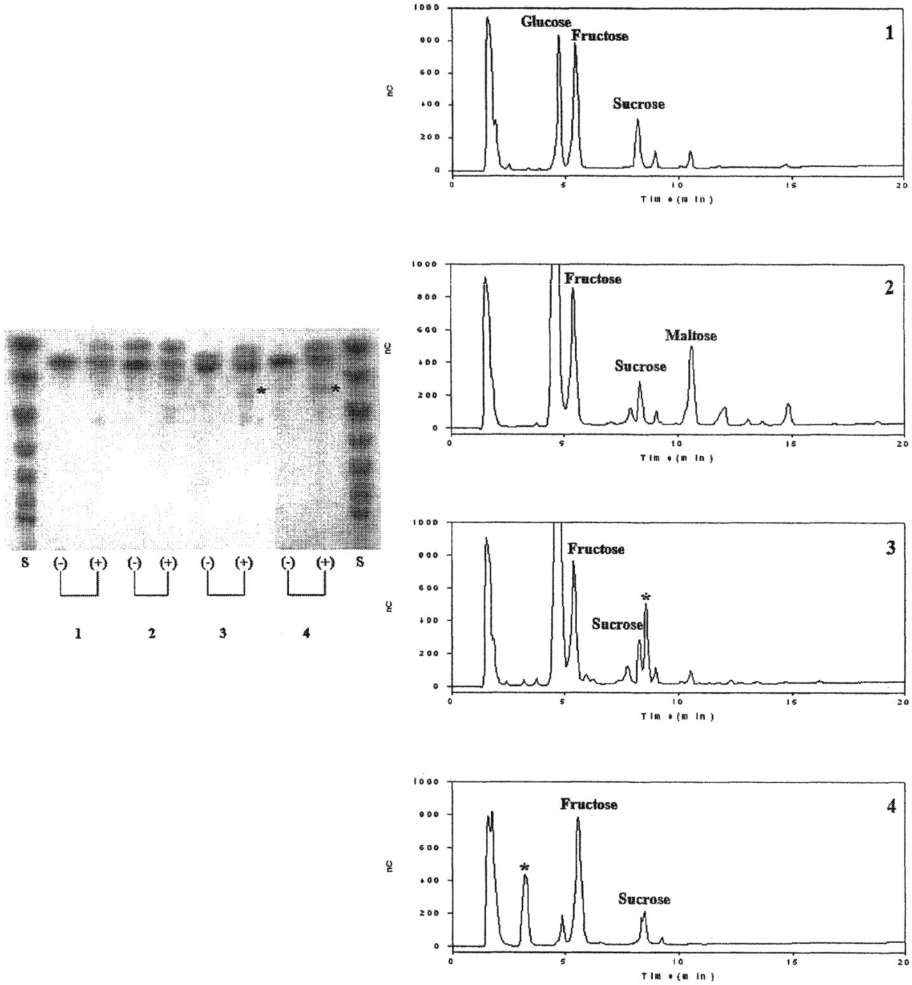

(A) (B)

Figure 5 TLC (A) and HPAEC (B) analyses of acceptor specificity of the transglucosylation reaction of DGAS on various monosaccharides such as glucose (2), galactose (3) and methyl-β-D-glucopyranoside (4)

(A) The lanes designated with (-) are the reactions without the addition of enzymes, whereas those marked with (+) are the samples reacted with DGAS. Lanes S are the maltooligosaccharide standards from glucose (G1) to maltoheptaose (G7). (1) is the reaction performed with sucrose only. Peaks and spots appeared as transglucosylation products were marked as star (*).

The acceptor specificity of the transglucosylation reaction of DGAS on various monosaccharides is also examined (Fig. 5). DGAS can not perform an efficient transglucosylation reaction with most monosaccharides, such as arabinose, fructose, fucose, mannose, rhamnose, ribose, sorbose, phenyl-α-D-glucopyranoside, and phenyl-β-D-glucopyranoside. However, some monosaccharides (galactose, xylose, methyl-α-D-glucopyranoside, and methyl-β-D-glucopyranoside) are effectively used as acceptors in the DGAS transglucosylation reaction. However, unlike with maltotriose, a variety of long maltooligosaccharides are not generated when galactose, xylose, methyl-α-D-glucopyranoside, or methyl-β-D-glucopyranoside is used as an acceptor. Only a couple of transglucosylation products are observed in the reaction, implying that the configuration of the transglucosylation product is important for use as an acceptor in consecutive transglucosylation reactions.

The industrial potential of AS for the synthesis of glucan, oligosaccharide, or glucoconjugates is further increased because sucrose is an inexpensive D-glucose donor compared with various other expensive nucleotide derivatives such as ADP- or UDP-glucose. To obtain the specific glucoconjugates with a maximum yield, a controlled reaction with minimized side reactions is necessary. Therefore, the adjustment of the enzymatic reaction of DGAS by temperature is useful for controlling the ratio of reaction products.

CONCLUSION

A novel thermostable AS gene was cloned from the moderate thermophile *D. geothermalis* and expressed in *E. coli.* DGAS shows remarkably thermostable characteristics by displaying its half-life of 6.8 h at 55°C. These results indicate that DGAS is the most attractive thermostable AS found to date. The most interesting aspect of DGAS is that two catalytic activities in DGAS, transglucosylation and hydrolysis, display totally different optimal reaction temperature profiles. The optimal temperature for transglucosylation was observed to be lower (25°C) than that for hydrolysis (45°C). Taken together, the present studies demonstrate the industrial potential of DGAS as a powerful enzyme to synthesize glucan, oligosaccharide, and glucoconjugates from an inexpensive D-glucose donor.

ACKNOWLEDGEMENT

This work was supported by the Korea Research Foundation Grant funded by the Korean Government (MOEHRD, Basic Research Promotion Fund) (KRF-2006-311-F00025)

REFERENCES

Albenne C, Skov L K, Mirza O, Gajhede M, Potocki-Veronese G, Monsan P, and Remaud-Simeon M (2002), 'Maltooligosaccharide disproportionation reaction: an intrinsic property of amylosucrase from *Neisseria polysaccharea*', *FEBS Lett*, 527, 67-70.

Albenne C, Skov L K, Mirza O, Gajhede M, Feller G, D'Amico S, Andre G, Potocki-Veronese G, van der Veen B A, Monsan P, and Remaud-Simeon M (2004), 'Molecular basis of the amylose-like polymer formation catalyzed by *Neisseria polysaccharea* amylosucrase', *J Biol Chem,* 279, 726-734.

Albenne C, Skov L K, Tran V, Gajhede M, Monsan P, Remaud-Simeon M, and Andre-Leroux G (2007), 'Towards the molecular understanding of glycogen elongation by amylosucrase', *Proteins*, 66, 118-126.

Buttcher V, Welsh T, Willmitzer L, and Kossmann J (1997), 'Cloning and characterization of the gene for amylosucrase from *Neisseria polysaccharea*: production of a linear α-1,4-glucan', *J Bacteriol,* 179, 3324-3330.

Buttcher V, Quanz M, and Willmitzer L (1999), 'Molecular cloning, functional expression and purification of a glucan branching enzyme from *Neisseria denitrificans*' *Biochim Biophys Acta,* 1432, 406-412.

Cote G L and Tao BY (1990), 'Oligosaccharide synthesis by enzymatic transglucosylation', *Glycoconjugate J,* 7, 145-162.

Emond S, Potocki-Veronese G, Mondon P, Bouayadi K, Kharrat H, Monsan P, and Remaud-Simeon M (2007), ' Optimized and automated protocols for high-throughput screening of amylosucrase libraries', *J Biomol Screen,* 12, 715-723.

Faijes M and Planas A (2007), 'In vitro synthesis of artificial polysaccharides by glycosidases and glycosynthases', *Carbohydr Res,* 342(12/13), 1581-1594.

Ferreira A C, Nobre M F, Rainey F A, Silva M T, Wait R, Burghardt J, Chung A P, and da Costa M S (1997), '*Deinococcus geothermalis* sp. nov. and *Deinococcus murrayi* sp. nov., two extremely radiation-resistant and slightly thermophilic species from hot springs', *Int J Syst Bacteriol,* 47, 939-947.

Filipkowski P, Duraj-Thatte A, and Kur J (2006), 'Novel thermostable single-stranded DNA-binding protein (SSB) from *Deinococcus geothermalis*', *Arch Microbiol,* 186, 129-137.

Kikuchi M, Ohnishi K, and Harayama S (1999), 'Novel family shuffling methods for the in vitro evolution of enzymes', *Gene,* 236, 159-167.

Lang M, Kamrat T, and Nidetzky B (2006), 'Influence of ionic liquid cosolvent on transgalactosylation reactions catalyzed by thermostable β-glycosylhydrolase CelB from *Pyrococcus furiosus*', *Biotechnol Bioeng,* 95, 1093-1100.

Lin J, Qi R, Aston C, Jing J, Anantharaman T S, Mishra B, White O, Daly M J, Minton K W, Venter J C, and Schwartz D C (1999), 'Whole-genome shotgun optical mapping of *Deinococcus radiodurans*', *Science,* 285, 1558-1562.

MacKenzie C R, Johnson K G, and McDonald I J (1977), 'Glycogen synthesis by amylosucrase from *Neisseria perflava*', *Can J Microbiol,* 23, 1303-1307.

Mirza O, Skov L K, Remaud-Simeon M, Potocki de Montalk G, Albenne C, Monsan P, and Gajhede M (2001), 'Crystal structures of amylosucrase from *Neisseria polysaccharea* in complex with D-glucose and the active site mutant Glu328Gln in complex with the natural substrate sucrose', *Biochemistry,* 40, 9032-9039.

Murata T, Honda H, Hattori T, and Usui T (2005), 'Enzymatic synthesis of poly-N-acetyllactosamines as potential substrates for endo-β-galactosidase-catalyzed hydrolytic and transglucosylation reactions', *Biochim Biophys Acta,* 1772, 60-68.

Nierman W C, Feldblyum T V, Laub M T, Paulsen I T, Nelson K E, Eisen J A, Heidelberg J F, Alley M R, Ohta N, Maddock J R, Potocka I, Nelson W C, Newton A, Stephens C, Phadke N D, Ely B, DeBoy R T, Dodson R J, Durkin A S, Gwinn M L, Haft D H, Kolonay J F, Smit J, Craven M B, Khouri H, Shetty J, Berry K, Utterback T, Tran K, Wolf A, Vamathevan J, Ermolaeva M, White O, Salzberg S L, Venter J C, Shapiro L, and Fraser C M (2001), 'Complete genome sequence of *Caulobacter crescentus*', *Proc Natl Acad Sci,* 98, 4136-4141.

Oh K W, Kim M J, Kim H Y, Kim B Y, Baik M Y, Auh J H, and Park C S (2005), 'Enzymatic characterization of a maltogenic amylase from *Lactobacillus gasseri* ATCC 33323 expressed in *Escherichia coli*', *FEMS Microbiol Lett,* 252, 175-181.

Park N Y, Baek N I, Cha J, Lee S B, Auh J H, and Park C S (2005), 'Production of a new sucrose derivative by transglucosylation of recombinant *Sulfolobus shibatae* β-glycosidase', *Carbohydr Res,* 340, 1089-1096.

Parkhill J, Achtman M, James K D, Bentley S D, Churcher C, Klee S R, Morelli G, Basham D, Brown D, Chillingworth T, Davies R M, Davis P, Devlin K, Feltwell T, Hamlin N, Holroyd S, Jagels K, Leather S, Moule S, Mungall K, Quail M A, Rajandream M A, Rutherford K M, Simmonds M, Skelton J, Whitehead S, Spratt B G, and Barrell B G (2000), 'Complete DNA sequence of a serogroup A strain of *Neisseria meningitidis* Z2491', *Nature*, 404, 502-506.

Pizzut-Serin S, Potocki-Veronese G, van der Veen B A, Albenne C, Monsan P, and Remaud-Simeon M (2005), 'Characterisation of a novel amylosucrase from *Deinococcus radiodurans*', *FEBS Lett,* 579, 1405-1410.

Potocki de Montalk G, Remaud-Simeon M., Willemot R M, Planchot V, and Monsan P (1999), 'Sequence analysis of the gene encoding amylosucrase from *Neisseria polysaccharea* and characterization of the recombinant enzyme', *J Bacteriol,* 181, 375-381.

Potocki de Montalk G, Remaud-Simeon M, Willemot R M, and Monsan P (2000a), 'Characterisation of the activator effect of glycogen on amylosucrase from *Neisseria polysaccharea*', *FEMS Microbiol Lett*, 186, 103-108.

Potocki de Montalk G, Remaud-Simeon M, Willemot R M, Sarcabal P, Planchot V, and Monsan P (2000b), 'Amylosucrase from *Neisseria polysaccharea*: novel catalytic properties', *FEBS Lett,* 471, 219-223.

Potocki-Veronese G, Putaux J L, Dupeyre D, Albenne C, Remaud-Simeon M, Monsan P, and Buleon A (2005), 'Amylose synthesized in vitro by amylosucrase: morphology, structure, and properties', *Biomacromolecules,* 6, 1000-1011.

Sarcabal P, Remaud-Simeon M, Willemot R, Potocki de Montalk G, Svensson B, and Monsan P (2000), 'Identification of key amino acid residues in *Neisseria polysaccharea* amylosucrase', *FEBS Lett*, 474, 33-37.

Skov L K, Mirza O, Henriksen A, Potocki de Montalk G, Remaud-Simeon M, Sarcabal P, Willemot R M, Monsan P, and Gajhede M (2001), 'Amylosucrase, a glucan-synthesizing enzyme from the α-amylase family', *J Biol Chem*, 276, 25273-25278.

Skov L K, Mirza O, Sprogoe D, Dar I, Remaud-Simeon M, Albenne C, Monsan P, and Gajhede M (2002), 'Oligosaccharide and sucrose complexes of amylosucrase. Structural implications for the polymerase activity', *J Biol Chem*, 277, 47741-47747.

Seo D H, Jung J H, Baik M Y, Yoo S H, and Park C S (2007a), 'Cloning and Expression of amylosucrase from *Deinococcus geothermalis* DSM 11300 (DGAS) in *E. coli*', *The 74th annual meeting of KoSFoST,* , Busan, Korean Society of Food Science and Technology.

Seo D H, Jung J H, Yoo S H, and Park C S (2007b), 'Characterization of a Novel Amylosucrase, a Glucan-synthesizing Enzyme, from *Deinococcus geothermalis* DSM 11300', *The 107th general meeting American Society for Microbiology*, Toronto, ASM press.

Tao B Y, Reilly P J, and Robyt J F (1988), '*Neisseria perflava* amylosucrase: characterization of its product polysaccharide and a study of its inhibition by sucrose derivatives', *Carbohydr Res,* 181, 163-174.

Thieme F, Koebnik R, Bekel T, Berger C, Boch J, Buttner D, Caldana C, Gaigalat L, Goesmann A, Kay S, Kirchner O, Lanz C, Linke B, McHardy A C, Meyer F, Mittenhuber G, Nies D H, Niesbach-Klosgen U, Patschkowski T, Ruckert C, Rupp O, Schneiker S, Schuster S C, Vorholter F J, Weber E, Puhler A, Bonas U, Bartels D, and Kaiser O (2005), 'Insights into genome plasticity and pathogenicity of the plant pathogenic bacterium *Xanthomonas campestris* pv. *vesicatoria* revealed by the complete genome sequence', *J Bacteriol,* 187, 7254-7266.

Turner P, Svensson D, Adlercreutz P, and Karlsson E N (2007), 'A novel variant of *Thermotoga neapolitana* β-glucosidase B is an efficient catalyst for the synthesis of alkyl glucosides by transglucosylation', *J Biotechnol*, 130, 67-74.

van der Veen B A, Skov L K, Potocki-Veronese G, Gajhede M, Monsan P, and Remaud-Simeon M (2006), 'Increased amylosucrase activity and specificity, and identification of regions important for activity, specificity and stability through molecular evolution', *FEBS J*, 273, 673-681.

SUBSTRATE SPECIFICITY, KINETIC MECHANISM AND OLIGOMERIC STATES OF CYCLOMALTODEXTRINASE FROM ALKALOPHILIC *BACILLUS* SP. I-5

Heeseob Lee

ABSTRACT

A Cyclomaltodextrinase (CDase) gene (cdase I-5) was isolated from alkalophilic *Bacillus* sp. I-5 that was isolated from Korean soil. The CDase I-5 gene encoded 583 amino acid residues for a 67,690 Da protein. Analysis of the kinetics of cyclomalto-dextrin (CD) hydrolysis by CDase I-5 indicated that ring-opening of the CD was the major limiting step, and that CDase I-5 preferentially degraded the linear maltodextrin chain by releasing the maltose unit. Three dimensional structure of CDase I-5 revealed that this enzyme forms a dodecamer, namely a hexamer of the dimer. Oligomeric states of CDase I-5 were modulated by salt concentration and pH of the reaction buffer. The mutagensis studies of this enzyme revealed that the dodecamerization of dimeric CDase I-5 was mediated by the protonation of H539 at the C-terminus.

Key words: cyclomaltodextrinase; kinetics; crystal structure; oligomeric states; dodecamer

INTRODUCTION

Cyclomaltodextrins (CDs) are a group of homologous cyclic nonreducing α-1, 4 linked D-glucose oligosaccharides obtained from starch by the action of cyclomaltodextrin glucanotransferase (CGTase) (Schenck et al., 1992). The most common CDs are α-, β-, or γ-CDs with six, seven and eight D-glucopyranose subunits, respectively. The polar hydroxyl groups are oriented toward the outside and make the CDs soluble in water, whereas the interior cavity is relatively hydrophobic. In aqueous solution, the hydrophobic environment of the cavity enables CDs to form inclusion complexes with many water-insoluble compounds (Clarke et al., 1988; Immel et al., 1996) that have many useful applications in the food and drug industries (Szejtli 1988). The enzyme that can hydrolyze CD is also important for the release of substances from CD inclusion complexes. CDs show varying degrees of resistance to hydrolysis by the common amylases. CDs are not readily hydrolyzed by exo-type amylases such as glucoamylase and β-amylase because CDs have no reducing ends.

Cyclomaltodextrinase (CDase; EC 3.2.1.54) hydrolyzes CDs much faster than starch and is clearly distinguished from α-amylases by the substrate preference (Bender 1986). Furthermore, CDase does not produce CDs from starch in contrast with cyclodextrin glucanosyltransferase (CGTase; EC 2.4.1.19) that both forms CDs from starch and has hydrolytic activity toward CDs (Kitahata, 1995). After the first report of a CDase from *Bacillus macerans* (DePinto et al., 1968), many enzymes denoted as CDase have been isolated from various microbial sources including *B. coagulans* (Kitahata et al., 1983), *Clostridium thermohydrosulfuricum* 39E (recently reclassified as *Thermoanaerobacter ethanolicus* 39E) (Podkovyrov et al., 1992), alkalophilic *Bacillus* sp. (Yoshida et al., 1991), *B. sphaericus* E-244 (Oguma et al., 1990; Oguma et al., 1993), alkalophilic *Bacillus* sp. I-5 (Kim et al., 1998), *Flavobacterium* sp. (Bender 1993), *B. sphaericus* ATCC7055 (Galvin et al., 1994), *Klebsiella oxytoca* M5a1 (Feederle et al., 1996; Fiedler et al., 1996), *Thermococcus* sp. B1001 (Hashimoto et al., 2001), *Thermotoga neapolitana* (Lunina et al., 2003), *Anoxybacillus flavithermus* (Turner et al., 2005), *Laceyella sacchari* (Turner et al., 2005), and *Archaeoglobus fulgidus* strain 7324 (LinksLabes et al., 2007). Most of these enzymes have an optimum temperature below 50°C and produce mainly maltose from CDs. Their molecular weights ranging from 62 to 90 kDa are higher than those of α-amylases.

The gene encoding CDase I-5 was cloned from an alkalophilic *Bacillus* sp. I-5 that was isloated from Korean soil and it encoded a protein consisting of 583 amino acids. The molecular mass of CDase I-5 was estimated about 67,690 daltons based on the summation of molecular mass of deduced amino acid residues (Kim et al., 1998; Lee 2002). In this review, the substrate specificities, reaction mechanisms, and quaternary structures of CDase I-5 will be discussed.

SUBSTRATE SPECIFICITY AND CATALYTIC PROPERTY OF CDASE I-5

Many of CDases show almost the same activities toward α-, β-, and γ-CDs, soluble starch, and amylose, and is active on branched CDs, pullulan, and maltose but not on glycogen. *Flavobacterium* CDase hydrolyzes linear maltohexaose, maltoheptaose and maltooctaose at higher rate than the corresponding CDs and shows a remarkable transglycosylation activity (Bender, 1993). The relative rates of hydrolysis of the various substrates differ among the enzyme sources, but are generally in the order of CD = maltooligosaccharides >> starch > pullulan.

CDase from *Bacillus* sp. I-5 hydrolyzes CD, pullulan, starch, and acarbose, a pseudotetrasaccharide and potent inhibitor of glucosidases, and displays a remarkable transglycosylation activity (Kim et al., 1998; Kim et al., 1999). CDase I-5 preferentially hydrolyzed β-CD. Starch and pullulan were hydrolyzed much slower than CDs (see Table 1). The main hydrolysis products from starch and CDs was maltose and glucose while that from pullulan was panose (Kim et al., 1998). The hydrolysis rate of CDase I-5 toward amylose and amylopectin revealed that amylose exhibited ~16 times higher k_{cat}/K_m value than amylopectin, indicating a unique specificity to amylose over amylopectin (Auh et al., 2006). This feature originated from its unusual quaternary structure in which a hydrolyzed product released from one active site on the assembly would be readily accepted into the other active sites of a cluster (Lee et al., 2002). Through the spatial arrangement of the active site in the supramolecular assembly, CDase I-5 of the dodecameric form would be more advantageous to discriminate the molecules in terms of their sizes, as compared to previously reported CD-degrading enzymes exhibiting this selectivity (Kamasaka et al., 2002). Therefore, a more linear shape of amylose molecules can be easily accessible to active site than amylopectin, resulting in a higher specificity on CDase I-5.

Table 1 Relative activities of CDase from alkalophilic *Bacillus* sp. I-5

Substrate	Relative activity (%)
α-CD	34
β-CD	100
γ-CD	79.1
amylose	32.9
soluble starch	8.4
Pullulan	3.0

(Adapted from Kim et al., 1998 and Lee et al., 2005)

ACTION MECHANISM OF CDASE I-5 ON CD

Hydrolysis pattern of α-, β-, and γ-CDs by CDase I-5 was studied in detail to elucidate the hydrolysis mechanism. The reaction products from various CDs and maltooligo-sccharides with respect to reaction time were analyzed. Main hydrolysis products by CDase I-5 were maltose and glucose. Generally, the product ratio of glucose to maltose was higher when substrates with odd numbers of glucose unit were used. The glucose to maltose ratio produced from α-CD was the same as that from maltohexaose. This was also true for β-CD and maltoheptaose (Kim et al., 2000).

It has been proposed that a multiple attack mechanism is an inherent property of the depolymerization enzymes (Takaku 1988; Suetsugu et al., 1974). It can be proposed that degradation of α-, β-, and γ-CDs follows a parallel-series of reactions and the kinetic parameters for the hydrolysis reaction were calculated based on the product formation (see Table 2). The rate constants for the hydrolysis of maltohexaose to maltose and maltotetraose (k_{4G6}) were higher than those of hydrolysis of maltohexaose to maltopentaose or maltotriose (k_{5G6}, k_{3G6}). The rate constant for the ring-opening reaction (k_d) was much lower than those for the reactions of hydrolyzing substrates to maltose (k_{4G6}, k_{3G5}, and k_{2G4}). A reaction constant for maltopentaose to maltotriose and maltose (k_{3G5}) was relatively greater than other reaction constants. However, the former reaction step (k_{5G6}) was so slow that k_{3G5} did not greatly influence the main degradation pathway of cyclodextrin. The results thus indicated that the major flow in the degradation of CD primarily depends on k_{4G6} and k_{2G4} (Kim et al., 2000).

The final reaction products from CDs and the resulting maltodextrins by CDase I-5 were maltose and glucose, which were the common products in the action of CDases (Kitahata et al., 1983; Oguma et al., 1990; Podkovyov, 1992). Maltose production as the major product could be confirmed by comparison of the reaction rate constants in Table 2. Each of reaction rate constants (k_{4G6}, k_{2G4}, and k_{3G5}) for producing maltose from maltodextrins was greater than those of others. Furthermore k_{4G6} was greater than k_{5G6} and k_{3G6} for the initial hydrolysis steps after ring opening reaction, which indicated that major flow of maltose production was the maltohexaose to maltotetraose and maltose (k_{4G6}) followed by the formation of maltose (k_{2G4}) from maltotetraose .

Despite the fact that CDase I-5 could not hydrolyze maltose, it produced glucose from α-, β-, and γ-CD. The product ratio of glucose to maltose varied depending on the number of glucose residues in the CDs. This suggests that CDase I-5 hydrolyzes the various α-1,4 glycosidic linkages after ring opening reaction. The same glucose to maltose ratio between α-CD and maltohexaose hydrolysis products indicates that both reactions follow the same reaction pathway after ring opening reaction of α-CD.

Table 2 First order rate constants for the ring opening of α-CD and for the hydrolysis of the various glycosidic bonds of the maltodextrins by CDase I-5

	k_d[a]	k_{5G6}[b]	k_{4G6}	k_{3G6}	k_{4G5}	k_{3G5}	k_{3G4}	k_{2G4}	k_{1G3}
Reaction rate constants ($\times 10^3$ min^{-1})	3.3± 0.28	2.2± 0.72	26.0± 1.26	2.4± 1.53	5.6± 0.39	33.0± 2.22	4.7± 0.07	21.5± 0.71	4.5± 0.23

[a] k_d is the rate constant for the ring opening of α-CD.
[b] Rate constants for the hydrolysis of individual gylcosidic bonds of the different maltodextrins are designated as follows: hydrolysis of the third bond of G6 is k_{3G6}, for the fourth bond is k_{4G6}, and for the fifth bond is k_{5G6}, and so forth.
(Adapted from Kim et al., 2000)

The rate constant (k_d) for the ring-opening reaction was much lower than those for the major hydrolytic reactions of the maltooligosaccharide to maltose (k_{4G6} and k_{2G4}), indicating that the ring opening of the CD is the rate limiting step in the degradation of CDs (Kim et al., 2000).

THREE-DIMENSIONAL STRUCTURES OF CDASE I-5

The crystal structure of CDase I-5 revealed that the monomeric structure contains a distinct N-domain in addition to a central $(\beta/\alpha)_8$-barrel domain and a C-domain (Fig. 1). The N-domain (residues 1-123) and the C-domain (residues 505-583) are composed exclusively of β-strands (Lee et al., 2002). Two molecules of CDase form a domain-swapped dimer in which the N-domain of one molecule is involved in extensive interactions with the $(\beta/\alpha)_8$-barrel domain of the adjacent molecule, as observed in the crystal structure of maltogenic amylase from *Thermus* strain (ThMA; Kim et al., 1999). In the dimeric structure, the C-domain is distinctively separated from the active site groove and is not involved in main-chain to main-chain hydrogen bond with either the N- or the $(\beta/\alpha)_8$-barrel domain. Instead, the interface between the C-domain and the $(\beta/\alpha)_8$-barrel domain consists predominantly of hydrophobic residues. The C-domain is found in the structures of all α-amylase family enzymes. Binding of raw starch is known to be the functional role of this domain in some amylases, such as cyclomaltodextrin glucanotransferase (CGTase) (Ohdan et al., 2000) and barley α-amylase (Søgaard et al., 1993; Tibbot et al., 2000). In all the known structures of α-amylase family enzymes in complex with an oligosaccharide at the active site, the bound sugar molecule is not in contact with the C-domain, and therefore it plays no direct role in the hydrolysis of substrate. Interestingly, the C-domain is critically involved in the supramolecular assembly of CDase.

The crystal packing of the CDase I-5 revealed an assembly composed of six copies of the dimeric units corresponding to the ThMA dimer. The dimeric units are related by the crystallographic two- and threefold symmetry axes of the cubic cell, resulting in a tightly packed hexameric assembly of the dimer. The predominant intermolecular interactions between the dimers are mediated by the C-domain of one molecule and the

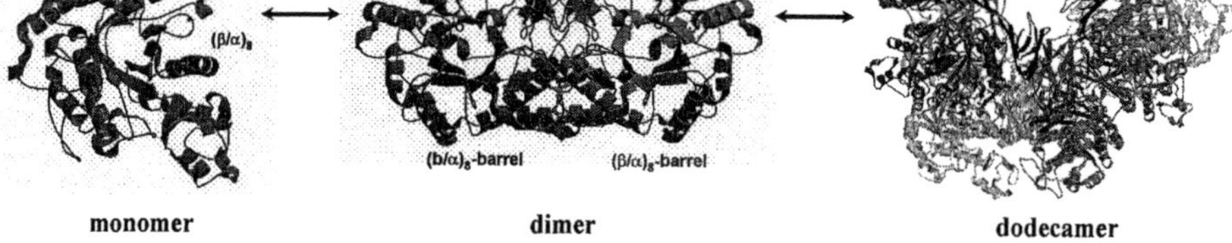

Figure 1 Three dimensional structure of CDase I-5
(Adapted from Lee et al., 2002)

N-domain of an adjacent molecule. Although the oligomerization states of CDase I-5 and ThMA are different, a superposition of CDase and of ThMA shows that the relative orientations of the C-domain and the N-domain with respect to the central domain are very similar in the two structures, both enzymes share 58% sequence identity.

All the twelve active sites are outwardly located on the dodecameric assembly. The hexamer formation does not shield any of the active sites from the access of the bulk solvent, and three active sites related by the crystallographic threefold axis are close to each other, thereby forming four identical clusters of three active sites according to the cubic symmetry of the supramolecular assembly. While the two active sites on the dimer are 180° away from each other facing the opposite direction, the clustered active sites face each other. The spatial arrangement of the active sites implied that the supramolecular assembly could confer a better enzyme activity than the dimeric form of ThMA and CDase, because a hydrolyzed product released from one active site on the assembly would be readily accepted into the other active sites of a cluster. Therefore it will spend less time before entering into another active site in comparison to the degradation of the substrate by the dimeric enzyme in which the same product released from one active site would travel a relatively long distance until it reaches the other active site (Lee et al., 2002).

FACTORS AFFECTING OLIGOMERIC STATES OF CDASE I-5

Enzymes in biological systems associate to form dimers or higher order oligomers. Oligomerization provides enzymes with many advantages such as high stability and control over accessibility and specificity of activie sites (Marianayagam et al., 2004; Bennet and Eisenberg, 2004). There are many factors affecting the dissociation/ association of the oligomeric protein, which are known to include pH, salt, and pressure. Recently, oligomeric states have been reported for the cyclodextrin-/pullulan-degrading enzymes such as CDase, maltogenic amylase (MAase; EC 3.2.1.133), and neo-pullulanase (NPase, EC 3.2.1.135). For example, 3D domain-swapped MAase from *Thermus* strain (ThMA) that exhibits increased substrate specificity via dimerization (Kim et al., 1999).

CDase I-5 existed as a dodecamer, which was consisted of a hexamer of dimeric units, and that the formation of the supramolecular assembly resulted in an increase in the catalytic efficiency compared with that of the dimeric unit of the enzyme (Lee et al., 2002). There were the exogenous and endogenous factors affecting the supramolecular assembly of CDase I-5. Dissociation/association of the CDase I-5 dodecamer was found to be dependent on pH and salt concentration. At pH 6.0, the enzyme preferentially dissociated into its dimeric units, which were enzymatically active; at pH 7.0, the enzyme existed predominantly in the dodecameric form, which had higher catalytic activity than the dimeric form. Conversely, CDase I-5 rapidly dissociated into dimeric units in the presence of KCl at pH 7.0. The association/dissociation process of CDase I-5 was examined in various oligomeric states in order to identify the mechanism and forces that contribute to the supramolecular assembly and function of the enzyme. In addition, the role of histidine residues at the interfaces in the formation of the dodecamer was investigated by site-directed mutagenesis.

Effect of KCl on the quaternary structure

The apparent molecular mass of the enzyme, calculated by comparing the elution time with those of standard proteins using gel permeation chromatography (Park, 2001), was

638 kDa, which was much larger than the molecular mass of the monomeric subunit (67.7 kDa). The result indicated that the major oligomeric state of CDase I-5 at pH 7.0 was dodecamer. However, the peak corresponding to dimer increased in the presence of 1 M KCl, while the area of the peak corresponding to dodecamer decreased, suggesting that the enzyme dissociated from dodecamers into dimers in the presence of salt (Lee et al., 2002). When the enzyme was treated with 1.0 M KCl, there was no significant change in the circular dichroism spectrum, while treatment with 1.0 M or 6.0 M urea produced significant changes (Lee et al., 2006). The results indicated that the secondary structure of CDase I-5 was not altered by KCl at concentrations of up to 1.0 M. Likewise, the ellipticity also showed that 1.0 M KCl did not affect the secondary structure of the enzyme, while urea and guanidine hydrochloride exerted a great influence. From these results, the secondary structure and peptide backbone of native CDase I-5 were stable and rigid at pH 7.0 in the absence or presence of KCl at concentrations up to 1.0 M.

The dissociation process induced by salts was too fast to monitor the inter-conversion of CDase I-5 by gel filtration chromatography between dodecamers and dimers. Therefore, the salt-induced dissociation of CDase was investigated using a stopped-flow apparatus. To characterize the changes in the quaternary structure of CDase I-5, the intrinsic fluorescence of CDase I-5 was measured at various concentrations of KCl and denaturants. Based on the crystal structure analysis of CDase I-5, the tryptophan residues of CDase I-5 at the 68, 68', 93, and 93' positions were possible candidates contributing to increased fluorescence intensity through dissociation upon exposure to solvent. The fluorescence intensity of CDase I-5 increased as the dodecameric enzyme dissociated into dimers upon the addition of 1.0 M KCl. Conversely, upon denaturation and unfolding of the protein by chemical modification, non-polar interior groups became exposed to the polar exterior phase, and the quenching of fluorescence was accompanied by a red shift and a decrease in intensity (Inouye et al., 2000). The intensity of fluorescence of CDase I-5 treated with 1.0 or 6.0 M urea at 25 °C was weak, and the wavelength of the spectral maximum was shifted to 355 nm.

To investigate the dissociation process of CDase I-5, changes in fluorescence intensity of the reaction mixture were monitored using an SFM-4 stopped-flow apparatus at different KCl concentrations (0–1.0 M KCl). The fluorescence intensity of CDase I-5 increased as the concentration of KCl increased. For a pseudo-first-order reaction, the rate constant of dissociation (k_d) from dodecamer into dimer was estimated at various concentrations of KCl using the Guggenheim method (Jonnalngadda and Gollapalli, 2000). The k_d values in the presence of 0.25 M and 1.0 M KCl were 5.96 s^{-1}

Table 3 Salt-induced dissociation rate constants (k_d) [a] of CDase I-5 determined by fast kinetic measurements

pH	Dissociation rate constant (s^{-1})			
	0.2 M KCl	0.5 M KCl	0.8 M KCl	1.0 M KCl
7.0	5.96 ± 0.06	6.36 ± 0.16	7.53 ± 0.12	7.99 ± 0.13
6.9	6.30 ± 0.11	7.03 ± 0.10	7.64 ± 0.06	8.79 ± 0.07
6.7	9.15 ± 0.17	10.46 ± 0.11	- [b]	11.81 ± 0.21
6.5	15.18 ± 0.13	18.38 ± 0.16	20.92 ± 0.15	21.92 ± 0.29

[a] Values for k_d were determined according to the Guggenheim method. Final concentrations after mixing were [CDase I-5] = 10 μM and [KCl] = 0.2–1.0 M.

[b] Not determined.

(Adapted from Lee et al., 2006)

and 7.99 s^{-1}, respectively (Table 3). The rate constants increased as the pH was lowered or the concentration of KCl was increased. The results suggested that the effect of salts on the oligomeric state of CDase I-5 correlated with the dissociation of the dodecameric form of the enzyme.

Stevens et al. (2000) reported that class Sigma glutathione *S*-transferase lost 60% of its catalytic activity and a single tryptophan residue per subunit became partly exposed when NaCl was added at concentrations up to 2 M. They reported that no significant change was detected either in the secondary structure of the protein according to far-UV circular dichroism data or in the size of the protein determined by size-exclusion HPLC. They suggested that the change might occur at or near the active site. However, in the case of CDase I-5, when the protein dissociated from dodecamers to dimers as shown by gel filtration chromatography, the activity on β-cyclodextrin decreased to 66%, but the activity on soluble starch increased by 160%. Large substrates such as soluble starch seemed to be able to access dimeric CDase more easily than the dodecameric form owing to less steric hindrance. These results suggested that the effect of salts on the oligomeric state of CDase I-5 correlated with the dissociation of the dodecameric form of the enzyme.

Effect of pHs on the quaternary structure

pH dependent dissociation/association

To investigate the effect of pH on the dissociation of dodecameric CDase I-5, sedimentation equilibrium analysis was performed at pH 5.0–8.5. The apparent molecular weight of CDase I-5 determined using analytical ultracentrifugation was plotted as a function of pH (Fig. 2). The results indicated that CDase I-5 existed as a monomer/dimer in the pH range of 5.0–6.0, while dodecameric CDase I-5 was predominant at pH 6.5–8.5. Dimeric CDase I-5 began to associate with a transition midpoint of pH 6.2, forming dodecameric CDase I-5 as a major form at pH values higher than 6.5 (Lee et al., 2006).

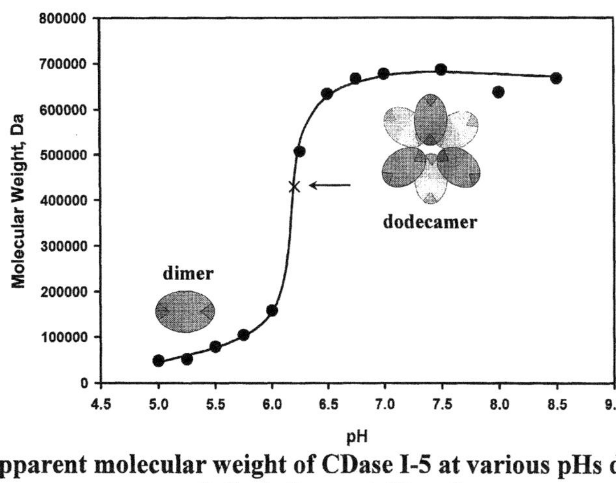

Figure 2 Apparent molecular weight of CDase I-5 at various pHs determined by analytical ultracentrifugation
(Adapted from Lee et al, 2006)

Table 4 Physicochemical properties of wild-type CDase I-5 at pH 6 and 7

Property	pH 6.0	pH 7.0
Transition to	Dissociation	Association
k (h^{-1})	0.0858	0.109
Oligomeric state	Dimer	Dodecamer
k_{cat} (s^{-1}) [a)]	8.5	78.2
K_m (mM) [a)]	0.889	0.454
k_{cat}/K_m ($s^{-1}\cdot mM^{-1}$) [a)]	9.5	172

(Adapted from Lee et al., 2006)

The reversibility of the association and dissociation processes of CDase I-5 was examined at pH 6.0 and 7.0. CDase I-5 was incubated in universal buffer (pH 6.0 or 7.0), and aliquots were taken at appropriate time intervals to determine the oligomeric state of the enzyme. Gel filtration chromatography was used to monitor the change of CDase I-5 from a dodecamer to a dimer. At pH 6.0, the peak corresponding to the dodecameric form decreased, while that corresponding to the dimer increased as the incubation time proceeded. In 72 h of incubation at 4 °C, dodecameric CDase I-5 was fully converted into the dimeric form. On the other hand, if the pH of the enzyme solution was elevated to 7.0 after dissociation at pH 6.0, the reverse was observed. The peak corresponding to the dimeric form of the enzyme shifted towards that corresponding to the dodecamer. The association process by which dimeric enzymes fully recovered their dodecameric form was completed in 106 h at 4 °C. These results indicated that separate dimers could form a dodecamer and that the dimer-dodecamer transition was a true association/ dissociation equilibrium process. The progress curve of the inter-conversion between dodecamer and dimer at pH 6.0 fitted a single exponential time course. Based on this observation, the kinetics of the dissociation process was analyzed in detail by calculating the peak area during the dissociation process. The rate of change in the peak area was estimated according to an equation of single exponential decay. The slope of the exponential line was considered to be the rate constant, giving a rate constant of 0.0858 h^{-1} for the dissociation of dodecamers to dimers (Table 4). The progress curve of the conversion of dimers to dodecamers at pH 7.0 also fitted a single exponential time course. The rate constant for the association of dimers to form dodecamers was determined as 0.109 h^{-1} (Lee et al., 2006). The kinetic parameters of CDase I-5 for β-cyclodextrin in either the dimeric or dodecameric state were compared by isothermal titration calorimetry at pH 6.0 and 7.0. The dodecameric form at pH 7.0 exhibited a k_{cat}/K_m value ~15 times larger than that of the dimeric form at pH 6.0 (Table 4; Lee et al., 2006).

Oligomerization states of certain proteins have been reported to be pH dependent (Kishimoto et al., 2000; Cabezon et al., 2000; Gordon-Smith et al., 2001). For example, bovine F_1-ATPase inhibitor protein, IF_1, forms tetramers at pH 8.0, while the protein is predominantly in the dimeric form below pH 6.5 (Cabezon et al., 2000; Gordon-Smith et al., 2001). The protonation of histidine residues appears to modify the structure of IF_1 and play an important role in the inter-conversion between dimers and tetramers given that the mutation of this residue to lysine abolishes the pH-dependent oligomerization without an alteration of enzyme activity (Cabezon et al., 2000). A 10 kDa light chain subunit of the cytoplasmic dynein complex LC8 shows a reversible monomer-dimer equilibrium at pH 7.0, but the dimers dissociate into monomers at lower pHs, with a transition midpoint at pH 4.8 (Barbar et al., 2001). This was explained by the titration of a histidine pair at the interface of the dimer. D-amino acid transaminase undergoes a reversible process of dissociation/association that is pH-dependent (Kishimoto et al., 2000), but this occurs at rates much slower than those of CDase I-5.

Amino acid residues affecting dissociation/association

Based on the information obtained about the three-dimensional structure of CDase I-5, the quaternary state of CDase I-5 was likely to be maintained by the intrinsic capability of the N-terminal and C-terminal regions of the enzyme to form a dodecamer at pH 7.0 and a dimer at pH 6.0. Crystallography of CDase I-5 has shown that a histidine residue in the C-terminal region (H539) and two of the four histidine residues in the N-terminal region (H49 and H89) are localized at the interfaces between dimeric units and are likely to be involved in the interaction between CDase I-5 molecules (Fig. 3). β-strand from K536 to L541 of a molecule is the major part contacting the adjacent β-strand from T50 to V54 of the other molecule in oligomerization. H539 is in the center of that contact region. The nitrogen (NE2) of the histidine residue forms a hydrogen bond to oxygen (OE1) in the side chain of Q516, of which the nitrogen (NE2) also forms hydrogen bond to side chain of D535. There are a total of six hydrogen bonds to support a sharp turn comprising from N533 to A537. Protonation of H539 may prevent the hydrogen bond to Q516 at lower pH, thereby destabilizing the region hold tightly by the hydrogen bond network from K536-T540 and leading to conformational change at the interface of a dimmer (Fig. 3). There are two hydrogen bonds at G538 and T540 to adjacent monomer and of which G538 form a hydrogen bond to the carbonyl oxygen of M51. Two residues at the N-terminus (H49 and H89) of a subunit were located close to

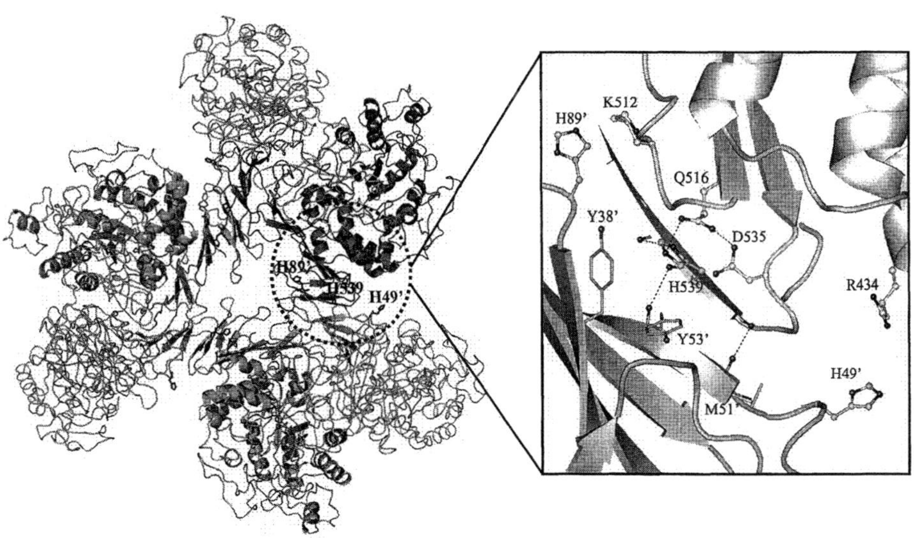

Figure 3 The three histidine residues at the interface of two CDase I-5 subunits constituting a dodecamer

Close view of the interface shows that H539 is involved in various hydrogen bondages. Amino acid residues in one subunit are primed and those in the other subunit are not. (Adapted from Lee et al., 2006)

the C-domain of the other CDase I-5 subunit. The isoelectric point of CDase I-5 (pI 7.8) suggested that a decrease in pH from 7.0 to 6.0 would increase the number of positively charged residues at the C-terminal region, particularly those arising from protonation of the histidinyl groups. These might destabilize the dodecameric structure of CDase I-5 by electrostatic repulsion of positively charged residues at low pH, resulting in the dissociation of dodecamers to dimers (Lee et al., 2006).

Single, double and triple mutations at three histidine residues (H49, H89, and H539) were constructed in various combinations. All mutant CDases purified from *E. coli* transformants carrying the mutant clones had specific activity toward β-cyclodextrin and optimal temperature and pH similar to those of wild-type CDase I-5. The dissociation rate constants of three mutants (H539V, H49V/H539V, and H49V/H89/H539V) were determined by gel permeation chromatography. The peak area corresponding to the dodecamer diminished with incubation time. The progress curves representing the dissociation of dodecamers to dimers fitted the equation of a single exponential decay. The dissociation rate constants of all mutants were increased compared with that of wild-type CDase I-5. The dissociation rate constants for H539V, H49V/H539V and H49V/H89V/H539V were 0.45 h^{-1}, 0.68 h^{-1}, and 1.36 h^{-1}, respectively (Table 5). The dissociation rate constant for H49V/H89V/H539V was about 16 times larger than that of wild-type CDase I-5. The mutation of histidine to valine showed the same effect even at pH 7 and above. This data indicated that the effect of pH on dissociation of the oligomer was mainly due to the protonation of a single residue rather than a global effect of pH on the protein. In agreement with the site-directed mutagenesis studies, H539 was most likely to be the target of this pH effect.

Wild-type and mutant CDases were stored in 50 mM sodium phosphate buffer (pH 7.0) at 4 °C, applied to gel permeation chromatography, and eluted with 50 mM sodium phosphate butter (pH 7.0) at a flow rate of 0.4 $mL \cdot min^{-1}$. One hundred microliters of the enzyme was applied to the column, and the absorbance of each eluent was measured at 280 nm. The proportion of dodecamers decreased as less protein was used. The dissociation constant (K_d) for the dodecamer was estimated. A very good fit to a line with a slope of 5.04 was obtained, and the K_d values for H49V/H89V/H539V and H49V/H539V were calculated as 1.79×10^{-30} M^5 and 4.63×10^{-32} M^5, respectively (Table 5). For wild-type CDase I-5, the enzyme was applied to a Superdex column at concentrations of up to 100 nM at pH 7.0, but no dissociation of the dodecameric enzyme was detected. The results indicated that the K_d value of wild-type CDase I-5 was much lower than those of the mutants. This result was confirmed by the sedimentation equilibrium and sedimentation velocity analytical ultracentrifugation analyses carried out at pH 7.0. In the sedimentation equilibrium analysis, the apparent

Table 5 Kinetic and equilibrium parameters of wild-type and mutant CDase I-5

Parameter	pH	Wild-type	Mutant		
			H539V	H49V/ H539V	H49V/H89V/ H539V
Dissociation rate constant, k_d (h^{-1})	6.0	0.0858	0.45	0.68	1.36
Equilibrium constant, K_d ($\times 10^{-30}$)	7.0	0.0	-[b]	0.046	1.79
Sedimentation coefficient (s)[a]	7.0	20	-[b]	-[b]	20, 5

[a] Apparent weight average sedimentation coefficient in Svedbergs.
[b] Not determined.
(Adapted from Lee et al., 2006)

molar masses of wild-type and mutant CDase were 736 kDa and 491 kDa, respectively. The data from a series of scans showed the common meniscus and the logical progression of the boundary and plateau regions. The sedimentation coefficient was calculated as described in the Materials and methods section. The apparent weight average sedimentation coefficients were 20 for wild-type and 20 and 5 for H49V/H89V/H539V. These results implied that the CDase mutant existed in a dimer/dodecamer equilibrium at pH 7.0 (Lee et al., 2006).

REFERENCES

Auh J H, Chae H Y, Kim Y R, Shim K H, Yoo S H, and Park K H (2006), 'Modification of rice starch by selective degradation of amylase using alkalophilic *Bacillus* cyclomaltodextrinase', *J Agric Food Chem*, 54, 2314-2319.

Barbar E, Kleinman B, Imhoff D, Li M, Hays T S, and Hare M (2001), 'Dimerization and folding of LC8, a highly conserved light chain of cytoplasmic dynein', *Biochemistry*, 40, 1596-1605.

Bender H (1986), 'Production, characterization, and application of cyclodextrins', *Adv Biotechnol Proc*, 6, 31-71.

Bender H (1993), 'Purification and characterization of a cyclodextrin-degrading enzyme from *Flavobacterium* sp.', *Appl Microbiol Biotechnol*, 39, 714-719.

Bennett M J and Eisenberg D (2004), 'The evolving role of 3D domain swapping in proteins', *Structure*, 12, 1339-1341.

Cabezon E, Jonathan P, Butler G, Runswick M J, and Walker J E (2000), 'Modulation of the oligomerization state of the bovine F_1-ATPase inhibitor protein, IF_1, by pH', *J Biol Chem*, 275, 25460-25464.

Clarke R J, Coates J H, and Lincoln S F (1998), 'Inclusion complexes of the cyclo-maltooligosaccharides (cyclodextrins)', *Adv Carbohydr Chem Biochem*, 46, 205-249.

DePinto J A and Campbell L L (1968), 'Purification and properties of the cyclo-dextrinase of *Bacillus macerans*', *Biochemistry*, 7, 121-125.

Duplay P, Bedouelle H, Fowler A, Zabin I, Saurin W, and Hofnung M (1984), 'Sequences of the *malE* gene and of its product, the maltose-binding protein of *Escherichia coli* K12', *J Biol Chem*, 256, 279-291.

Feederle R, Pajatsch M, Kremmer E, and Böck A (1996), 'Metabolism of cyclodextrins by *Klebsiella oxytoca* M5a1: purification and characterisation of cytoplasmically located cyclodextrinase', *Arch Microbiol*, 165, 206-212.

Fiedler G, Pajatsch M, and Böck A (1996), 'Genetics of a novel starch utilisation pathway present in *Klebsiella oxytoca*', *J Mol Biol*, 256, 279-291.

Galvin N M, Kelly C T, and Fogarty W M (1994), 'Purification and properties of the cyclodextrinase of *Bacillus sphaericus* ATCC7055', *Appl Microbiol Biotechnol*, 42, 46-50.

Gordon-Smith D J, Carbajo R J, Yang J C, Videler H, Runswick M J, Walker J E, and Neuhaus D (2001), 'Solution structure of a C-terminal coiled-coil domain from bovine IF_1: the inhibitor protein of F_1-ATPase', *J Mol Biol*, 308, 325-339.

Hashimoto Y, Yamamoto T, Fujiwara S, Takagi M, and Imanaka T (2001), 'Extra-cellular synthesis, specific recognition, and intracellular degradation of cyclomaltodextrins by the hyperthermophilic archaeon *Thermococcus* sp. strain B1001', *J Bacteriol*, 183, 5050-5057.

Immel S and Lichtenthaler F W (1996), 'Per-O-methyl α- and β-CD: cyclodextrins with. inverse hydrophobicity', *Starch*, 48, 225-232.

Inouye K, Hiromi T, and Hiroshi O (2000), 'States of tryptophanyl residues and stability of recombinant human matrix metalloproteinase 7 (matrilysin) as examined by fluorescence', *J Biochem*, 128, 363-369.

Jonnalngadda S B and Gollapalli N R (2000), 'Kinetics of reduction of toluidine blue with sulfite-kinetic salt effect in elucidation of mechamism', *J Chem Edu*, 77, 506-509.

Kamasaka H, Sugimoto K, Takata H, Nishimura T, and Kuriki T (2002), '*Bacillus stearothermophilus* neopullulanase selective hydrolysis of amylase to maltose in the presence of amylopectin', *Appl Environ Microbiol*, 68, 1658-1664.

Kim T J, Shin J H, Oh J H, Kim M J, Lee S B, Ryu S, Kwon K, Kim J W, Choi E H, Robyt J F, and Park K H (1998), 'Analysis of the gene encoding cyclomaltodextrinase from alkalophilic *Bacillus* sp. I-5 and characterization of enzymatic properties', *Arch Biochem Biophys*, 353, 221-227.

Kim J S, Cha S S, Kim H J, Kim T J, Ha N C, Oh S T, Cho H S, Cho M J, Kim M J, Lee H S, Kim J W, Choi K Y, Park K H, and Oh B H (1999), 'Crystal Structure of a Maltogenic Amylase : Provides Insights into a Catalytic Versatility', *J Biol Chem*, 274, 26279-26286.

Kim M J, Park W S, Lee H S, Kim T J, Shin J H, Yoo S H, Cheong T K, Ryu S R, Kim J C, Kim J W, Moon T W, Robyt J F, and Park K H (2000) 'Kinetics and inhibition of cyclomaltodextrinase from Alkalophilic *Bacillus* sp. I-5', *Arch Biochem Biophys*, 373, 110-115.

Kishimoto K, Yasuda C, and Manning J M (2000), 'Reversible dissociation/ association of D-amino acid transaminase subunits: properties of isolated active dimers and inactive monomers', *Biochemistry*, 39, 381-387.

Kitahata S, Taniguchi M, Beltran S D, Sugimoto T, and Okada S (1983), 'Purification and some properties of cyclodextrinase from *Bacillus coagulans*', *Agric Biol Chem*, 47, 1441-1447.

Kitahata S (1995) 'Cyclomaltodextrin glucanotransferase' in The Amylase research society of Japan (Ed.), *Enzyme chemistry and molecular biology of amylases and related enzymes*, Tokyo, CRC Press, 6-17.

Lee H S, Kim M S, Cho H S, Kim J I, Kim T J, Choi J H, Park C, Lee H S, Oh B H, and Park K H (2002), 'Cyclomaltodextrinase, neopullulanase, and maltogenic amylase are nearly indistinguishable from each other', *J Biol Chem*, 227, 21891-21897.

Lee H S, Kim J S, Shim K H, Kim J W, Park C S, and Park K H (2005), 'Quaternary structure and enzymatic properties of cyclomaltodextrinase from alkalophilic *Bacillus* sp. I-5', *Biologia*, 60 (Suppl. 16), 73-77.

Lee H S, Kim J S, Shim K, Kim J W, Inouye K, Oneda H, Kim Y W, Cheong K A, Cha H, Woo E J, Auh J H, Lee S J, Kim J W, and Park K H (2006), 'Dissociation/association properties of a dodecameric cyclomaltodetrinase: Effects of pH and salt concentration on the oligomeric state', *FEBS J*, 273, 109-121.

Marianayagam N J, Sunde M, and Matthews J M (2004), 'The power of two: protein dimerization in biology', *Trends Biochem Sci*, 29, 618-625.

Oguma T, Kikuchi M, and Mizusawa K (1990), 'Purification and some properties of cyclodextrin-hydrolyzing enzyme from *Bacillus sphaericus*', *Biochim Biophys Acta*, 1036, 1-5.

Oguma T, Matsuyama A, Kikuchi M, and Nakano E (1993), 'Cloning and sequence analysis of the cyclomaltodextrinase gene from *Bacillus sphaericus* and expression in *Escherichia coli* cells', *Appl Microbiol Biotechnol*, 39, 197-203.

Ohdan K, Kuriki T, Takata H, and Okada S (2000), 'Cloning of the cyclodextrin glucanotransferase gene from alkalophilic *Bacillus* sp. A2-5a and analysis of the raw starch-binding domain', *Appl Microbiol Biotechnol*, 53, 430-434.

Park K H (2001), 'The multisubstrate specificity and the quaternary structure of cyclodextrin-/pullulan-degrading enzymes', *J appl Glycosci*, 48, 293-299.

Park K H, Lee H S, Kim T J, Cheong K A, Nguyen V D, Min M J, Cho H Y, Kim Y W, Park C S, Oh B H, and Kim J W (2002), 'N- and C-terminal region mediated oligomerization of the cyclodextrin-/pullulan degrading enzymes', *Biologia*, 57 (Suppl. 11), 87-92.

Podkovyrov S M and Zeikus J G (1992), 'Structure of the gene encoding cyclomalto-dextrinase from *Clostridium thermohydrosulfuricum* 39E and characterization of the enzyme purified from *Escherichia coli*', *J Bacteriol*, 174, 5400-5405.

Puyet A and Espinosa M (1993), 'Structure of the maltodextrin-uptake locus of *Streptococcus pneumoniae*: Correlation to the *Escherichia coli* maltose regulon', *J Mol Biol*, 230, 800-811.

Schenck F W and Hebeda R E (1992), 'Starch hydrolysis products-worldwide technology, production and application', Weinheim/New York, VCH, 319-333.

Søgaard M, Kadziola A, Haser R, and Svensson B (1993), 'Site-directed mutagenesis of histidine 93, aspartic acid 180, glutamic acid 205, histidine 290, and aspartic acid 291 at the active site and tryptophan 279 at the raw starch binding site in barley alpha-amylase 1', *J Biol Chem*, 268, 22480-22484.

Stevens J M, Armstrong R N, and Dirr H W (2000), 'Electrostatic interactions affecting the active site of class Sigma glutathione S-transferase', *Biochem J*, 347, 193-197.

Szejtli J (1988), 'Chemical and physical properties', in Szejtli J, *Cyclodextrin technology*, Dordrecht, The Netherlands, Kluwer academic publisher, 1-20.

Suetsugu N, Koyama S, Takeo K, and Kuge T (1974), 'Kinetic studies on the hydrolyses of alpha-, beta-, and gamma-cyclodextrins by Taka-amylase A', *J Biochem*, 76, 57-63.

Takaku H (1988), 'Anomalously linked oligosaccharides mixture', in The Amylase Research Society of Japan (Ed.), *Handbook of amylase and related enzymes: Their sources, isolation methods, properties and applications*, New York, Pergamon, 215-217.

Tibbot B K, Wong D W, and Robertson G H (2000), 'A functional raw starch-binding domain of barley alpha-amylase expressed in *Escherichia coli*', *J Protein Chem*, 19, 663-669.

Yoshida A, Iwasaki Y, Akiba T, and Horikoshi K (1991), 'Purification and properties of cyclomaltodextrinase from alkalophilic *Bacillus* sp.', *J Ferment Bioeng*, 71, 226-229.

PART II

FUNCTIONS AND APPLICATIONS OF CARBOHYDRATE-ACTIVE ENZYMES

ENZYMATIC MODIFICATION OF STARCH FOR FOOD INDUSTRY

Kwan-Hwa Park, Jin-Hee Park, Suyong Lee, Sang-Ho Yoo, and Jung-Wan Kim

ABSTRACT

Various enzymatic modifications of starch have been attempted for the novel applications to the food industry. The major targets of molecular modification of starch by enzymes include the amylose content, the molecular mass, and the structure of amylopectin chains. The main approaches are the indirect *in vitro* method using the carbohydrate hydrolyzing enzymes from microorganisms or the direct *in vivo* method suppressing or over-expressing the enzymes in the transgenic plants. In this article, we address the microbial enzymes with potentials in modifying starch and starch-based foods by hydrolysis, debranching, and/or disproportionation reactions. Maltogenic amylase from various bacteria has shown that the enzyme could hydrolyze amylose readily, but hardly attack amylopectin. The size discrimination of maltogenic amylase can be explained by the geometry of the enzyme's active site, limiting the molecular size and shape of the substrate. Thus, maltogenic amylase has a great potential in producing starch with different amylose content. Chain distribution of amylopectin can be engineered by 4-α-glucanotransferase that disproportionates the side chains of glucan, which eventually alters the side chain length. Molecular size of starch can also be controlled by the selective hydrolysis reaction such as 4-α-glucanotransferase that preferentially cleaves the α-1,4-glycosidic linkage of the glucan segment between the amylopectin clusters. As a result, the apparent mass can be reduced to the level of an amylopectin cluster. Microbial debranching enzymes play a role in shaping glycogen in the cells. We employed debranching enzymes to modify amylopectin. The enzymatic modification of starch molecules directly affected properties of the modified starch especially in freeze-thaw stability of gels and retardation of retrogradation during storage.

Key words: carbohydrate enzymes; starch; glycogen; enzymatic modification; food industry

INTRODUCTION

Starch is a major energy source on earth, providing up to 80% of the calories consumed by humans. Recently, ethanol production from starch also becomes increasing importance due to the shortage of fossil fuels. Starch comprises about 70% of the dry weight of cereal seeds and is composed of amylopectin (~75%) and amylose (~25%). Amylopectin is a highly branched homopolysaccharide with 4~5% of α-1,6-glucosidic linkage while amylose is a planar polymer exclusively composed of glucose linked by α-1,4-glucosidic bond. Branching of amylopectin takes place every 24-30 glucose units and the molecule is soluble in water. Its counterpart in animals is glycogen with more branching that is present every 8 to 12 glucose units. The α-1,4-glucosidic bonds of amylose promote the formation of α helix structure. Amylose is less readily hydrolyzed than amylopectin. Strands of amylopectin form double helical structures with each other or with amylose.

The relatively simple chemical component of starch implies that few enzymes are involved in the biosynthesis of starch in nature. Even though the structure and self-assembly process of starch are not well understood, the enzymatic system for the synthesis and degradation of starch needs to be very rapid.

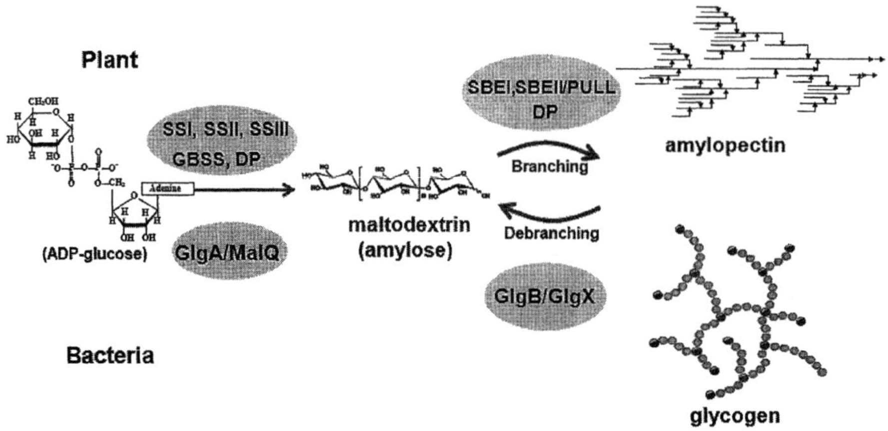

Figure 1 Comparison of the relevant enzymes for the synthesis of starch and glycogen in plants and bacteria, respectively

SS represents starch synthase; GBSS, granule bound starch synthase; SBE, starch branching enzyme; PULL, pullulanase; DP, disproportionating enzyme; GlgA, glycogen synthase; GlgB, branching enzyme; GlgX, glycogen debranching enzyme; MalQ, 4-α-glucanotransferase. The structure and physical properties of starches produced by transgenic plants are discussed intensively by Jobling et al. (2004).

As shown in Figure 1, the process of starch biosynthesis includes sugar activation forming ADP-glucose by ADP-glucose pyrophosphorylase, chain elongation by soluble starch synthase (E.C.2.4.1.21; ADP-glucose; 1,4-α-D-glucan-4-α-D-glucosyltransferase) and granule bound starch synthase, shaping by starch branching enzymes (E.C. 2.4.1.18; 1,4-α-D-glucan-6-α-(1,4- α-glucano)-transferase) and debranching enzymes (Preiss, 2004). The process is very similar to the biosynthesis of glycogen in animals and microorganisms (Henrissat et al., 2002; Preiss, 2006). Only difference is the glucose activation; ADP-glucose for plants and bacteria, UDP-glucose for animals and fungi. The degradation of starch or glycogen is carried out by glycosidases such as α-amylase and glycogen phosphorylases. The enzymes involved in biosynthesis and degradation of starch and glycogen include various members of families GT5 (starch or glycogen synthases), GT35 (glycogen phosphorylases), GH13 (α-amylase, pullulanase, cyclomaltodextrinase *etc.*), GH15 (glucoamylases), GH31 (α-glucosidase), GH57 (4-α-glucanotransferase), and GH77 (amylomaltase and 4-α-glucanotransferase).

Modification of starch can be manipulated in plants by mutation or by transgenic technology to generate novel starch. The technology may include suppression or over-expression of the enzymes involved in starch biosynthesis and degradation. The amylose content of starch can be controlled genetically by suppressing the enzyme such as granule-bound starch synthase that is responsible for amylose biosynthesis. In potato, down-regulation of two starch branching enzymes (SBEI and SBEII) led to the production of starches with higher amylose levels (Jobling, 2004). The enzymes involved in amylopectin biosynthesis are starch synthases, starch branching enzymes, and debranching enzymes. Amylopectin chain length can be engineered by suppression or over-expression of soluble starch synthase. In addition, various attempts have been made to engineer the structure of amylopectin by suppression or over-expression of starch branching enzyme. The resulting amylopectin molecules contained less α-1,6-glucosidic branch points with longer α-1,4-glucosidic backbone chains than the regular molecules. The production of highly branched starch such as phytoglycogen could be achieved by suppressing the debranching enzyme in rice (Nakamura, 2002).

Modern enzyme technology has been adopted for the modification of starch *in vitro* (Fig. 2). Many of the recent developments for the utilization of carbohydrate enzymes are associated with improving their texture, functionality, and nutritional quality of starch. These can be achieved by modifying starch in chain length, branch point formation, phosphate substitution, debranching, and disproportionation (Blennow, 2004). The catalytic properties and their potential applications of the enzymes in food industry will be discussed briefly in this chapter.

Starch-modifying Enzymes

Starch is one of the most abundant natural compounds and accumulates as a complex granular structure, which composed of α-1,4 linked and α-1,6 branched glucans. Various enzymes acting on starch have been classified into glycosyl hydrolases (EC 3.2.1.X) and glycosyl transferase (EC 2.4.X.Y) on the basis of the catalytic reaction, substrate specificity, and sequence similarity (IUBMB, 1992).

Coutinho and Henrissat (1999) classified all the enzymes that act on starch into several families (CAZY website: http://afmb.curs-mrs.fr/CAZY). Families 13, 57, and 77 include most of the important enzymes that have been used for modifying starch.

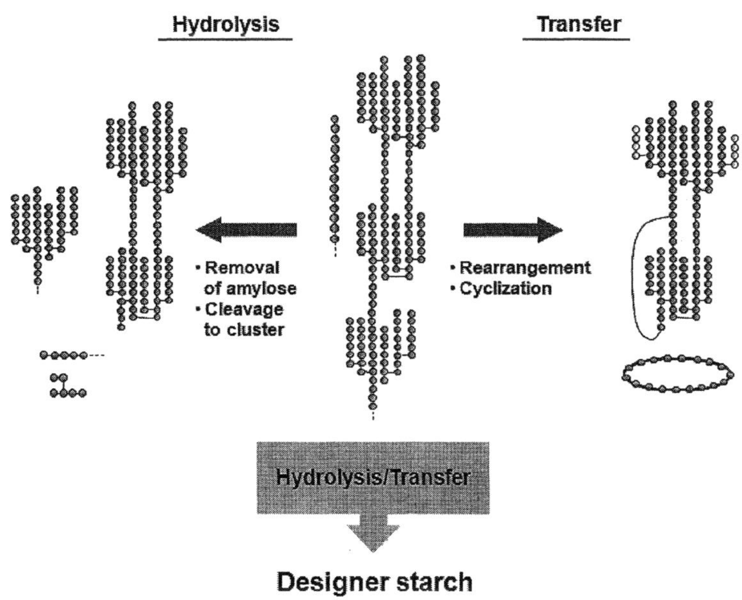

Figure 2 Principles of starch modification by the hydrolyzing and transferring reactions of carbozymes

Notably, family 13 includes enzymes such as α-amylase (EC 3.2.1.1), pullulanase (EC 3.2.1.41), isoamylase (EC 3.2.1.68), glucan branching enzyme (EC 2.4.1.18), and cyclodextrin glycosyltransferase (EC 2.4.1.19). Family 77 possesses amylomaltase, 4-α-glycanotransferase (EC 2.4.1.25), and so on.

In food industry, most starch modification has been achieved by acid or enzyme treatment to obtain hydrolysis products. Conversion of starch to low DE includes the processes for liquefaction and saccharification. An alternative useful starch modification is to design starch with novel structure, in which side chain distribution, molecular mass, or amylose content is changed by chemical, physical, and/or enzymatic method. Enzymatic method is one of the promising processes for the modification of starch, which can be achieved by using the carbohydrate active enzymes from microorganisms, plants, and animals.

Interestingly, similar catalytic properties of the enzymes have been found among the enzymes involved in the starch and glycogen biosynthesis of plants and bacteria (Fig. 1). Moreover, ADP-glucose is used as a common starting material for the synthesis of starch or glycogen in plants and bacteria, respectively. Consequently, the relevant enzymes from bacteria can be introduced to the modification process of starch.

Debranching enzymes in plants and bacteria

Debranching enzymes hydrolyze α-1,6-glucosidic linkage in polyglucan such as amylopectin, glycogen, and pullulan. The enzymes are widely distributed in nature including animal, plant, and bacteria (Okada et al., 1988; Kuriki et al., 1988). Isoamylase and pullulanase are originated from plants and bacteria and possess hydrolytic activity on α-1,6-glucosidic linkage, exclusively (Fig. 3). In contrast, amylo-1,6-glucosidase, the debranching enzyme from mammalians and yeasts, have two distinct activities of a transferase (1,4-α-D-glucan-4-α-glycosyltransferase; EC 2.4.1.25) and a glycosidase (dextrin 6-α-glucosidase; EC 3.2.1.33; Gillard et al., 1980). The enzyme transfers maltotriosyl groups from the branches of phosphorylase limit glycogen to the non-reducing end glucose of the main chain to form an α-1,4-glucosidic linkage. Then the same enzyme further hydrolyzes the remaining glucosyl residue at the

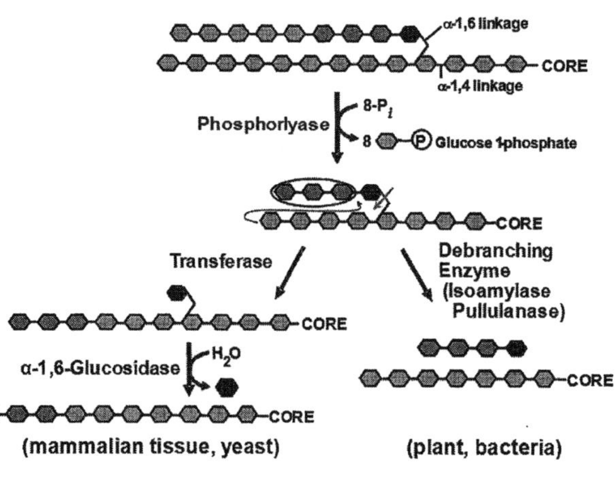

Figure 3 Breakdown of glycogen in mammalian tissues, plants, and bacteria

branch point to produce glucose and limit maltodextrins (Liu et al., 1991). Unlike mammalian debranching enzymes, those originated from bacteria and plants hydrolyze α-1,6-glucosidic linkage at the branch points of glycogen or glycogen to release maltooligosaccharides and limit maltodextrin (Ball and Morell, 2003). It is interesting debranching enzymes of mammalian tissue can act best on maltotetraosyl branch chain that is provided by glycogen phosphorylase, while bacterial debranching enzyme can directly act on the branch chains of amylopectin or glycogen (Park et al., 2008). Therefore, debranching enzymes from bacteria are likely to be practically applicable for modifying starch.

Disproportionating enzymes in plants and microorganisms

Disproportionating enzyme catalyzes the transfer of a certain fragment of glucan to the C-4 of the acceptor. Disproportionating enzymes found in plants resemble 4-α-glucanotransferase, a disproportionating enzyme (amylomaltase) of microorganisms. One of the microbial disproportionating enzyme, MalQ, has been investigated intensively to understand the maltose utilization system in *E.* coli (Pugsley and Dubreuil, 1988; Dippel and Boos, 2005; Boos and Shuman, 1998; Palmer et al., 1976). The enzyme catalyzes the transfer of glucosyl and maltodextrinyl residues from non-reducing end of maltodextrin to acceptor molecules (Fig. 4).

The starch degradation pathway in plants is compared with that of maltose utilization in *E. coli* (Fig. 5). In the cytosol of plants, degradation of starch may occur in a process similar to the maltose utilization system of *E. coli*. Delatte et al. (2006) and Steichen et al. (2008) proposed that degradation of glucans released from starch granules proceeded via the hydrolysis reactions catalyzed by a series of amylolytic enzymes such as α- and β-amylase to produce maltose in plants. The maltose molecules are exported to the cytosol for further elongation reaction by the cytosol disproportionating enzyme and then the resulting maltodextrin (heteroglycan in Fig. 5) can be converted to hexose phosphate by glucan phosphorylase. Similarly, maltose taken up through the cell membrane is elongated by a disproportionating enzyme, MalQ, to produce maltodextrin that can be further degraded to hexose phosphate by maltodextrin phosphorylase, MalP (corresponding to glucan phosphorylase in Fig. 5) in *E. coli* (Dippel and Boos, 2005).

$$(\alpha\text{-1,4-glucan})_a + (\alpha\text{-1,4-glycan})_b \longrightarrow (\alpha\text{-1,4-glucan})_{a-x} + (\alpha\text{-1,4-glycan})_{b+x}$$

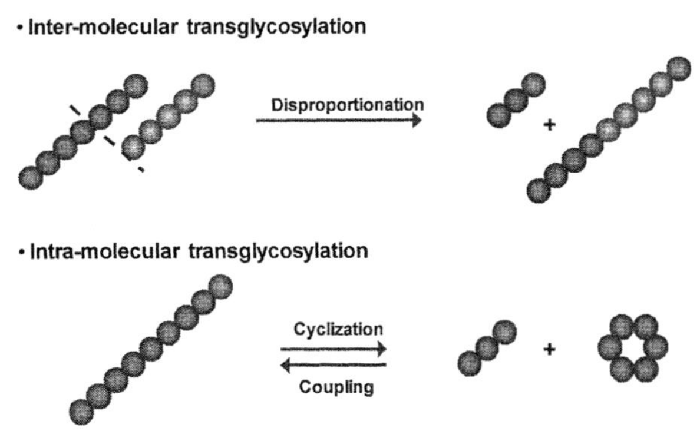

Figure 4 Modification of starch by transglycosylation reactions

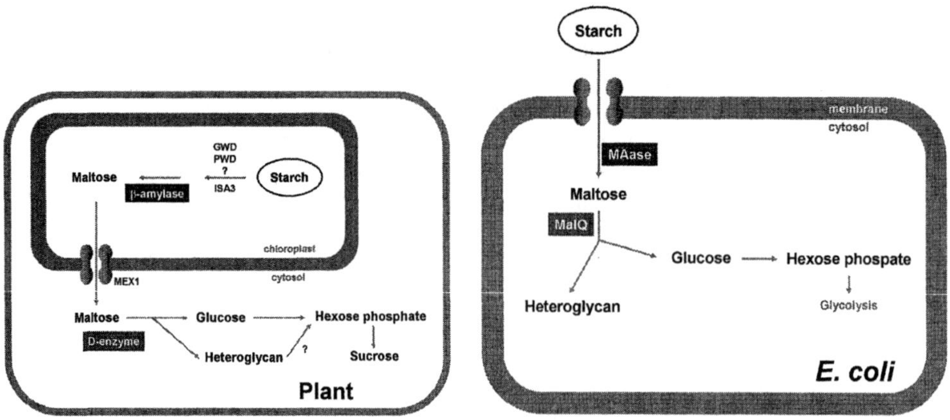

Figure 5 Comparison of the starch degradation processes in plants and *E. coli*
GWD, glucan-water kinase; PWD, phosphoglucan-water dikinase; ISA3, isoamylase3; MEX1, maltose transporter (Reprinted, with permission, from the *Annual Review of Plant Biology,* Volume 56 ©2005 by Annual Reviews, www.annualreviews.org).

Maltogenic amylase from bacteria

Maltogenic amylase (EC 3.2.1.33) along with cyclomaltodextrinase (EC 3.2.1.54), and neopullulanase (EC 3.2.1.135) is capable of hydrolyzing various substrates such as maltooligosaccharides, starch, pullulan, and cyclodextrin. More than 20 cyclomalto-dextrinase, maltogenic amylase, and neopullulanase have been isolated only from bacteria and they share 40-86% sequence identity with each other (Park et al., 2000). Their multi-substrate specificity for cyclodextrins, pullulan, and soluble starch make them easily be distinguished from other members of GH13 such as α-amylase, isoamylase, and pullulanase.

This group of enzymes hydrolyzes cyclodextrins and starch mainly to maltose, and pullulan to panose by cleaving α-1,4-glucosidic linkages, whereas α-amylases lack the activity on cyclodextrins and pullulan (Fig. 6). Maltogenic amylase has been reported to hydrolyze acarbose, a strong inhibitor of α-amylase, to a pseudotrisaccharide and a glucose. They are also capable of transferring a sugar moiety released from the substrates to a receptor sugar molecule by forming mainly α-1,6-glucosidic linkage as well as α-1,3- and α -1,4-glucosidic linkages when the substrates are present in excess. These catalytic properties of the enzymes have made it possible to use them for the production of highly branched oligosaccharide (or isomaltooligosaccharide) mixtures from liquefied starch.

Branched oligosaccharides have several properties that are attractive to consumers of modern society with great concern for healthcare. Branched oligosaccharides are better sweetener with softer and milder taste than sucrose. They are beneficial in preventing dental caries since oral microbial flora do not hydrolyze them, and also helpful for the diabetes patient and people on a diet because they are not digested readily in human body (Glor, 1988; Kaneko et al., 1992; Kaneko et al., 1995). They are also known to promote the growth of intestinal bacteria such as *Bifidobacteria* (Park et al., 1990). Kweon et al. (1994) reported that branched oligosaccharides could improve shelf life of foods by lowering water activity and thereby preventing the growth of spoilage microorganisms. They also could retard retrogradation of starchy foods, when

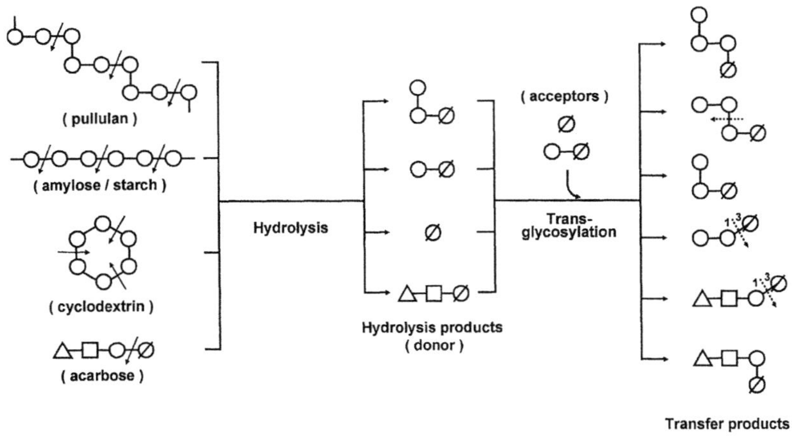

Figure 6 Action pattern of maltogenic amylase
(with permission of Park et al., 2000, *BBA-Protein Structure and Molecular Enzymology*, 1478, 2, 165-185)

added as a food component. Using maltogenic amylase from *B. licheniformis* (BLMA) or *B. stearothermophilus* (BSMA), a mixture with 58.5% of various branched oligosaccharides was produced from 30% (w/v) liquefied rice or corn starch (Lee et al., 1995). The mixture contained the most branched DP4, followed by branched DP5, panose, and isomaltose (Table 1). The content of branched oligosaccharides could be increased even higher (85.7%) by removing glucose and maltose through yeast fermentation, improving its efficacy as a humectant in bread (Yoo et al., 1995). Recently,

Table 1 Compositions of various branched oligosaccharide mixtures

Components		BSMA[a]	BSMA Fremented[a]	BSMA + α-GTase[a]	Brand A[a,b]	Brand B[a,b]
DP1	Glucose	9.7	-	10.8	42.1	22.0
DP2	Maltose	14.1	0.9	11.4	4.6	21.0[c]
	BDP2[d]	3.0	7.3	8.2	24.4	
DP3	Maltotriose	6.2	7.0	2.7	0.0	24.0[c]
	BDP3[d]	16.0	28.5	23.8	15.7	
DP4	Maltotetraose	2.4	2.4	1.3	0.0	33.0[c,e]
	BDP4[d]	23.8	28.6	25.0	9.0	
≥DP5	≥Maltopentaose	10.0	4.0	6.4	3.2	
	≥BDP5[d]	14.7	21.3	10.4	1.0	
Total amount		57.5	85.7	67.4	50.1	65.0

a : the content was expressed in weight percentage.
b : commercial product (declared composition)
c : both maltooligosaccharides and branched oligosaccharides
d : branched oligosaccharide with a degree of polymerization
e : included maltooligosaccharides larger than maltotetraose

a very efficient process for the production of branched oligosaccharides has been developed, using BSMA and 4-α-glucanotransferase from *Thermotoga maritima* (Lee et al., 2002a). The cooperative action of the enzymes promoted the formation of branched oligosaccharides to the final concentration of 68% with relatively larger branched compounds compared with the products obtained by the reaction without 4-α-glucanotransferase. Time course analysis of the reaction suggested that 4-α-glucanotransferase transferred donor sugar molecules to the hydrolysis products such as maltose and maltotriose to form various branched molecules. BSMA hydrolyzed maltopentaose and maltohexaose most readily into maltose and maltotriose, simultaneously transferring them to acceptor molecules to form larger branched oligosaccharides (Fig. 7). The contents and the preparation process of these branched oligosaccharides and other commercial mixtures are compared in Table 1 and Figure 8, respectively.

Their unique enzyme characteristics were found to be derived from the protein structure. Alignment of the primary structure of maltogenic amylase in comparison with those of other amylolytic enzymes revealed 4 common conserved regions that are located in the main catalytic domain (Table 2, Fig. 9). The spacing between each region was also conserved among the enzymes. In addition, maltogenic amylase has a unique domain, the domain N, which is consisted of approximately 130 amino acid residues. This unique N-terminal domain as seen in the tertiary structure apparently contributes to the formation of a dimer. The tertiary structure of maltogenic amylase comprises three domains; a N-terminal domain, a central $(\beta/\alpha)_8$-barrel domain, and a C-terminal domain.

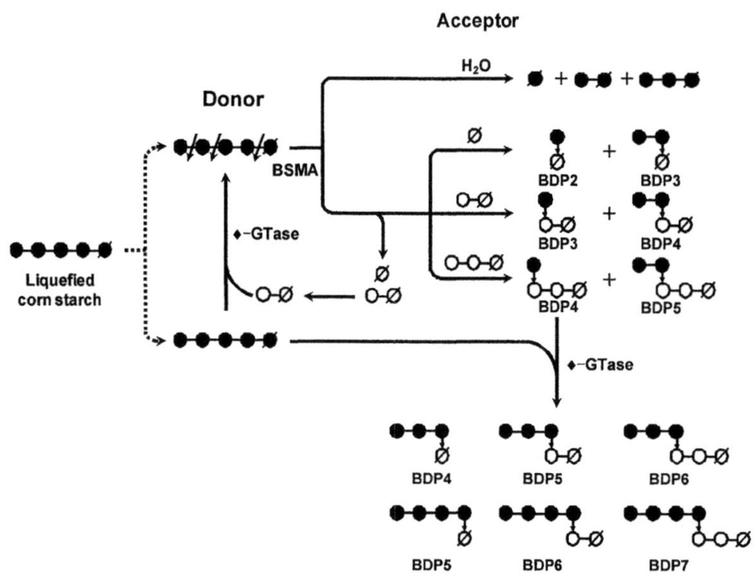

Figure 7 Cooperative action modes of BSMA and 4-α-glucanotransferase for the production of branched oligosaccharides

Open and closed circles represent nonreducing glucopyranosyl residue in acceptor and donor molecules, respectively; open circle with a slash, reducing glucose residues. BSMA with hydrolyzing and transglycosylation activities and 4-α-glucanotransferase with disproportionation activity promoted the production of various branched oligosaccharides (with permission of Lee et al., 2002, *J Agric Food Chem*, 50(10), 2812-2817).

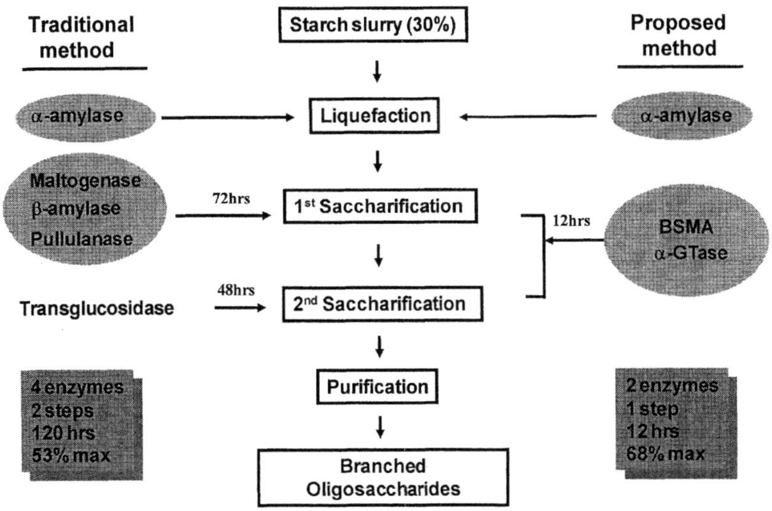

Figure 8 Comparison of the procedures for the branched oligosaccharide mixture (with permission of Lee et al., 2002, *J Agric Food Chem*, 50(10), 2812-2817)

Table 2. Comparison of amino acid sequences in the regions conserved among various amylases

Amylolytic enzymes[a]	Amino acids of conserved domains			
	I	II	III	IV
BSMA	DAVFNH	GWRLDVANE	EIWH	LLGSHD
BLMA	DAVFNH	GWRLDVANE	EIWH	LLDSHD
NPL	DAVFNH	GWRLDVANE	EIWH	LLGSHD
TVA II	DAVFNH	GWRLDVANE	EIWH	LLDSHD
Pullulanase	DVVYNH	GFRFDLASV	EGWD	YVSKHD
CDase I-5	DAVFNH	GWRLDVANE	EVWH	LLDSHD
BLA	DVVINH	GFRLDAVKH	EYWQ	FVDNHD
BMCGTase	DFAPNH	GIRFDAVKH	EWFL	FIDNHD
BSCGTase	DFAPNH	GIRMDAVKH	EWFL	FIDNHD
Consensus sequence	DAVINH	GFRLDAAKH	EVID	FVDNHD

a: BSMA, MAase from *Geobacillus stearothermophilus* (Genbank ID:1255196); BLMA MAase from *B. licheniformis* (gi:39564); NPL, neopullulanase from *G. stearothermophilus* (gi:541633); TVA II, *Thermoactinomyces vulgaris* amylase II (gi:13537293); pullulanase, pullulanase from *Klebsiella pneumoniae* (gi:110591424); CDase I-5, Cyclodextrinase from alkalophilic *Bacillus* I-5 (gi:); BLA, thermostable α-amylase from *B. licheniformis* (gi:113813); BMCGTase, cyclodextrin glucanotransferase from *Paenibacillus macerans* (gi:39625); BSCGTase, cyclodextrin glucanotransferase from *G. stearothermophilus* (gi:39833).

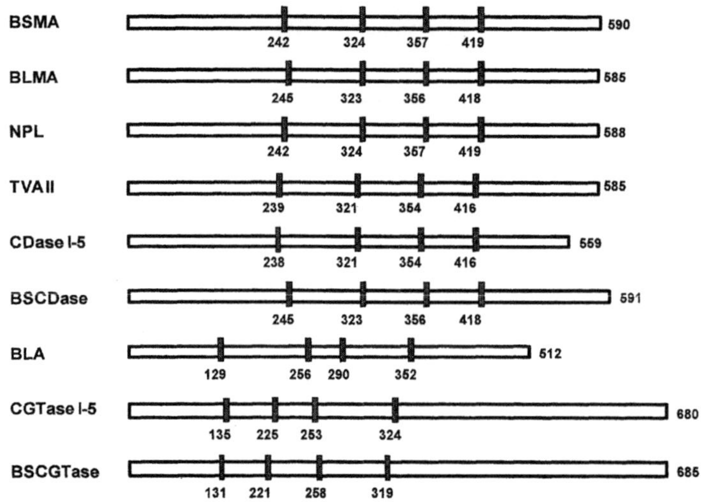

Figure 9 Comparison of the spacings between the regions conserved among various amylolytic enzymes

For the abbreviations refer to the footnote of Table 2.

The N-terminal domain is composed exclusively of β-strands (Kim et al., 1999; Fig. 10). They suggested that the protein is present in equilibrium of monomeric and dimeric form in solution via the N-terminal domain (Fig. 10). When they form a dimer, the N-terminal domain of one subunit covers partly the top of the have preference to smaller and thin substrates such as cyclodextrins and short linear maltodextrins. On the other hand, when the enzyme is present in monomeric form, the catalytic activity toward larger and bulky substrates such as starch and pullulan increases dramatically due to the easier accessibility to the active site. Therefore, the substrate specificity of the enzymes is likely to be modulated by monomer-dimer equilibrium. The tertiary structures of maltogenic amylases from *Bacillus* and *Thermus* showed that the enzymes exist as oligomers consisted of basic dimeric units. Oligomerization in a similar way has also been reported for cyclomaltodextrinase (Kim et al., 1992; Kim et al., 1999; Lee et al., 2002b), TVAII (Kamitori et al., 1999), and neopullulanase (Hondoh et al., 2003).

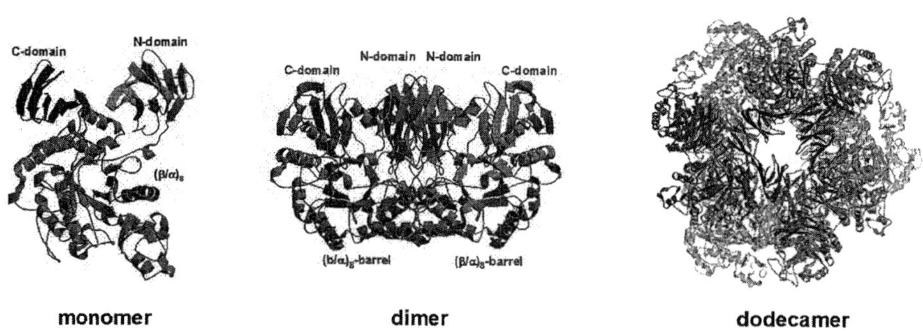

Figure 10 Tertiary structure of cyclodextrinase from *Bacillus* sp. I-5
(with permission of Lee et al., 2002, *J Biol Chem*, 277, 24, 21891-21897)

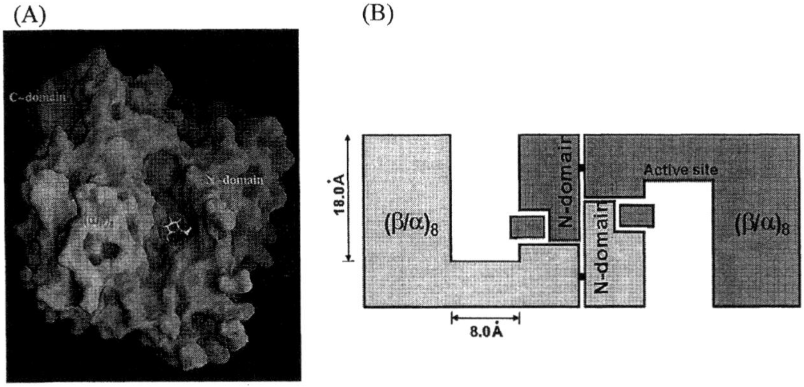

Figure 11 Surface model (A) and schematic diagram (B) of dimeric maltogenic amylase

(with permission of Park et al., 2000, *BBA-Protein Structure and Molecular Enzymology*, 1478, 2, 165-185)

Cyclomaltodextrinase I-5 from an alkalophilic *Bacillus* sp. forms a dodecamer that is a hexamer of the dimeric form (Lee et al., 2002b). The intermolecular interactions between the dimers are mediated by the C-terminal domain of one molecule and the N-terminal domain of an adjacent molecule (Fig. 11). Since the active sites of cyclomaltodextrinase I-5 are outwardly oriented on the dodecameric assembly, the hexamer formation does not shield the active sites from a substrate molecule. The specific activity of cyclomaltodextrinase I-5 was found to be about two times higher than that of dimeric *Thermus* maltogenic amylase (ThMA). Furthermore, cyclomaltodextrinase I-5 exhibited an exceptionally higher preference toward amylase than amylopectin.

The dimer interface at the top of the barrel forms a narrow and deep groove that is ~17 Å in length, ~8 Å in width, and ~18 Å in depth, distinctively different from the wide and shallow active site cleft of α -amylase. Three invariant catalytic residues are located at the bottom of the groove. The shape of the active site is closely related with the substrate specificity. We have proposed that the geometry of the active site in the homodimer is able to discriminate the molecular size of the substrate. Small substrate like amylose can access the deep, narrow groove whereas large substrate such as amylopectin was unable to bind with the catalytic site (Kim et al., 2000).

Subsite binding affinity of the enzyme was examined by using maltogenic amylase and maltooligosaccharides labeled with ^{14}C (G3*-G7*). The patterns of the products obtained by the reactions revealed that maltogenic amylase split the first glycosidic bond from the reducing end, producing maltose as a major product (Fig. 12; Park et al., 2007c).

Structured-starch by maltogenic amylase

One of the approaches modifying the structure of starch is the enzymatic removal of amylose, thereby changing the composition of amylase and amylopectin in starch. Many efforts have been made to find enzymes that are capable of hydrolyzing amylase selectively to reduce the content of amylose in starch.

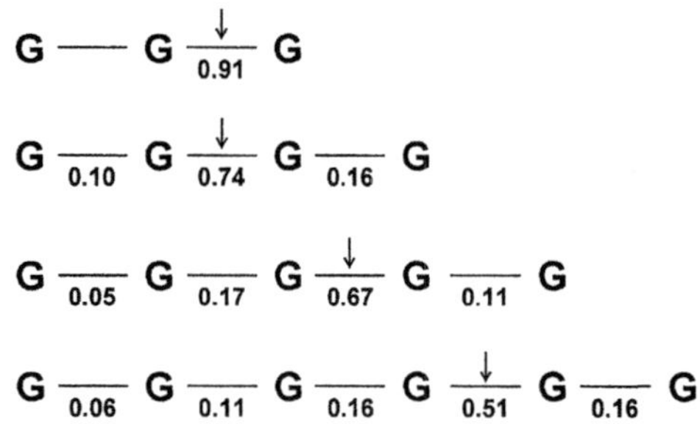

Figure 12 Subsite binding affinity of maltogenic amylase
(with permission of Park et al., 2005, *BBA-Protein and Proteomics*, 1751, 2, 170-177)

Recently, the substrate specificity of maltogenic amylase toward amylose and amylopectin was analyzed (Kamasaka et al., 2002; Auh et al., 2006). The results have shown that maltogenic amylase hydrolyzed amylose to produce maltose, but it hardly attacked amylopectin. Specific activity of maltogenic amylase toward amylose was 30 times higher than that toward amylopectin. The unique action pattern of maltogenic amylase can be used to produce low-amylose starch or amylose-free starch by specifically degrading amylose molecules. Maltogenic amylases from various bacteria were examined for various starches from rice, tapioca and kudzu. As shown in Table 3, when starches from various sources were incubated with BSMA, amylose greatly decreased without significant change in the content of amylopectin. The decrease of amylose was explicit in the starches from roots than those from grains, which could be due to the difference in the structure of the starches. However, the structural differences among starches still remain to be elucidated.

Modification of rice starch by maltogenic amylase/cyclomaltodextrinase

Cyclomaltodextrinase belongs to the subclass of cyclodextrin degrading enzyme along with maltogenic amylase and neopullulanase (Park et al., 2000). The distinct action pattern of alkalophilic *Bacillus* I-5 cyclomaltodextrinase on starch structure was examined for starch modification (Auh et al., 2006). Analysis of the hydrolysis reaction carried out by the enzyme revealed that the k_{cat}/K_m value on amylose was 14.6 s^{-1} $(mg/ml)^{-1}$, whereas that for amylopectin was 0.92 $s^{-1}(mg/ml)^{-1}$.

Table 3 Amylose and amylopectin content in kudzu, tapioca, rice powder, and rice starch treated by BSMA

	Amylose content		Amylopectin remained (%)
	Control	BSMA treatment	
Kudzu	18.8	2.6	95.5
Tapioca	20.6	3.9	98.2
Rice powder	19.1	8.6	89.1
Rice starch	23.6	9.2	90.3

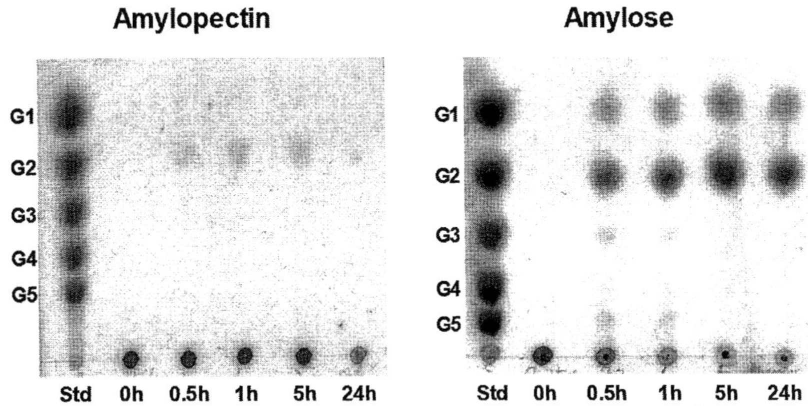

Figure 13 TLC analysis of amylose and amylopectin hydrolyzed by cyclomaltodextrinase I-5
(with permission of Auh et al., 2006, *J Agric Food Chem*, 54, 6, 2314-2319)

This high preference toward amylose can be applied for modifying rice starch to produce low-amylose starch. Amylose was selectively hydrolyzed mainly to maltose by cyclomaltodextrinase in the presence of amylopectin (Fig. 13, Table 4). There was little difference in the side chain length distribution between the control and amylopectin treated with cyclomaltodextrinase.

Applications of amylose-low starch

Maltogenic amylase/cyclomaltodextrinase can be employed in the production of amylose-free or amylose-varied starch. A partially degraded starch by the enzymes can be widely used in the food industry. The ratio of amylose to amylopectin in rice starch greatly influences the taste of cooked rice. Furthermore, low-amylose rice (5-15% of amylose) is suitable for frozen cooked rice.

Amylose is known to be responsible for the short-term retrogradation involving in aggregating double helices of amylose and amylopectin, that eventually contributes the synergic effects on retrogradation (Miles et al., 1985). The maltogenic amylase treatment significantly retarded the retrogradation of cooked rice, since a substantial amount of amylose was degraded by the action of the enzyme (Auh et al., 2006).

Table 4 Change in the content of amylose and amylopectin in starch according to the reaction time during the hydrolysis by cyclomaltodextrinase I-5
(with permission of Auh et al., 2006, *J Agric Food Chem*, 54, 6, 2314-2319)

Reaction time (min)	Amylopectin (%)	Amylose (%)
0	71	28.5
10	72	12
30	72	10
60	72	9

Amylose-free and short chain amylopectin starches have improved freeze-thaw stability, which can be applied in the food products that needs to be stored at chilled temperature. The repeated heating and cooling of 4-α-glucanotransferase treated starch solution showed excellent thermoreversibility (Kaper et al., 2005; Lee et al., 2006).

Implication of maltogenic amylase in nature

The catalytic properties of maltogenic amylase and its relationship to the structure have well been characterized *in vitro*, making it possible to apply the enzyme to modification of starch. Maltogenic amylase has been found only in bacteria, indicating its specific role in the life cycle or survival of the cells. Maltogenic amylase is distinguished from typical α-amylase not only by the catalytic properties including multi-substrate specificity but also by the location in the cell. The enzyme does not have the signal sequence for secretion and is likely to be confined in the cell, suggesting its role different from that of α-amylase in the cell.

Immunolocalization study showed that the enzyme was located close to the cell membrane during vegetative growth, allowing rapid degradation of maltodextrin and cyclodextrin to maltose as they are transported via their specific transporters, MdxE and CycB, respectively (Kamionka and Dahl, 2001; Schönert et al., 2006). Thus, maltogenic amylase seems to be involved in utilization of linear maltodextrins and β-cyclodextrin in the *Bacillus* cell. Furthermore, maltogenic amylase was localized solely to the core of endospore during sporulation. Since there is no further uptake of maltodextrins from the environment during the cellular differentiation, the enzyme is likely to be involved in breakdown of endogenous glycogen with cooperative action of a debranching enzyme, pullulanase, to generate glucose as the energy source (Fig. 14).

These results correlated very well with those the expression study of the maltogenic amylase gene in *B. subtilis* (Kim et al., 2004). The promoter of the gene (*yvdF*) was induced by maltose, β-cyclodextrin, or starch at late exponential growth phase, but

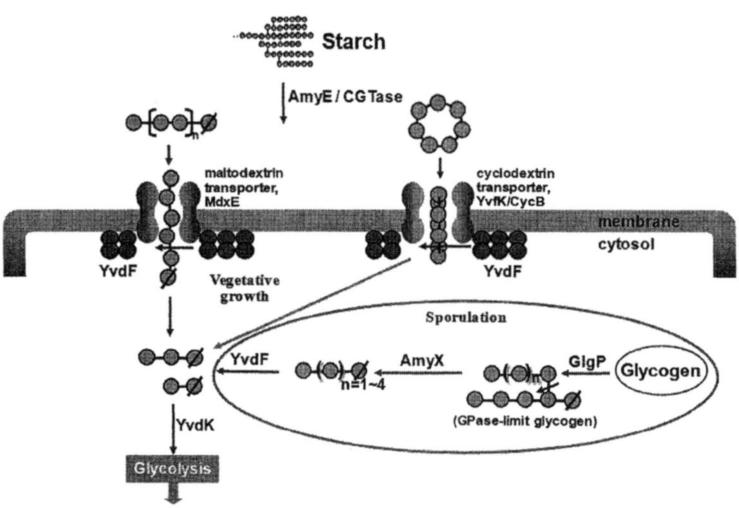

Figure 14 A proposed model for sugar utilization and glycogen breakdown by Maltogenic amylase in *Bacillus* sp.

AmyE represents α-amylase; CGTase, cyclodextrin glucanotransferase; YvdF, maltogenic amylase; AmyX, pullulanase; GlgP and GPase, glycogen phosphorylas

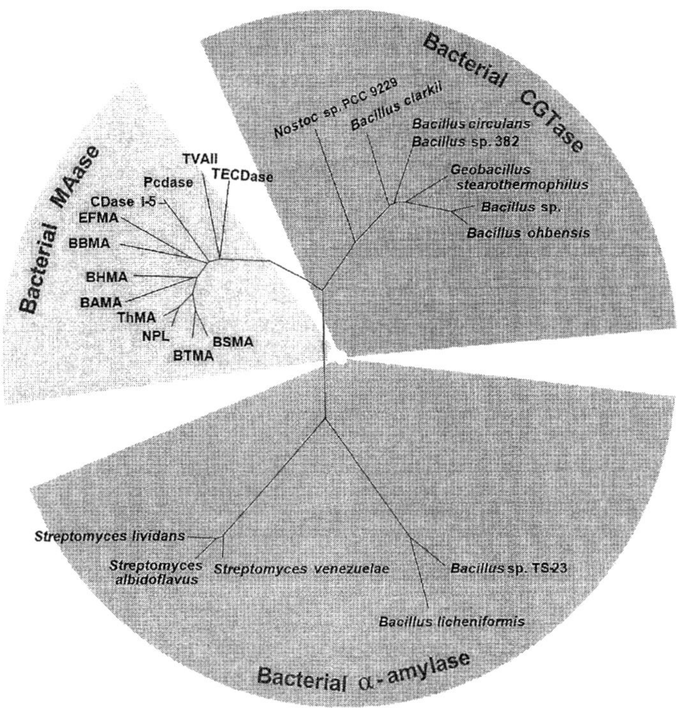

Figure 15 A phylogenetic tree of maltogenic amylase (MAase), cyclodextrin glucanotransferase (CGTase), and α-amylase

Phylip format tree outputs from the CLUSTAL X analysis were visualized with TreeView PPC based on the distance matrix using the neighbor-joining method. The unrooted phylogenetic tree was built from entire sequences of the following enzymes: ThMA, MAase from *Thermus* sp. IM6501 (gi:3089607); BAMA, MAase from *B. acidopullulyticus* (gi:3960830); BBMA, MAase from *B. subtilis* (gi:6689858); EFMA, MAase from *Enterococcus faecalis* (gi:29375914); BTMA, MAase from *Bacillus thermoalkalophilus* (gi:51038505); BSMA, MAase from *B. stearothermophilus* (gi:1255196); TVAII, α-amylase II from *Thermoactinomyces vulgaris* (gi:1171687); CDase I-5, cyclodextrinase (CDase) from alkalophilic *Bacillus* sp. I-5 (gi:1236529); NPL, neopullulanse from *B. stearothermophilus* (gi:13182951); CGTase from *Nostoc* sp. PCC 9229 (gi:20258046), *B. clarkii* (gi:126364303), *B. circulans* (gi: 39420), *Bacillus* sp. 38-2 (gi:216248), *Bacillus* sp. (gi:3298517), *Geobacillus stearothermophilus* (gi:4099127), and *B. ohbensis* (gi:27263167); α-amylases from *Aspergillus kawachii* (gi:2570150), *B. licheniformis* (gi:99030348), *Bacillus* sp. TS-23 (gi:722279), *Streptomyces albidoflavus* (gi:80685), *Streptomyces lividans* (gi:167508809), and *Streptomyces venezuelae* (gi:153159) (adopted from Kim et al., 2007).

repressed by glucose, fructose, sucrose, or glycerol in the culture medium. The promoter was not detected in the *spo0A* mutant, which was defective in sporulation and the knock-out mutation of *yvdF* promoted sporulation by two-folds, implying its role during the process.

The phylogenetic relationship among the four distinct groups of eubacterial maltogenic amylases, archaeal maltogenic amylases, 4-α-glucanotransferases, and α-amylases was shown as a phylogenetic tree (Fig. 15). The eubacterial maltogenic amylases are placed close to the cluster of archaeal maltogenic amylase. The action patterns on various carbohydrate substrates are also useful for understanding the phylogenetic relationship between these groups of maltogenic amylase, 4-α-glucanotransferases and α-amylase.

The enzymes are often found as dimeric or oligomeric assembly consisted of dimeric unit in solution (Table 5). The oligomeric states of the enzymes have been determined by ultracentrifugation analysis and size exclusion column chromatography. Crystal structures of ThMA, neopullulanase from *B. stearothermophilus*, and TVII showed that the N-terminal domain is involved in dimerization (Kamitori et al., 1999; Kim et al., 1999; Hondoh et al., 2003). Recently, TpMA from *Thermoplasma volcanium* and SMMA from *Staphylothermus marinus* were found to be dimer in solution, too (Kim et al., 2007). As mentioned above for maltogenic amylases, the substrate accessibility for the catalytic site depends on the size of substrate and the geometry of the catalytic and substrate binding sites of the enzymes. It is interesting to notify that thermostability and optimal temperature of the enzymes range from 25°C for *Nostoc punctiforme* to 105°C for *Staphylothermus marinus.* The enzymes with the variety of the optimal temperature would make various reaction conditions for the modification of starch possible.

Starch treated with 4-α-Glucanotransferase

The enzyme, 4-α-glucanotransferase from *Thermus scotoductus* catalyzes the transfer of α-glucan chains from one α-glucan molecule to the non-reducing end of another (Takaha and Smith, 1999; Park et al., 2007b). This intermolecular glucan transfer reaction is called 'disproportionation'. In addition, the enzyme catalyzes the intramolecular glucan transfer reaction which creates cyclic glucans such as cycloamylose (Tachibana et al., 2000; Bhuiyan et al., 2003; Park et al., 2007b). The action pattern of 4-α-glucanotransferase is shown in Figure 16. The transfer products from maltose to maltoheptaose were produced by transferring glucose unit. Maltose was the smallest donor molecule for the disproportionation reaction.

Maize granular starch was modified by *Thermus scotoductus* 4-α-glucanotransferase and the physicochemical properties of the products were characterized by Park et al. (2007a). The average molecular weight of amylopectin decreased rapidly from 4.4×10^8 Da to 7.1×10^5 Da, indicating that the inner chains (C or B chains) of amylopectin were cleaved in addition to the trimming of outer chains. As the result, small amylopectin clusters with shortened branch chains could be produced. The chain length distribution of branches in rice starch and maize starch was examined by the 4-α-glucanotransferase treatment. In the case of rice starch, which contains ~20% of amylose, the number of longer branch chains with DP>25 significantly increased as well as that of shorter branch chains. The increase could be attributed to shortened amylose. The formation of amylopectin clusters with rearranged branch chains can be responsible for the formation of thermoreversible gel. The 4-α-glucanotransferase-modified starches displayed a smaller proportion of branch chains with DP 7-20 and a larger proportion of branch chains of DP>20 than the control (Fig. 17). The results indicated that the fragments of amylose are transferred to the branch chains of amylopectin through disproportionation reactions of 4-α-glucanotransferase (Park et al., 2007a). The enzyme also demonstrated the capability to produce cycloamylose with DP 19-35 from rice and maize starch (Fig. 18).

The gelatinization and pasting temperatures of 4-α-glucanotransferase-modified starch were decreased, whereas the peak, setback, and the final viscosity was lowered. Also, 4-α-glucanotransferase-modified starch exhibited a slower retrogradation rate. The enzyme treatment changed the dynamic rheological properties of the starch, leading to decreases in its elastic (G′) and viscous (G″) moduli (Park et al., 2007a).

Table 5 Major oligomeric state of CD-/pullulan-degrading and related enzymes

Enzyme	Origin	Molecular mass[a]	Optimal temp. (°C)	Sequence identity[b]	N-terminal segment[c]	Major oligomeric state	Ref.
CDase	*B. sphaericus* ATCC7055	91.2-95	40	N.A.[d]	N.D.[e]	Dimer	A
	B. sphaericus E-244	72	45	57	120-130	Dimer	B-D
	B. stearothermophilus K-12481	67	60	N.A.	120-130	Dimer	E
	T. ethanolicus 39E	66	65	48	N.A.	NA	F,G
	Alkalophilic *Bacillus* sp.	67	50	51	120-130	Dimer	H
	Alkalophilic *Bacillus* sp. I-5	65	45	100	120-130	Tetramer/octamer	I
	B. coagulans	62	50	N.A.	N.A.	NA	J
	Flavobacterium sp.	62	N.D.	N.A.	N.A.	NA	K
	Xanthomonas campestris K-11151	55	55	N.A.	N.A.	NA	L
	Thermotoga maritima	55	85	34	N.A.	NA	M
MAase	*B. licheniformis*	67	50	45	121	Dimer	N
	B. stearothermophilus ET1	69	55	54	121	Dimer	O
	B. subtilis SUH4-2	69	40	46	121	Dimer	P
	Thermus sp. IM6501	68	60	69	124	Dimer	Q
	Nostoc punctiforme	55	25-30	39	N.A.	Trimer	U.D.[f]
	Pyrococcus furiosus	76	90	30	190	Dimer	R
	Thermoplasma volcanium	71	75	30	170	High oligomer	S
	Staphylothermus marinus	82.4	100	32	188	Dimer	U.D.
TVAII	*T.vulgaris* R-47	71	40	48	121	Dimer	T
Neo-	*B. stearothermophilus*	62	60	53	N.A.	N.A.	U,V
pullulanase	*Bacteroides thetaiotaomicron*	70	N.D.	29	N.A.	N.A.	W,X
	K. pneumoniae	66	N.D.	N.A.	N.A.	N.A.	Y
	Alkalophilic *Bacillus* sp. KSM-1876	68.6	40	51	N.A.	N.A.	†
	Bacillus polymyxa	58	N.D.	28	N.A.	N.A.	‡

[a] Molecular mass of monomer (kDa); [b] Primary structure of CDase from Alkalophilic *Bacillus* sp. I-5 was used as a template; [c] Number of amino acids.
A, (Galvin et al., 1994); B-D, (Bender, 1977; Oguma et al., 1991; Oguma et al., 1993); E, (Abe et al., 1996); F,G, (Saha and Zeikus, 1990; Podkovyrov et al., 1993); H, (Yoshida et al., 1991); I, (Kim et al., 1998), J, (Kitahata et al., 1983); K, (Bender, 1993); L, (Abe et al., 1994); M, (Nelson et al., 1999); N, (Kim et al., 1992); O, (Cha et al., 1998); P, (Cho et al., 2000); Q, (Kim et al., 1999); R, (Yang et al., 2004); S, (Kim et al., 2007); T, (Tonozuka et al., 1995); U,V, (Kuriki et al., 1988; Takata et al., 1992); W,X, (Smith and Salyers, 1991; D'Elia and Salyers, 1996); Y, (Bloch, 1986); †, (Igarashi et al., 1992); ‡, (Yebra et al., 1999); d, information not available; e, not determined; f, unpublished data.

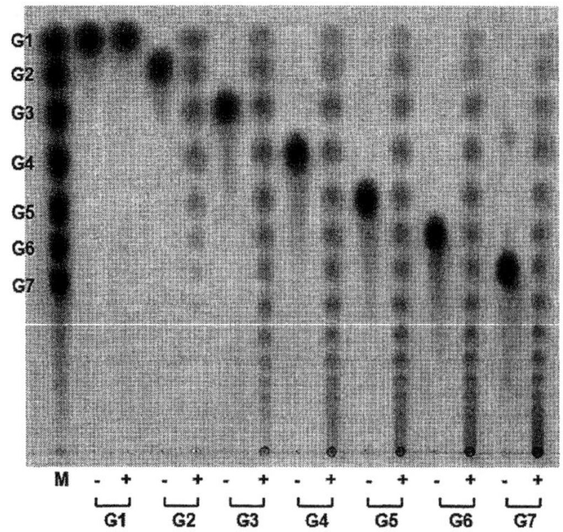

Figure 16 TLC analysis of reaction pattern of 4-α-glucanotransferase on maltooligosaccharides

Lane M was spotted with maltooligosaccharide standards; reactions using glucose (G1) to maltoheptaose (G7) with (+) or without (-) the enzyme (with permission of Park et al., 2007, *Carbohydr Polym*, 67, 2, 164-173).

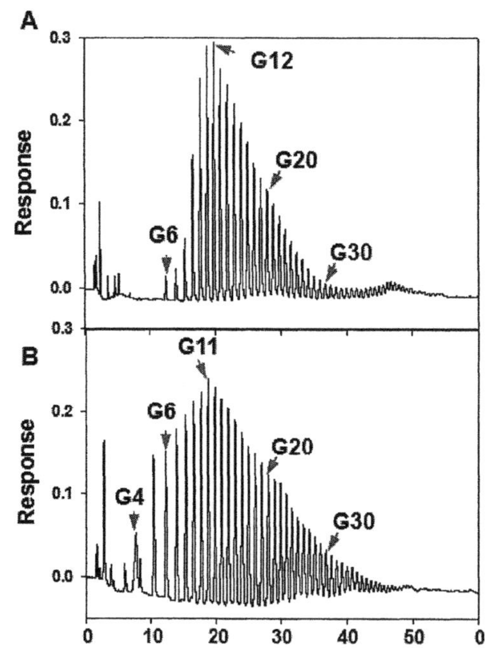

Figure 17 HPAEC analysis of the chain length distribution of branch in rice starch

Rice starch (control, A) and that modified with 4-α-glucanotransferase (B) were treated with isoamylase (with permission of Park et al., 2007, *Carbohydr Polym*, 67, 2, 164-173).

Amylopectin re-shaped by debranching enzyme

As mentioned above, mammalian debranching enzymes possess a bifunctional activity of glucan transferase and amylo-1,6-glucosidase. In contract, bacterial debranching enzymes catalyze the hydrolysis of α-1,6-glucosidic linkage at branch points of polyglucan.

TreX, a debranching enzyme originated from *Sulfolobus solfataricus* exhibited hydrolyzing activity toward α-1,6-glucosidic linkages of amylopectin, glycogen, pullulan, and other branched substrates. The enzyme showed high specificity for the hydrolysis of the side chains with DPs ranging from 3 to 9 or longer (Park et al., 2008). This activity may facilitate debranching of relatively long branched side chain from amylopectin or glycogen, rearranging the side chain of the molecules.

GlgX, a debranching enzyme from *E. coli* has high specificity for the outer chains with DP4 in glycogen (Dauvillee et al., 2005), indicating that the enzyme may be involved in reducing the frequency of short external chains in the molecule. The side chain length distribution of glycogen in *E. coli* revealed that the number of the side chains with DP4 dramatically increased in the GlgX knock-out mutant whereas that of the wild type was relatively high (Dauvillee et al., 2005).

The substrate specificities of various debranching enzymes were also investigated regarding the branch chain length (Sakano et al., 1991). Plant pullulanase cleaves the short branch chain such as maltosyl- and maltotriosyl residues more easily than microbial pullulanase (Walker, 1968). Based on our tests for the specificity, the debranching enzymes from various bacteria showed different specificity toward the chain length of branch. Thus, we proposed a model that the specific activity of debranching enzymes can possibly be attributed to shaping of amylopectin or glycogen.

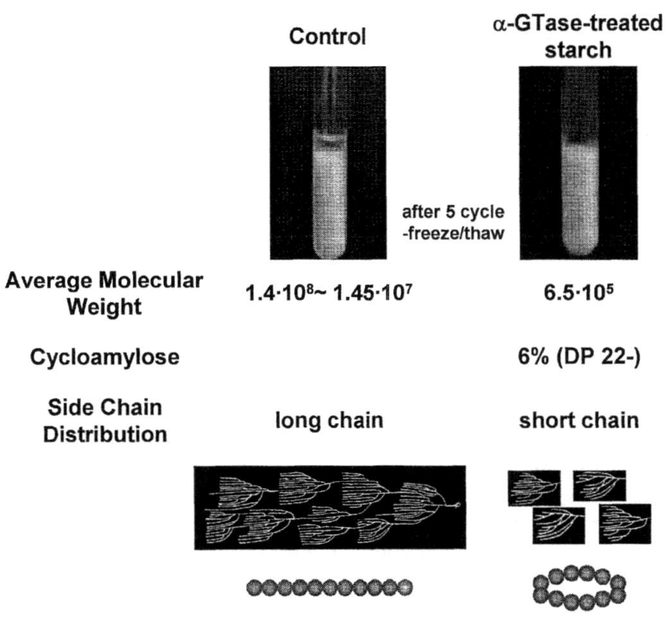

Figure 18 Molecular changes and freeze-thaw stability of starchmodified by 4-α-glucanotransferase

During starch biosynthesis in plants, debranching enzyme, a pullulanase (R-enzyme), may cleave preferentially the short branch chain (maltosyl- and maltotriosyl-), resulting in the production of amylopectin with long branch chains. The amylopectin synthesized in plants can be reconstructed using various microbial debranching enzymes with specificities for various branch chain lengths (Fig. 19).

Highly branched tapioca or rice starch modified by the combined action of -Glucanotransferase-/Maltogenic amylase or BE/Maltogenic amylase

Tapioca or rice starch can be modified using BE, 4-α-glucanotransferase, or maltogenic amylase. By the branching enzyme treatment, the molecular weight of the starches decreased from 3.1×10^8 to 1.7×10^6, the number of shorter branch chains (DP 6-12) increased, the number of longer branch chain (DP>25) decreased, and the content of amylose decreased from 18.9% to 0.75% (Le et al., 2008). The results indicated that most of the long chains of amylose were cleaved and branched by the formation of α - 1,6-glucosidic linkage. To prepare highly branched tapioca starch, BE-treated tapioca starch was treated further with maltogenic amylase. The analysis of branch chains of the products revealed that a series of small peaks appeared newly between linear maltooligosaccharides. These peaks were identified as extra-branched maltooligosaccharides. Maltogenic amylase transferred sugar moieties of the shorter branch chains of amylopectin by forming α-1,6-glucosidic linkages. Likewise, rice starch was modified by combination reaction of 4-α-glucanotransferase and maltogenic amylase (unpublished data). A schematic diagram of enzymatic modification of tapioca starch by branching enzyme and maltogenic amylase is shown in Figure 20.

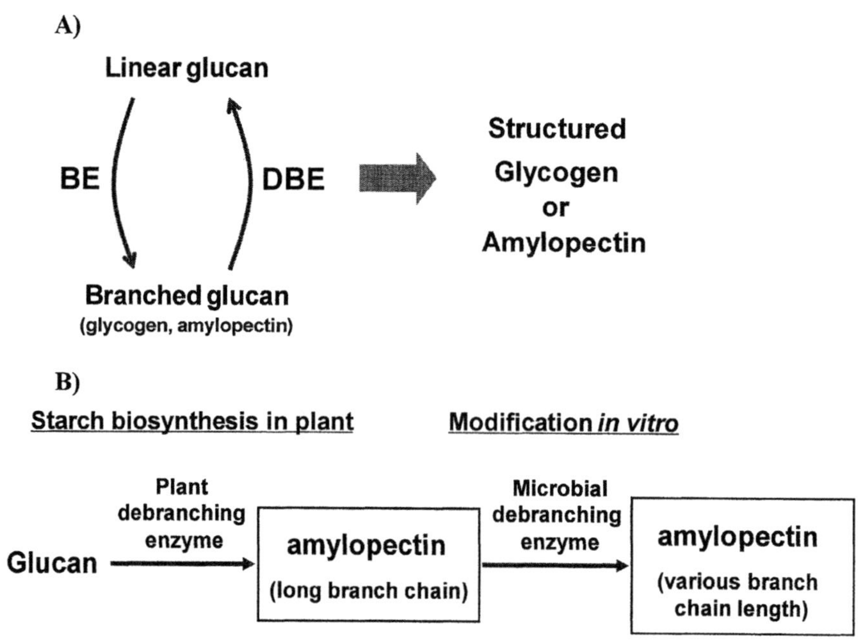

Figure 19 Shaping of starch by debranching enzyme *in vivo* (A) and *in vitro* (B)

The susceptibility to enzymatic digestion of the highly-branched tapioca/rice starch was determined using human pancreatic α-amylase and glucoamylase from *A. niger*. The highly-branched tapioca/rice starch gave significantly lowered susceptibility to digest enzymes, composed to native starches.

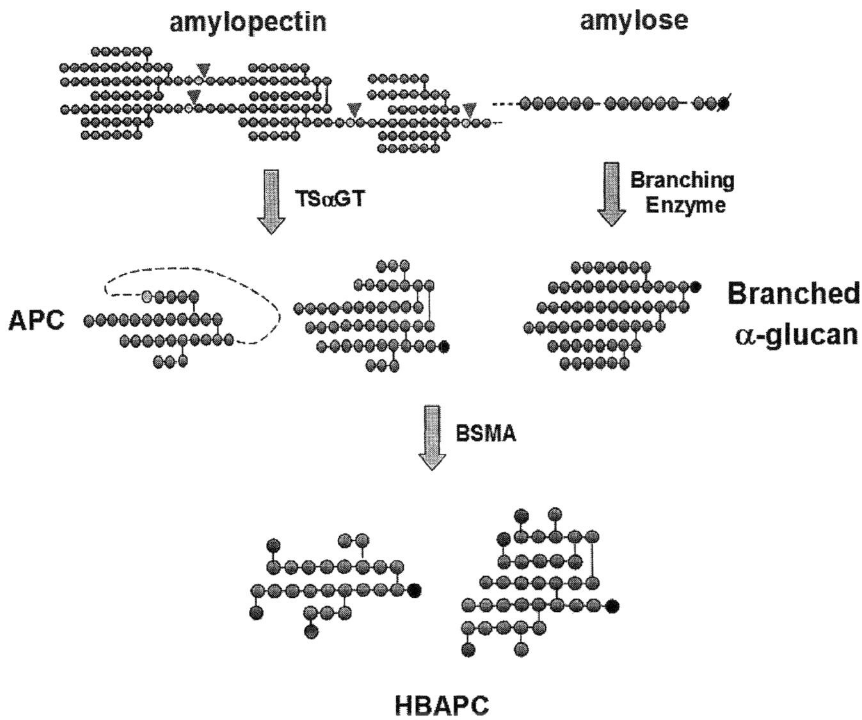

Figure 20 Schematic diagram of enzymatic modification of tapioca starch with branching enzyme and maltogenic amylase

The reducing ends of glucan chain are shown by black circle (with permission of Lee et al., 2008, *J Agric Food Chem*, 56, 1, 126-131).

AKNOWDGEMENT

This work was supported by a grant from the Korea Health21 R&D project, Ministry of Health and Welfare, Republic of Korea (AD050376).

REFERENCES

Abe J, Nakazono O, and Hizuruki S (1996), 'Characterization of cyclodextrinase from *Bacillus stearothermophilus* K-12481', *J Appl Glycosci,* 43, 43155-43159.

Abe J, Onitsuka N, Nakano T, Shibata Y, Hizukuri S, and Entani E (1994), 'Purification and characterization of periplasmic alpha-amylase from *Xanthomonas campestris* K-11151', *J Bacteriol,* 176(12), 3584-3588.

Auh J H, Chae H Y, Kim Y R, Shim K H, Yoo S H, and Park K H (2006), 'Modification of rice starch by selective degradation of amylose using alkalophilic *Bacillus* cyclomaltodextrinase', *J Agric Food Chem,* 54(6), 2314-2319.

Ball S G and Morell M K (2003), 'From bacterial glycogen to starch: understanding the biogenesis of the plant starch granule', *Annu Rev Plant Biol,* 54207-54233.

Bender H (1977), 'Cyclodextrin-glucanotransferase von *Klebsiella pneumoniae.* 2. bedeutung des enzyms für den metabolismus der cyclodextrine bei *Klebsiella pneumoniae* M5a1', *Arch. Microbiol,* 11349-11366.

Bender H (1993), 'Purification and characterization of a cyclodextrin-degrading enzyme from *Flavobacterium* sp', *Appl Microbiol Biotechnol,* 39(6), 714-719.

Bhuiyan S H, Kitaoka M, and Hayashi K (2003), 'A cycloamylose-forming hyperthermostable 4-α-glucanotransferase of Aquifex aeolicus expressed in Escherichia coli', *J Mol Catal, B Enzym,* 22(1-2), 45-53.

Blennow A (2004), 'Starch bioengineering'. IN ELIASSON, A. C. (Ed.) *Starch in food: Structure, function and applications.* Cambridge, Woodhead Publishing Limited.

Bloch M (1986), 'Comparison of the malA regions of *Escherichia coli* and *Klebsiella pneumoniae*', *J Bacteriol,* 168(3), 1220-1227.

Boos W and Shuman H (1998), 'Maltose/maltodextrin system of Escherichia coli: transport, metabolism, and regulation', *Microbiol Mol Biol Rev,* 62(1), 204-229.

Cha H J, Yoon H G, Kim Y W, Lee H S, Kim J W, Kweon K S, Oh B H, and Park K H (1998), 'Molecular and enzymatic characterization of a maltogenic amylase that hydrolyzes and transglycosylates acarbose', *Eur J Biochem,* 253251–253262.

Cho H Y, Kim Y W, Kim T J, Lee H S, Kim D Y, Kim J W, Lee Y W, Lee S B, and Park K H (2000), 'Molecular characterization of a dimeric intracellular maltogenic amylase of *Bacillus subtilis* SUH4-2', *BBA/Protein Structure and Molecular Enzymology,* 1478(2), 333-340.

Coutinho P M, and Henrissat B (1999), 'Carbohydrate-active enzymes: an integrated database approach' In: Gilbert H J, Davies G J, Henrissat B,, and Svensson B (Ed.) *Recent Advances in Carbohydrate Bioengineering,* Cambridge, Royal Society of Chemistry, 3-12

D'Elia J N and Salyers A A (1996), 'Contribution of a neopullulanase, a pullulanase, and an alpha-glucosidase to growth of *Bacteroides thetaiotaomicron* on starch', *J Bacteriol,* 178(24), 7173-7179.

Dauvillee D, Kinderf I S, Li Z, Kosar-Hashemi B, Samuel M S, Rampling L, Ball S, and Morell M K (2005), 'Role of the *Escherichia coli glgX* Gene in Glycogen Metabolism', *J Bacteriol,* 187(4), 1465-1473.

Delatte T, Umhang M, Trevisan M, Eicke S, Thorneycroft D, Smith S M, and Zeeman S C (2006), 'Evidence for distinct mechanisms of starch granule breakdown in plants', *J Biol Chem,* 281(17), 12050-12059.

Dippel R and Boos W (2005), 'The maltodextrin system of *Escherichia coli*: metabolism and transport', *J Bacteriol,* 187(24), 8322-8331.

Galvin N M, Kelly C T, and Fogarty W M (1994), 'Purification and properties of the cyclodextrinase of *Bacillus sphaericus* ATCC 7055', *Appl Microbiol Biotechnol,* 42(1), 46-50.

Gillard B K, White R C, Zingaro R A, and Nelson T E (1980), 'Amylo-1, 6-glucosidase/4-alpha-glucanotransferase. Reaction of rabbit muscle debranching enzyme with an active site-directed irreversible inhibitor, 1-S-dimethylarsino-1-thio-beta-D-glucopyranoside', *J Biol Chem,* 255(18), 8451-8457.

Glor E B (1988), 'Degradation of starch and its hydrolytic products by oral bacteria', *J Dent Res,* 67(1), 75-81.

Henrissat B, Deleury E, and Coutinho P M (2002), 'Glycogen metabolism loss: a common marker of parasitic behaviour in bacteria?', *Trends Genet,* 18(9), 437-440.

Hondoh H, Kuriki T, and Matsuura Y (2003), 'Three-dimensional Structure and Substrate Binding of *Bacillus stearothermophilus* Neopullulanase', *J Mol Biol,* 326(1), 177-188.

Igarashi K, Ara K, Saeki K, Ozaki K, Kawai S, and Ito S (1992), 'Nucleotide sequence of the gene that encodes a neopullulanase from an alkalophilic *Bacillus*', *Biosci Biotechnol Biochem,* 56(3), 514-516.

Jobling S (2004), 'Improving starch for food and industrial applications', *Curr Opin Plant Biol,* 7(2), 210-218.

Kamasaka H, Sugimoto K, Takata H, Nishimura T, and Kuriki T (2002), '*Bacillus stearothermophilus* neopullulanase selective hydrolysis of amylose to maltose in the presence of amylopectin', *Appl Environ Microbiol,* 68(4), 1658.

Kamionka A, and Dahl M K (2001), '*Bacillus subtilis* contains a cyclodextrin-binding protein which is part of a putative ABC-transporter', *FEMS Microbiol Lett,* 204(1), 55-60.

Kamitori S, Kondo S, Okuyama K, Yokota T, Shimura Y, Tonozuka T, and Sakano Y (1999), 'Crystal structure of *Thermoactinomyces vulgaris* R-47 α-amylase II (TVAII) hydrolyzing cyclodextrins and pullulan at 2.6 Å resolution', *J Mol Biol,* 287(5), 907-921.

Kaneko T, Kohmoto T, Kikuchi H, Fukui F, Shiota M, Yatake T, Takaku H, and Ilno H (1992), 'Digestibility of isomaltoologosaccharides by rats and effects on serum lipids', *Nippon Nogeikagaku Kaishii*, 66, 1211-1220.

Kaneko T, Yokoyama A, and Suzuki M (1995), 'Digestibility characteristics of isomaltooligosaccharides in comparison with several saccharides using the rat jejunum loop method', *Biosci Biotechnol Biochem,* 59(7), 1190-4.

Kaper T, Talik B, Ettema T J, Bos H, van der Maarel M, and Dijkhuizen L (2005), 'Amylomaltase of *Pyrobaculum aerophilum* IM2 produces thermoreversible starch gels', *Appl Environ Microbiol,* 71(9), 5098.

Kim D Y, Cha C H, Oh W S, Yoon Y J, and Kim J W (2004), 'Expression of the promoter for the maltogenic amylase gene in *Bacillus subtilis* 168', *J Microbiol,* 42(4), 319-327.

Kim I C, Cha J H, Kim J R, Jang S Y, Seo B C, Cheong T K, Lee D S, Choi Y D, and Park K H (1992), 'Catalytic properties of the cloned amylase from *Bacillus licheniformis*', *J Biol Chem,* 267(31), 22108-22114.

Kim J S, Cha S S, Kim H J, Kim T J, Ha N C, Oh S T, Cho H S, Cho M J, Kim M J, Lee H S, Kim J W, Choi K Y, Park K H, and Oh B H (1999), 'Crystal structure of a maltogenic amylase provides insights into a catalytic versatility ', *J Biol Chem,* 274(37), 26279-26286.

Kim J W, Kim Y H, Lee H S, Yang S J, Kim Y W, Lee M H, Kim J W, Seo N S, Park C S, and Park K H (2007), 'Molecular cloning and biochemical characterization of the first archaeal maltogenic amylase from the hyperthermophilic archaeon *Thermoplasma volcanium* GSS1', *BBA-Proteins and Proteomics,* 1774(5), 661-669.

Kim T J, Park C S, Cho H Y, Cha S S, Kim J S, Lee S B, Moon T W, Kim J W, Oh B H, and Park K H (2000), 'Role of the glutamate 332 residue in the transglycosylation activity of *Thermus* maltogenic amylase', *Biochemistry (Washington),* 39(23), 6773-6780.

Kim T J, Shin J H, Oh J H, Kim M J, Lee S B, Ryu S, Kwon K, Kim J W, Choi E H, and Robyt J F (1998), 'Analysis of the gene encoding cyclomaltodextrinase from alkalophilic *Bacillus* sp. I-5 and characterization of enzymatic properties', *Arch Biochem Biophys,* 353(2), 221-227.

Kitahata S, Taniguchi M, Beltran S D, Sugimoto T, and Okada S (1983), 'Purification and some properties of cyclodextrinase from *Bacillus coagulans*', *Agric Biol Chem,* 47(7), 1441-1447.

Kuriki T, Okada S, and Imanaka T (1988), 'New type of pullulanase from *Bacillus stearothermophilus* and molecular cloning and expression of the gene in *Bacillus subtilis*', *J Bacteriol,* 170(4), 1554-1559.

Kweon M R, Park C S, Auh J H, Cho B M, Yang N S, and Park K H (1994), 'Phospholipid hydrolysate and antistaling amylase effects on retrogradation of starch in bread', *J Food Sci*, 59(5), 1072-1076

Le Q T, Lee C K, Kim Y W, Lee S J, Zhang R, Withers, S G, Kim Y R, Auh J H, and Park K H (2008), 'Amylolytically-resistant tapioca starch modified by combined treatment of branching enzyme and maltogenic amylase',*Carbohydr Polym*, In press

Lee C K, Le Q T, Kim Y H, Shim J H, Lee S J, Park J H, Lee K P, Song S H, and Auh J H (2008), 'Enzymatic synthesis and properties of highly branched rice starch amylose and amylopectin cCluster', *J Agric Food Chem,* 56(1), 126-131.

Lee H S, Auh J H, Yoon H G, Kim M J, Park J H, Hong S S, Kang M H, Kim T J, Moon T W, and Kim J W (2002a), 'Cooperative action of α-glucanotransferase and maltogenic amylase for an improved process of isomaltooligosaccharide (IMO) production', *J Agric Food Chem,* 50(10), 2812-2817.

Lee H S, Kim M S, Cho H S, Kim J I, Kim T J, Choi J H, Park C, Lee H S, Oh B H, and Park K H (2002b), 'Cyclomaltodextrinase, neopullulanase, and maltogenic amylase are nearly indistinguishable from each other', *J Biol Chem,* 277(24), 21891-21897.

Lee K Y, Kim Y R, Park K H, and Lee H G (2006), 'Effects of α-glucanotransferase treatment on the thermo-reversibility and freeze-thaw stability of a rice starch gel', *Carbohydr Polym,* 63(3), 347-354.

Lee S J, Yoo S H, Kim M J, Kim J W, Seok H M, and Park K H (1995), 'Production and characterization of branched oligosaccharides from liquefied starch by the action of *Bacillus licheniformis* amylase, *Starch,* 47(4), 127-134.

Liu W, Madsen N B, Braun C, and Withers S G (1991), 'Reassessment of the catalytic mechanism of glycogen debranching enzyme', *Biochemistry,* 30(5), 1419-1424.

Miles M J, Morris V J, Orford P D, and Ring S G (1985), 'The roles of amylose and amylopectin in the gelation and retrogradation of starch', *Carbohydr Res,* 135(2), 271-281.

Nakamura Y (2002), 'Towards a better understanding of the metabolic system for amylopectin biosynthesis in plants: Rice endosperm as a model tissue', *Plant Cell Physiol,* 43(7), 718-725.

Nelson K E, Clayton R A, Gill S R, Gwinn M L, Dodson R J, Haft D H, Hickey E K, Peterson J D, Nelson W C, and Ketchum K A (1999), 'Evidence for lateral gene transfer between Archaea and Bacteria from genome sequence of *Thermotoga maritima*', *Nature,* 399323-329.

Oguma T, Kikuchi M, and Mizusawa K (1991), 'Some culture conditions for the production of cyclodextrin-hydrolyzing enzyme from *Bacillus sphaericus*', *Agric Biol Chem,* 55(6), 1661-1662.

Oguma T, Matsuyama A, Kikuchi M, and Nakano E (1993), 'Cloning and sequence analysis of the cyclomaltodextrinase gene from *Bacillus sphaericus* and expression in *Escherichia coli* cells', *Appl Microbiol Biotechnol,* 39(2), 197-203.

Okada G, Unno T, and Sawai T (1988), 'A Simple purification procedure for and further characterization of a glucodextranase from *Arthrobacter globiformis* I42', *Agric Biol Chem,* 52(9), 2169-2176.

Palmer T N, Ryman B E, and Whelan W J (1976), 'The action pattern of amylomaltase from *Escherichia coli*', *Eur J Biochem,* 69(1), 105-115.

Park J H, Park K H, and Jane J L (2007a), 'Physicochemical properties of enzymatically modified maize starch using 4-alpha-glucanotransferase', *Food Sci Biotechnol,* 16(6), 902-909.

Park J H, Kim H J, Kim Y H, Cha H, Kim Y W, Kim T J, Kim Y R, and Park K H (2007b), 'The action mode of *Thermus aquaticus* YT-1 4-α-glucanotransferase and its chimeric enzymes introduced with starch-binding domain on amylose and amylopectin', *Carbohydr Polym,* 67(2), 164-173.

Park J H, You J Y, Shim O H, Shin O H, Shin H K, Lee S J, and Park K H (1990), 'Growth effect of branched oligosaccharides on principal intestinal bacteria', *Kor J Appl Micro Biotech,* 20, 237-242

Park J T, Park H S, Kang H K, Hong J S, Cha H, Woo E J, Kim J W, Kim M J, Boos W, Lee S, and Park K H (2008), 'Oligomeric and functional properties of a debranching enzyme (TreX) from the archaeon *Sulfolobus solfataricus* P2', *Biocatal Biotransformation,* 2676-2685.

Park K H, Kim T J, Cheong T K, Kim J W, Oh B H, and Svensson B (2000), 'Structure, specificity and function of cyclomaltodextrinase, a multispecific enzyme of the α-amylase family', *BBA-Protein Structure and Molecular Enzymology,* 1478(2), 165-185.

Park S H, Cha H, Kang H K, Shim J H, Woo E J, Kim J W, and Park K H (2005), 'Mutagenesis of Ala290, which modulates substrate subsite affinity at the catalytic interface of dimeric ThMA', *BBA-Proteins and Proteomics,* 1751(2), 170-177.

Park S H, Kang H K, Shim J H, Woo E J, Hong J S, Kim J W, Oh B H, Lee B H, Cha H, and Park K H (2007c), 'Modulation of substrate preference of *Thermus* maltogenic amylase by mutation of the residues at the interface of a dimer', *Biosci Biotechnol Biochem,* 71(6), 1564-1567.

Podkovyrov S M, Burdette D, and Zeikus J G (1993), 'Analysis of the catalytic center of cyclomaltodextrinase from *Thermoanaerobacter ethanolicus* 39E', *FEBS Lett,* 317(3), 259-262.

Preiss J (2004), 'Plant starch synthesis', In Eliasson, A. C. (Ed.) *Starch in food: Structure, function and applications.* Cambridge, Woodhead Publishing Limited.

Preiss J (2006), 'Bacterial glycogen inclusions: enzymology and regulation of synthesis', *Microbial Monographs,* Berlin, Springer.

Pugsley A P, and Dubreuil C (1988), 'Molecular characterization of malQ, the structural gene for the *Escherichia coli* enzyme amylomaltase', *Mol Microbiol,* 2(4), 473-479.

Saha B C, and Zeikus J G (1990), 'Characterization of thermostable cyclodextrinase from *Clostridium thermohydrosulfuricum* 39E', *Appl Environ Microbiol,* 56(9), 2941-2943.

Sakano Y, Nagahata N, and Fujimoto D (1991), 'Comparison of action patterns of debranching amylases', *Denpun Kagaku* 38, 187-192.

Schönert S, Seitz S, Krafft H, Feuerbaum E A, Andernach I, Witz G, and Dahl M K (2006), 'Maltose and maltodextrin utilization by *Bacillus subtilis*', *J Bacteriol,* 188(11), 3911-3922.

Smith A M, Zeeman S C, and Smith S M (2005), 'Starch degradation', *Annu Rev Plant Biol*, 56, 73-98.

Smith K A, and Salyers A A (1991), 'Characterization of a neopullulanase and an alpha-glucosidase from *Bacteroides thetaiotaomicron* 95-1', *J Bacteriol,* 173(9), 2962-2968.

Steichen J M, Petty R V, and Sharkey T D (2008), 'Domain characterization of A 4 alpha-glucanotransferase essential for maltose metabolism in photosynthetic leaves', *J Biol Chem*, 283(30), 20797-20804.

Tachibana Y, Takaha T, Fujiwara S, Takagi M, and Imanaka T (2000), 'Acceptor specificity of 4-α-glucanotransferase from *Pyrococcus kodakaraensis* KOD1, and synthesis of cycloamylose', *J Biosci Bioeng,* 90(4), 406-409.

Takaha T, and Smith S M (1999), 'The functions of 4-alpha-glucanotransferases and their use for the production of cyclic glucans', *Biotechnol Genet Eng Rev,* 16257-16280.

Takata H, Kuriki T, Okada S, Takesada Y, Iizuka M, Minamiura N, and Imanaka T (1992), 'Action of neopullulanase. Neopullulanase catalyzes both hydrolysis and transglycosylation at alpha-(1-4)-and alpha-(1-6)-glucosidic linkages', *J Biol Chem,* 267(26), 18447-18452.

Tonozuka T, Mogi S, Shimura Y, Ibuka A, Sakai H, Matsuzawa H, Sakano Y, and Ohta T (1995), 'Comparison of primary structures and substrate specificities of two pullulan-hydrolyzing α-amylases, TVA I and TVA II, from *Thermoactinomyces vulgaris* R-47', *BBA-Protein Structure and Molecular Enzymology,* 1252(1), 35-42.

Walker G J (1968), 'Metabolism of the reserve polysaccharide of streptococcus mitis some properties of a pullulanase', *Biochem J,* 108, 33-40.

Yang S J, Lee H S, Park C S, Kim Y R, Moon T W, and Park K H (2004), 'Enzymatic analysis of an amylolytic enzyme from the hyperthermophilic archaeon *Pyrococcus furiosus* reveals its novel catalytic properties as both an alpha-amylase and a cyclodextrin-hydrolyzing enzyme', *Appl Environ Microbiol,* 70(10), 5988-5995.

Yebra M J, Blasco A, and Sanz P (1999), 'Expression and secretion of *Bacillus polymyxa* neopullulanase in *Saccharomyces cerevisiae*', *FEMS Microbiol Lett,* 17041–17049.

Yoo S H, Kweon M R, Kim M J, Auh J H, Jung D S, Kim J R, Yook C, Kim J W, and Park K H (1995), 'Branched oligosaccharides concentrated by yeast fermentation and effectiveness as a low sweetness humectant', *J Food Sci,* 60(3), 516-521.

Yoshida A, Iwasaki Y, Akiba T, and Horikoshi K (1991), 'Purification and properties of cyclomaltodextrinase from alkalophilic *Bacillus* sp.', *J Ferment Bioeng,* 71(4), 226-229.

GLYCOSYLATION OF CARBOXYLIC GROUP: A NEW REACTION OF SUCROSE PHOSPHORYLASES

Koji Nomura, Kazuhisa Sugimoto, Hiromi Nishiura, Takashi Kuriki

ABSTRACT

We found a new reaction of sucrose phosphorylases; transglycosylation of carboxyl group. Sucrose phosphorylases from two different sources were tested for glycosylation of carboxylic acid compounds. *Streptococcus mutans* sucrose phosphorylase showed remarkable transglycosylating activity, especially under acidic conditions. *Leuconostoc mesenteroides* sucrose phosphorylase exhibited very weak transglycosylating activity. When benzoic acid and sucrose were used as an acceptor and a donor molecule, 1-*O*-benzoyl α-D-glucopyranoside was produced which was identified by 1D- and 2D-NMR analyses of the purified product and its acetylated product. *S. mutans* sucrose phosphorylase showed broad acceptor-specificity. The sucrose phosphorylase catalyzed transglycosylation to various carboxylic compounds such as short-chain fatty acids, hydroxy acids, dicarboxylic acids, phenolic carboxylic acids, and acetic acid.

Key words: sucrose phosphorylase; transglycosylation; carboxylic acid; benzoic acid; acetic acid

INTRODUCTION

We have developed systems to produce glucose polymers with liner (Yanase et al., 2007), branched (Takata et al., 1996; Kakutani et al., 2007), and cyclic structures (Takaha et al., 1996) at industrial level (Fujii et al., 2003). We have also improved enzymes used for the systems based on the concept of α-amylase family (Takata et al., 1992; Kuriki, 1992) as a rational tool for designing and engineering the enzymes (Kuriki et al., 1996; Kuriki et al., 2006). Thus, exploring and application of new transglycosylation reactions are the core competence of our research group. From the physiological viewpoint, glycosylation is an important factor of various bioactive compounds. Indeed, glycosylation have been used for improving physicochemical and biological properties of many compounds. For example, glycosylation of hesperidin greatly improved its solubility in water, and glycosylation of arbutin significantly improved its inhibitory effect on human tyrosinase (Sugimoto et al., 2003). There are many reports on enzymatic glycosylation of aglycones having glycosyl residues, alcoholic OH group, and phenolic OH group (Sugimoto et al., 2004; Sugimoto et al., 2005). However, there had been no report on glycosylation of carboxylic groups in various aglycones using transglycosylating reaction of carbohydrate active enzymes before our publication (Nomura et al., 2004). In this article, we review our first report for glycosylation of carboxylic compounds by sucrose phosphorylase, an α-amylase family enzyme (Sugimoto et al., 2007). Detailed mechanism and the structure of the products using benzoic acid as a model of carboxylic compounds is also described (Sugimoto et al., 2007).

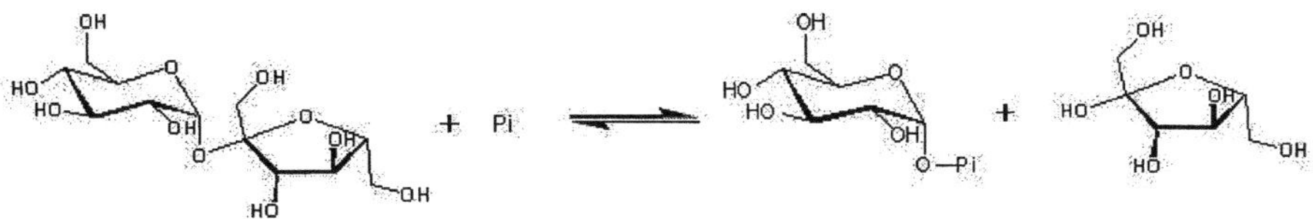

Figure 1 Reaction catalyzed by sucrose phosphorylase

BACKGROUND OF OUR INTEREST FOR THE REACTIONS CATALYZED BY SUCROSE PHOSPHORYLASE

Sucrose phosphorylase catalyzes the reversible conversion of sucrose and inorganic phosphate to α-D-glucose-1-phosphate and D-fructose (Mieyal and Abeles, 1972) (Fig. 1). In the phosphorolytic reaction, the enzyme catalyzes the transfer of the glucosyl moiety of sucrose to inorganic phosphate to form α-D-glucose-1-phosphate and D-fructose. Water, methanol, ethanol, 1-2-cyclohexanediol, ethylene glycol, and poly-phenol compounds such as chatechins and hydroquinone have also been reported to act as acceptors in place of inorganic phosphate (Mieyal and Abeles, 1972; Kitao et al., 1993; Kitao et al., 1994). The acceptor specificity of *Leuconostoc mesenteroides* sucrose phosphorylase was extensively studied (Kitao et al., 1994). They described that the enzyme could not transfer the glucose moiety of sucrose to benzoic acid. We reconsidered this conclusion based on the catalytic mechanism of glycosyl transfer reaction (Kuriki and Imanaka, 1999) (Fig. 2). The p*K*a of benzoic acid is known to be 4.2. Therefore, essentially all of carboxylic moiety of benzoic acid was dissociated at the pH of 7.5, which was the optimum pH of the reaction for *L. mesenteroides* sucrose phosphorylase. Hence, we employed *Streptococcus mutans* sucrose phosphorylase, which had significant enzymatic activity even at pH 4.0 (Fujii et al., 2006), to detect the transglycosylation of the carboxyl group by the enzyme.

Table 1 Effect of pH on efficiency of glucosylation by sucrose phosphorylases

Enzyme (initial pH of reaction)	Transfer ratio (%)
Leuconostoc mesenteroides SPase	
3.9	0.0
5.1	10.0
6.1	4.0
7.1	0.5
Streptococcus mutans SPase	
3.9	55.0
5.1	8.0
5.9	1.0
7.1	0.5

Sucrose phosphorylases from *S. mutans* and *L. mesenteroides* were used, and benzoic acid was used as an acceptor molecule (14). The reaction mixture was analyzed by HPLC. Transfer ratio was expressed as the percentage of the peak area of the transfer product against the total peak area of the transfer product and unreacted benzoic acid.

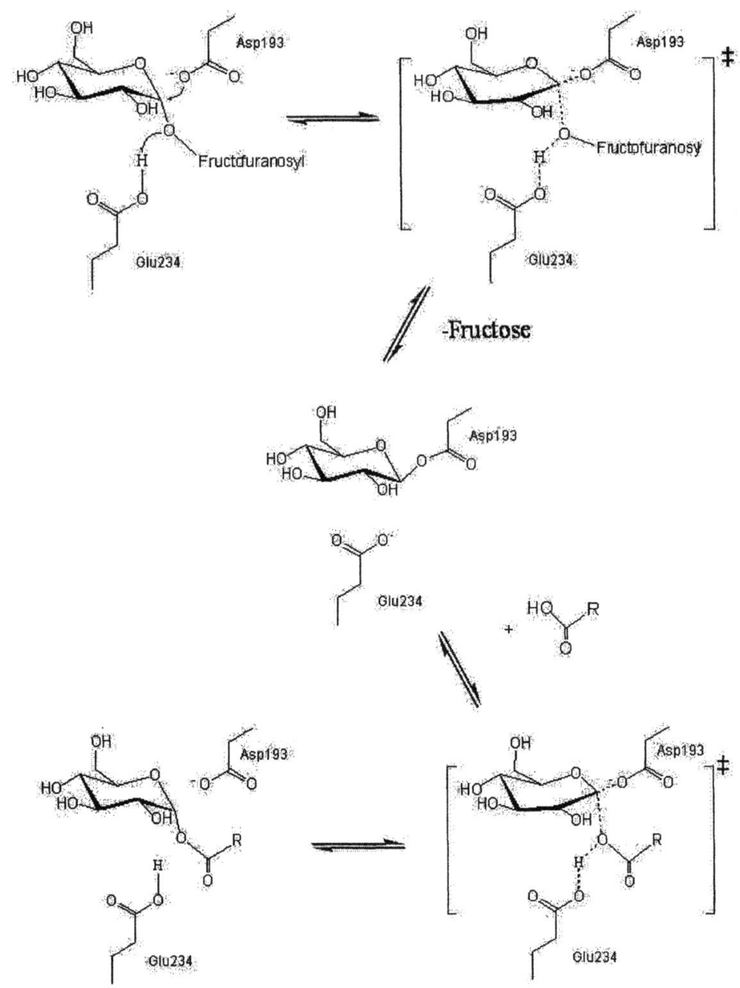

Figure 2 Possible catalytic mechanism of sucrose phosphorylase on carboxylic acid

GLYCOSYLATION OF BENZOIC ACID BY SUCROSE PHOSPHORYLASES

Benzoic acid was used as a model of carboxylic compounds to examine this reaction in detail (Sugimoto et al., 2007), because it's easily detected with its UV absorbing property. We examined the glycosylation reaction of *S. mutans* and *L. mesenteroides* sucrose phosphorylases at several pH values. As we expected, we found that both of these enzymes catalyze the transglycosylation reaction to benzoic acid, particularly under acidic conditions (Table 1). It is known so far that UDP-glucuronosyltransferase catalyzes the transfer of D-glucuronic acid to carboxylic acid using UDP-glucuronic acid as a donor molecule (Clarke and Burchell, 1994). Our results are the first to show that other carbohydrate active enzymes catalyze the transglycosylation reaction on carboxylic groups without nucleotide activated sugars. The optimum pH and the pH-activity profile of the transglycosylation activity of sucrose phosphorylase from *S.*

mutans toward benzoic acid (Table 1) were different from those of the phosphorolytic activity (Fujii et al., 2006). This is also consistent with our hypothesis that the undissociated carboxyl group is essential for the acceptor of the transglycosylation catalyzed by sucrose phosphorylases. As mentioned above, the p*K*a of benzoic acid is known to be 4.2, and the concentration of the undissociated carboxyl group of benzoic acid around neutral pH is very low. This is also quite reasonable from the view point of the proposed catalytic mechanism of phosphorolysis by sucrose phosphorylase from *Bifidobacterium adolescentis* that protonated phosphate group is necessary for binding to the catalytic domain of the enzyme (Mirza et al., 2006). The transfer efficiency of the transglycosylation reaction of sucrose phosphorylase from *L. mesenteroides* was much lower than that of sucrose phosphorylase from *S. mutans* at pH 3.9. The difference between these two enzymes with regard to activity and stability under acidic conditions (Fujii et al., 2006; Kitaoka et al., 1994) were most likely to be the major causes of their different transfer efficiencies. These results suggest that *S. mutans* sucrose phosphorylase is more appropriate to the transglycosylation reaction on carboxylic compounds than *L. mesenteroides* sucrose phosphorylase. Therefore, we used *S. mutans* sucrose phosphorylase in further studies.

We examined the transglycosylation reaction to benzoic acid by the enzyme and the formation of the products in the reaction mixture in detail. The reaction mixture containing sucrose and benzoic acid used as donor and acceptor molecules, respectively, was incubated with the enzyme. The pH of the solution was adjusted between 4.6 - 4.8 with 5 N HCl during the reaction. HPLC analysis of the reaction mixture revealed that three compounds, 1, 2 and 3 (Fig. 3) were produced. In the initial part of the reaction, benzoic acid was decreased, and compound 1 was initially produced and increased during the first reaction period. However, another two compounds, 2 and 3, appeared and gradually increased as the reaction continued. At the end of the reaction, the relative amount of compounds 1, 2 and 3 were approximately 25 %, 25 % and 20 %, respectively, and the total amount of the glucosylated products reached close to 70 %. During the reaction, the pH of the reaction mixture was adjusted between 4.6 and 4.8 with hydrochloric acid, because a change in the pH of the reaction mixture to a higher pH was observed with the decrease of the unreacted benzoic acid. When the reaction was performed without the pH control, several products other than three, compounds 1, 2, and 3 were observed.

These three compounds were purified to determine the structures. In the process of the purification, the rapid interconversion between compounds 2 and 3 was observed. Therefore, we also obtained their acetylated products and determined their structures. As the results of the spectroscopic analyses of the purified products, structures of these three compounds identified as follows. Compound 1, the initial product of the enzyme reaction, was identified as 1-*O*-benzoyl α-D-glucopyranoside (Fig. 3). Compounds 2 and 3 were identified as 2-*O*-benzoyl α-D-glucopyranose and 2-*O*-benzoyl β-D-glucopyranose, respectively (Fig. 3). From the result of the production pattern of these compounds during the reaction, we predicted that 1-*O*-benzoyl α-D-glucopyranoside was produced initially by the enzyme reaction, and thereafter the other two compounds were produced by the non-enzymatic structural change of 1-*O*-benzoyl α-D-glucopyranoside.

We examined the products produced from purified 1-*O*-benzoyl α-D-glucopyranoside in an aqueous solution over time. In an aqueous solution, the amount of 1-*O*-benzoyl α-D-glucopyranoside decreased with time, and 2-*O*-benzoyl α-D-glucopyranose and 2-*O*-benzoyl β-D-glucopyranose were produced spontaneously. With prolonged

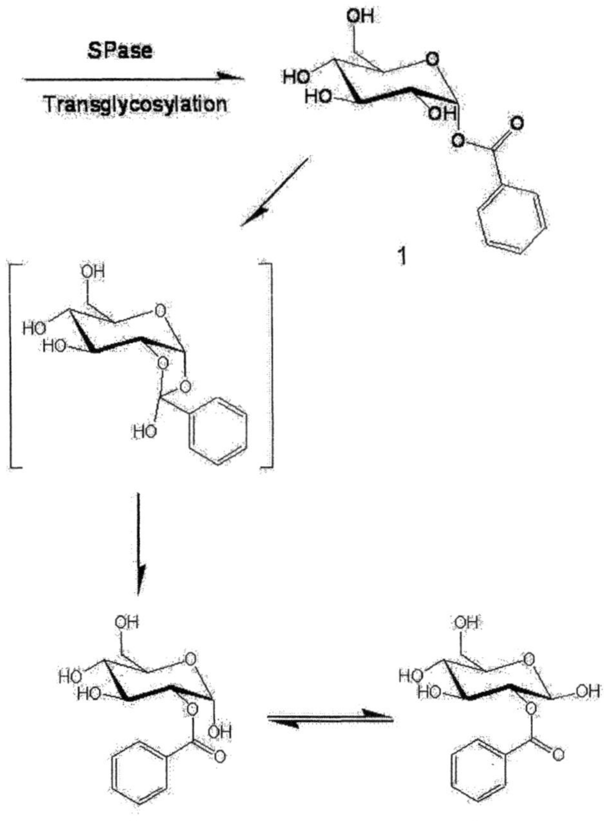

Figure 3 Proposed scheme of the production of benzoic acid glucoside and its isomers in the reaction mixture of sucrose phosphorylase when sucrose and benzoic acid are used as the substrates

incubation, 1-*O*-benzoyl α-D-glucopyranoside disappeared, and another six compounds appeared sequentially. It is well-known that 1-*O*-acyl β-D-glucopyranuronate are produced as a major product *in vivo* metabolite for many carboxylate drugs and that those compounds were converted to isomeric glucuronides by intramolecular acyl migration in aqueous solution under physiological conditions (Fenselau, 1994; Spahn-Langguth and Benet, 1992). The initial product, 1-*O*-benzoyl α-D-glucopyranoside was synthesized by transglycosylation reaction of sucrose phosphorylase, and it was converted to 2-*O*-benzoyl α-D-glucopyranose by an intramolecular acyl migration reaction, probably via the orthoacid ester intermediate, and that 2-*O*-benzoyl β-D-glucopyranose was produced by mutarotation from its α anomer (Fig. 3). Furthermore, other isomeric benzoyl glucoses were observed in the reaction mixture, at higher pH values were also produced in the same manner. We detected a small amount of benzoic acid especially in aqueous solutions at higher pH. We considered that the hydrolysis of the benzoyl glucose also occurred in the aqueous solution.

The acceptor specificity of the enzyme was examined by the HPLC analyses of the reaction products. The enzyme was incubated with sucrose and several carboxylic compounds as donor and acceptor molecules, respectively. The enzyme catalyzed the transfer of the glucosyl moiety of sucrose not only to benzoic acid but also to short chain fatty acids, dicarboxylic acids, hydroxy acids and aromatic carboxylic acids. Particularly, when acetic acid, propionic acid, butyric acid, valeric acid, malonic acid, fumaric acid, lactic acid and benzoic acid were used as acceptor molecules, the conversion ratio for each glucosylated product was more than 50 %, showing this enzyme has wide acceptor specificity. We expect that the *S. mutans* sucrose phosphorylase should be a very useful enzyme for forming many carboxylic glucosides and glucoses compounds occurring in nature.

GLYCOSYLATION OF ACETIC ACID BY SUCROSE PHOSPHORYLASE

Acetic acid is the main component of vinegar. It is known that acetic acid has several physiological activities including an enhancing effect on calcium absorption (Kishi et al., 1999), and on the prevention of hypertension (Kondo et al., 2001). However, solutions of acetic acid at high concentration are difficult to drink because of a strong sour taste. We anticipated the improvement of the strong sour taste of acetic acid by glycosylation. Therefore, the glycosylation of acetic acid and the properties of the glycoside were examined (Nomura et al., 2008).

Sucrose phosphorylase from *S. mutans* was incubated with sucrose and acetic acid as donor and acceptor molecules, respectively. New peaks and spots other than acetic acid were detected by HPLC and TLC analyses, respectively. The effect of pH and the concentrations of sucrose and acetic acid on the transglycosylation reaction of the enzyme were examined in detail. When the reaction was performed with 40 % sucrose and 0.4 M acetic acid at pH 5.0 at 37°C, more than 80 % of acetic acid supplied to the reaction was glucosylated and the yield of the glucose transfer products was maximized. We isolated the initial product and determined the structure of the purified product by spectroscopic analyses. We concluded the structure of the initial product of the reaction is 1-*O*-acetyl α-D-glucopyranoside (Fig. 4).

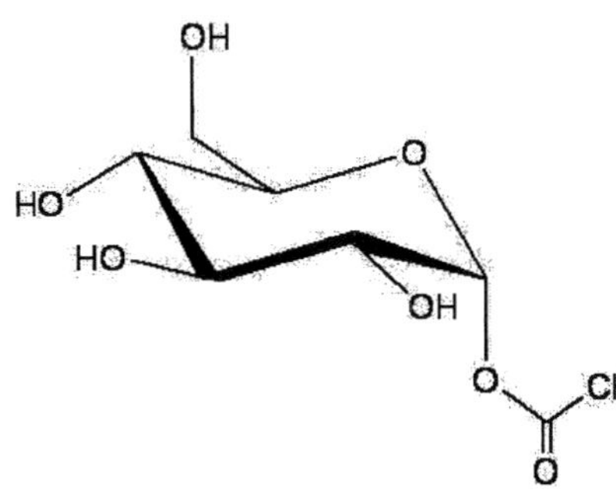

Figure 4 Structure of the transglycosylation product, 1-*O*-acetyl-α-D-glucopyranoside

The sensory test of the solutions of acetic acid and acetic acid glucosides were carried out. Aqueous solutions of several concentrations were prepared, and the intensity of the acidic taste of them was estimate by panels of professional tasters. The acidic taste of acetic acid was markedly reduced by glycosylation. The threshold value of the sour taste of the acetic acid glucosides was more than 1.0 M, whereas that of acetic acid was 10^{-2} M. Thus, the threshold value for acetic acid glucosides was approximately 100 times greater than that for acetic acid. While acetic acid glucosides were not very sour, they were slightly sweet and bitter (Nomura et al., 2008).

CONCLUSION

We have found that sucrose phosphorylases catalyze transglycosylation reaction on carboxylic compounds, and various α-glucosides can be synthesized from sucrose and carboxylic compounds. Enzymatic synthesis of mono-acyl glucoses using the trans-glycosylation reaction of the enzymes is more convenient method than chemical synthesis because we can synthesize them without using protection groups. *S. mutans* sucrose phosphorylase catalyzes the transglycosylation reaction with high transfer efficiency and wide acceptor specificity. These characteristics of the enzyme are suitable for synthesizing glycosides of various carboxylic compounds in high yield. Development of the functional glycosides by using the sucrose phosphorylases is now in progress.

REFERENCES

Clarke D J and Burchell B (1994), 'The uridine diphosphate glucuronosyltransferase multigene family: function and regulation' in Kauffman F C (ed.), *Handbook of experimental pharmacology*, vol. 112, Conjugation-deconjugation reactions in drug metabolism and toxicity, Springer-Verlag, Budapest, 3-43.

Fenselau C (1994), 'Acyl glucuronides as chemically reactive intermediates' in Kauffman F C (ed.), *Handbook of experimental pharmacology*, vol. 112, Conjugation-deconjugation reactions in drug metabolism and toxicity, Berlin, Springer-Verlag, 367-389.

Fujii K, Takata H, Yanase M, Terada Y, Ohdan K, Takaha T, Okada S, and Kuriki T (2003), 'Bioengineering and application of novel glucose polymer', *Biocatal Biotransform*, 21(4/5), 167-172.

Fujii K, Iiboshi M, Yanase M, Takaha T, and Kuriki T (2006), 'Enhancing the thermal stability of sucrose phosphorylase from *Streptococcus mutans* by random mutagenesis', *J Appl Glycosci*, 53(2), 91-97.

Kakutani R, Adachi Y, Kajiura H, Takata H, Kuriki T, and Ohno N (2007), 'Relationship between structure and immunostimulating activity of enzymatically synthesized glycogen', *Carbohydr Res*, 342(16), 2371-2379.

Kishi M, Fukaya M, Tsukamoto Y, Nagasawa T, Takehana K, and Nishizawa N (1999), 'Enhancing effect of dietary vinegar on the intestinal absorption of calcium in ovariectomized rats', *Biosci Biotechnol Biochem*, 63(5), 905-910.

Kitao S, Ariga T, Matsudo T, and Sekine H (1993), 'Syntheses of catechin-glucosides by transglycosylation with *Leuconostoc mesenteroides* sucrose phosphorylase', *Biosci Biotech Biochem*, 57(12), 2010-2015.

Kitao S and Sekine H (1994), 'α-D-Glucosyl transfer to phenolic compounds by sucrose phosphorylase from *Leuconostoc mesenteroides* and production of α-arbutin', *Biosci Biotech Biochem*, 58(1), 38-42.

Kitaoka K, Takahashi H, Hara K, Hashimoto H, Sasaki T, and Taniguchi H (1994), 'Purification and characterization of sucrose phosphorylase from *Leuconostoc mesenteroides* ATCC 12291 cells, and disaccharides synthesis by the enzyme', *J Appl Glycosci*, 41(2), 165-172.

Kondo S, Tayama K, Tsukamoto Y, Ikeda K, and Yamori Y (2001), 'Antihypertensive effects of acetic acid and vinegar on spontaneously hypertensive rats', *Biosci Biotechnol Biochem*, 65(12), 2690-2694.

Kuriki T (1992), 'Can protein engineering interconvert glucanohydrolases/glucanotransferases, and their specificities?', *Trends Glycosci Glycotechnol*, 4(20), 567-572.

Kuriki T, Kaneko H, Yanase M, Takata H, Shimada J, Handa S, Takada T, Umeyama H, and Okada S (1996), 'Controlling substrate preference and transglycosylation activity of neopullulanase by manipulating steric constraint and hydrophobicity in active center', *J Biol Chem*, 271(29), 17321-17329.

Kuriki T and Imanaka T (1999), 'The concept of the α-amylase family: Structural similarity and common catalytic mechanism', *J Biosci Bioeng*, 87(5), 557-565.

Kuriki T, Takata H, Yanase M, Ohdan K, Fujii K, Terada Y, Takaha T, Hondoh H, Matsuura Y, and Imanaka T (2006), 'The concept of the α-amylase family: A rational tool for interconverting glucanohydrolases/glucanotransferases, and their specificities', *J Appl Glycosci*, 53, 155-161.

Mieyal J J and Abeles R H (1972), 'Disaccharide phosphorylases' in Boyer P D (ed.), *The enzymes*, vol. 7, 3rd ed., New York, Academic Press, 515-532.

Mirza O, Skov L K, Sprogøe D, van den Broek L A, Beldman G, Kastrup J S, and Gajhede M (2006), 'Structural rearrangements of sucrose phosphorylase from *Bifidobacterium adolescentis* during sucrose conversion', *J Biol Chem*, 281(46), 35576-35584.

Nomura K, Sugimoto K, Takii H, Ueyama R, Nishiura H, Nishimura T, and Kuriki T (2004), *Japanese published patent application*, 2006-180875 (submitted on December 2, 2004).

Nomura K, Sugimoto K, Nishiura H, Ohdan K, Nishimura T, Hayashi H, and Kuriki T (2008), 'Glucosylation of acetic acid by sucrose phosphorylase', *Biosci Biotech Biochem*, 72(1), 82-87.

Spahn-Langguth H and Benet L Z (1992), 'Acyl glucuronides revisited: is the glucuronidation process a toxification as well as a detoxification mechanism?', *Drug Metab Rev*, 24(1), 5-48.

Sugimoto K, Nishimura T, Nomura K, Sugimoto K, and Kuriki T (2003), 'Syntheses of arbutin-α-glycosides and a comparison of their inhibitory effects with those of α-arbutin and arbutin on human tyrosinase', *Chem Pharm Bull*, 51(7), 798-801.

Sugimoto K, Nishimura T, Nomura K, Sugimoto K, and Kuriki T (2004), 'Inhibitory effects of α-arbutin on melanin synthesis in cultured human melanoma cells and a three-dimensional human skin model', *Biol Pharm Bull*, 27(4), 510-514.

Sugimoto K, Nomura K, Nishimura T, Kiso T, Sugimoto K, and Kuriki T (2005), 'Syntheses of α-arbutin-α-glycosides and their inhibitory effects on human tyrosinase', *J Biosci Bioeng*, 99(3), 272-276.

Sugimoto K, Nomura K, Nishiura H, Ohdan K, Nishimura T, Hayashi H, and Kuriki T (2007), 'Novel transglucosylating reaction of sucrose phosphorylase to carboxylic compounds such as benzoic acid', *J Biosci Bioeng*, 104(1), 22-29.

Takaha T, Yanase M, Takata H, Okada S, and Smith S M (1996), 'Potato D-enzyme catalyzes the cyclization of amylose to produce cycloamylose, a novel cyclic glucan', *J Biol Chem*, 271(6), 2902-2908.

Takata H, Kuriki T, Okada S, Takesada Y, Iizuka M, Minamiura N, and Imanaka T (1992), 'Action of neopullulanase. Neopullulanase catalyzes both hydrolysis and transglycosylation at α-(1→4)- and α-(1→6)-glucosidic linkages', *J Biol Chem*, 267(26), 18447-18452.

Takata H, Takaha T, Okada S, Hizukuri S, Takagi M, and Imanaka T (1996), 'Cyclization reaction catalyzed by branching enzyme', *J Bacteriol*, 178(6), 1600-1606.

Yanase M, Takaha T, and Kuriki T (2007), 'Developing and engineering enzymes for manufacturing amylose', *J Appl Glycosci*, 54(2), 125-131.

STRATEGY FOR CONVERTING AN INVERTING GLYCOSIDE HYDROLASE INTO A GLYCOSYNTHASE

Motomitsu Kitaoka, Yuji Honda, Masafumi Hidaka, and Shinya Fushinobu

ABSTRACT

We found a novel inverting xylanolytic enzyme belonging to GH8, reducing end xylose-releasing exo-oligoxylanase (Rex, EC. 3.2.1.156), that hydrolyzed xylooligosaccharides (X_3 or larger) to release X_1 at their reducing end. Rex hydrolyzed α-X_2F into X_2 only in the presence of X_1, clearly proving the Hehre-resynthesis hydrolysis mechanism. A library of mutant Rex at the catalytic base (Asp263) was constructed by saturation mutagenesis. Among them, D263C showed the highest level of X_3 production, and D263N exhibited the fastest consumption of α-X_2F. However, F^- releasing activities of the mutants were much less than that of wild type. Next, Y198 of Rex that forms a hydrogen bond with the nucleophilic water was substituted with phenylalanine, causing a drastic decrease in the hydrolytic activity and a small increase in F^- releasing activity from α-xylobiosyl fluoride in the presence of xylose. Y198F of Rex accumulates much more product during the glycosynthase reaction than D263C. We here conclude that an inverting glycosidase is effectively converted into glycosynthase by mutating a residue holding the nucleophilic water molecule with the general base residue while keeping the general base residue intact.

Key words: glycosnythase; inverting glycoside hydrolase; reducing-end-xylose releasing exo-oligoxylanase; transglycosylation; xylobiosyl fluoride

BACKGROUND

Enzymes that hydrolyze glycosyl linkages (Glycoside hydrolases, GH) are generally categorized into two types, retaining and inverting enzymes, based on changes in the anomeric configurations during the reactions (Henrissat, 1991; Henrissat and Bairoch, 1993; Henrissat and Bairoch, 1996; Davies and Henrissat, 1995; Sinnott, 1990). Typical reaction mechanisms of both types are similar using two acidic residues acting as a general acid (a proton donor) and a general base (a nucleophile) as illustrated in Figure 1. The retaining GH reaction proceeds with the following steps: (1) the general acid residue donates a proton to the glycosyl oxygen atom and the base residue directly attacks the anomeric center in concert, producing a covalent-bound intermediate at the base residue with Walden inversion; (2) the intermediate undergoes another inverting hydrolysis resulting the anomeric retention during the overall reaction. The reaction of the inverting GH differs in the nucleophilic reagent attacking the anomeric center; in this case, a water molecule activated by the base reisdue attacks the anomeric center to hydrolyze the glycoside with anomeric inversion.

Many retaining GHs are utilized in the production of various glycosides using their transglycosylation activity. However, none of inverting GH shows the transglycosylation activity. The difference in the occurance of transglycosylation is due to the difference in their mechanism (see Figure 1). In the case of a retaining GH, if the glycosyl-enzyme intermediate is attacked by another alcohol instead of water, transglycosylation occurs. It should be noted that no water molecule participate in the reaction. Considering the transglycosylation reaction by an inverting GH, a dehydration step must be postulated because the reaction is initiated by the hydration of the glycosyl

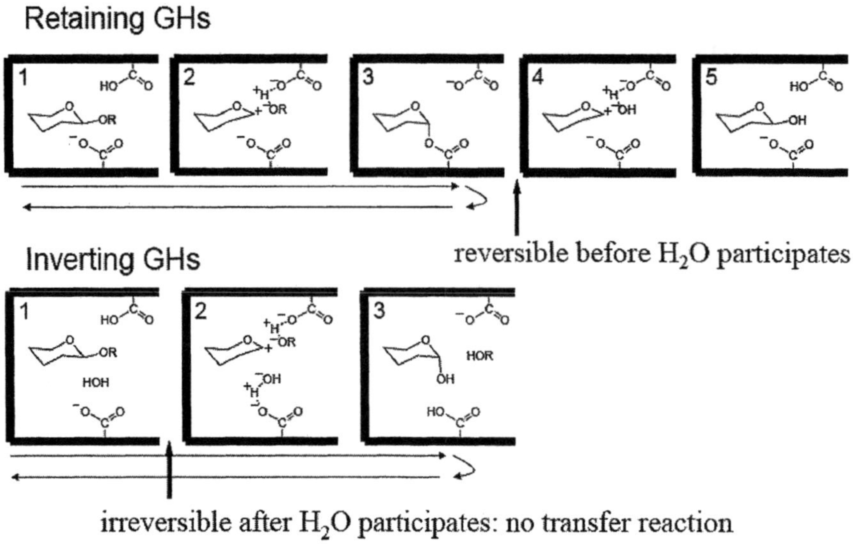

Figure 1 Typical reaction mechanisms of retaining and inverting GHs
If the reaction proceeds to stage 3 and returning to stage 1with another alcohol, glycosyl transfer reaction occurs (indicated with narrow allows). In the case of retaing enzyme, a water molecule, which shows a very high concentration in the reaction solution, firstly participate in the reaction at stage 4. Thus, the reactions until stage 3 are reversible, making transferring reaction possible. Contrarily water molecule participates in the reaction of inverting GHs at stage 2, suggesting the irreversible reaction from stage 1 to 2.

linkage. Since the enzymatic reaction must be performed in water, presence of the abundant molecules of water makes the dehydration step impossible. Thus usages of the inverting GHs are limited in hydrolysis, not in the production of glycosides by transglycosylation.

Hehre and co-workers reported that β-amylase, an inverting GH, hydrolyzed β-maltosyl fluoride, the opposite anomer of the glycosides to be hydrolyzed, into maltose and F^- (Hehre et al., 1979). Later, various inverting enzymes were found to hydrolyze the "wrong" glycosyl fluorides (Konstantinidis et al., 1993; Becker et al., 2000; Hehre, 2000; Williams and Withers, 2000), suggesting that the reaction is common among inverting GHs. Instead of simple hydrolysis, the reaction mechanism consists of two steps. In the first step, a new glycoside of the correct anomer forms from the wrong glycosyl fluoride and an acceptor by Walden inversion. The new glycoside is then hydrolyzed with anomeric inversion at the same site on the enzyme before it is released from the active center, which is the normal reaction of an inverting GH (Hehre, 2000; Williams and Withers, 2000). The mechanism was confirmed by analyzing the reaction of trehalase, which hydrolyzed an α-1,α-1 linkage with anomeric inversion (Hehre et al., 1982). This mechanism was later named the "Hehre resynthesis-hydrolysis mechanism" (Williams and Withers, 2000).

The mutant GH enzymes that catalyze the synthesis of glycosides from the glycosyl fluoride of the opposite anomer are called glycosynthase. In 1998, Withers's group reported the first glycosynthase, in which a mutant retaining enzyme, GH 1 β-glucosidase from *Agrobacterium* sp., with a mutation at its nucleophilic residue (E358), catalyzed synthesis of various β-glucosides using α-glucosyl fluoride as a donor and various *p*-nitrophenyl-β-glycosides as acceptors (Mackenzie et al., 1998). Then, various retaining GHs have been converted into glycosynthases by substituting their nucleophilic residues (Malet and Planas, 1998;, Trincone et al., 2000; Fort, 2000; Mayer et al., 2000; Nashiru, 2001; Okuyama et al., 2002; Hrmova et al., 2002; Ducros et al., 2003; van Lieshout et al., 2004; Drone et al., 2005; Sugimura et al., 2006; Kim et al., 2006; Hommalai et al., 2007).

Williams and Withers commented that the glycosynthase technique was developed by mimicking the Hehre resynthesis-hydrolysis mechanism of the inverting GHs (Williams and Withers, 2000). However, no glycosynthase mutant derived from inverting GH has yet been reported. Thus we have attempted to convert an inverting GH into glycosynthase, which is expected to expand the usage of inverting GHs into the synthesys of glycosides.

REDUCING-END-XYLOSE RELEASING EXO-OLIGOXYLANASE (Rex)

Xylanase belonging to GH8

GH8 is a family of inverting hydrolase whose members are mainly cellulase, chitinase, and chitosanase. Recently, a GH8 endo β-1,4 xylanase (pXyl) was found in a culture supernatant of *Pseudoalteromonas haloplanktis* and characterized (Collins et al., 2002). The 3D structures of two GH8 enzymes were found to have $(\alpha/\alpha)_6$ barrel structure (clan GH-M) (Guerin et al., 2002; Van Petegem et al., 2003). The GH8 xylanase, pXyl, has the highest amino acid identity (32.6%) with the protein encoded by the BH2105 gene (GenBank accession number:BAB05824) of *Bacillus halodurans* C-125, an alkalophilic bacterium whose genomic sequence is available (Takami et al., 2000; Collins et al., 2002). We planned to express the BH2105 protein on *Escherichia coli* and characterize the properties of the enzyme.

Characterization of recombinant BH2105

The recombinant BH2105 protein was expressed in *E. coli* and purified to yield a 45-kDa protein on SDS-PAGE. This sequence information suggested that the enzyme has no signal peptide. This enzyme did not hydrolyze birch wood xylan as well as any other polymeric substrates for GH8 enzymes (chitosan, lichenan, curdlan, and carboxymethyl cellulose). Various pentasaccharides (xylopentaose, cellopentaose, laminaripentaose, chitopentaose and chitosanpentaose) were examined for the substrate and the enzyme showed hydrolytic activity only on xylopentaose (X_5), producing initially X_1 and X_4, and finally X_1 and X_2. Then hydrolysis of xylooligosaccharides (Xn: n = 2 to 6) were examined. In the initial stage, the enzyme released X_1 and X_{n-1} from X_n and the final products were X_1 and X_2, when $n \geq 3$. It hardly hydrolyzed X_2. On the other hand, the enzyme exhibited no activity on 4-nitrophenyl-β-xyloside and 4-nitrophenyl-β-xylobioside at all, suggesting the possibility of hydrolysis at the reducing end. The enzyme had no transglycosidation activity as usual for inverting enzymes.

When β-1,4 linked D-glucose and D-xylose-based trisaccharides (Shintate et al., 2003) were examined as the substrate, the enzyme hydrolyzed only G-X-X, X-X-G and G-X-G at the linkage of the reducing end side, judging from the products, with much slower rate than X_3 (Figure 2). The reducing end specificity is completely different from

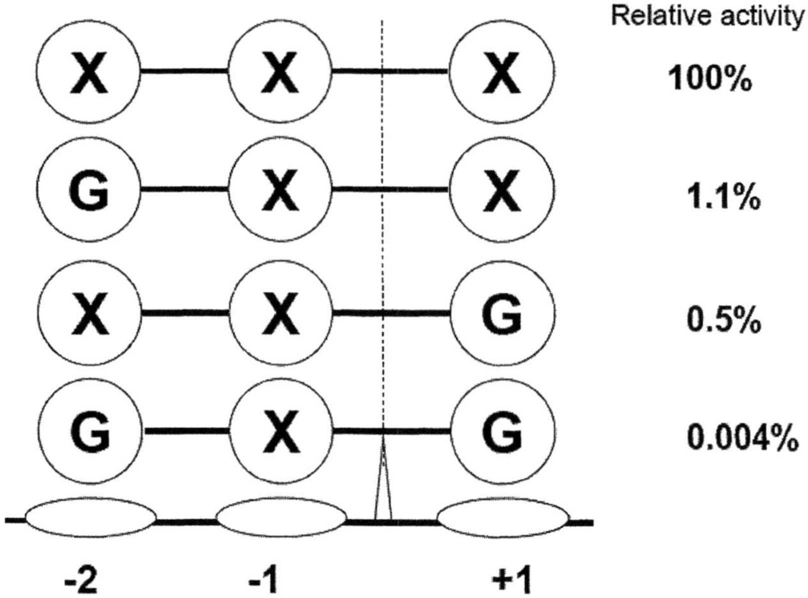

Figure 2 Reaction pattern of Rex towardβ-1,4 linked D-glucose and D-xylose-based trisaccharides

Rex did not hydrolyze the other trisaccharides (G-G-G, G-G-X, X-G-G, and X-G-X). (Adapted from Honda and Kitaoka, 2004.)

that of β-xylosidase, which liberates xylose from the non-reducing end. Judging from the activity, the enzyme is specific to homo xylooligosaccharides.

Anomeric hydroxyl group recognition by the enzyme

The anomeric composition of the degradation products of X3 by the BH2105 protein were analyzed by HPLC (Honda and Kitaoka, 2004). As shown in Figure 3, the enzyme produced β-anomer of X_1 and α-anomer of X_2 from X_3 in the reaction for 1 min. Furthermore, the α-anomer of X3 was the predominant anomer remaining in the reaction. This result strongly suggests that the enzyme hydrolyzed only the β-anomer of X_3 at the linkage of the reducing-end side with anomeric inversion to form α-X_2 and β-X_1.

Name of the enzyme

We determined that the GH8 glycosidase from *B. halodurans* C-125 has novel enzymatic properties as listed below: (1) the enzyme releases xylose from the reducing end of a xylooligosaccharide in an exo-splitting manner; (2) the enzyme strictly recognizes the β-anomeric hydroxyl group at the reducing end of the substrate. Since the enzyme reaction is clearly different from that of β-xylosidase, we propose that the name of the enzyme be reducing-end-xylose releasing exo-oligoxylanase (Rex). It has been registered to be a new enzyme as EC 3.2.1.156 in the Enzyme Nomenclature.

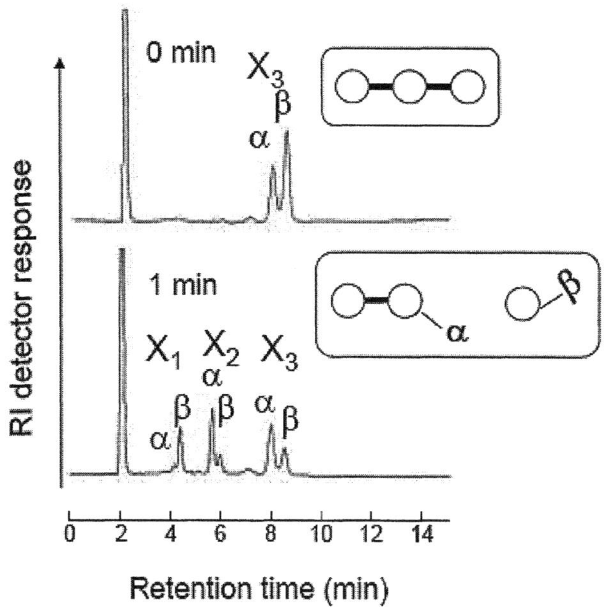

Figure 3 Anomer analysis of the hydrolytic products from X_3
The enzymatic reaction toward 50 mM X_3 was carried out in 25 mM sodium phosphate buffer (pH 7.1) at 25°C with an enzyme concentration of 5.5 μM. After incubation for 1 min, an aliquot (10 μl) of the reaction solution was immediately loaded onto a TSK-GEL AMIDE-80 column (4.6 × 250 mm, Tosoh, Japan), and eluted with acetonitrile-water (7:3 v/v) at a flow rate of 1.5 ml/min at 25°C, separating the xylooligosaccahrides anomers. The initial substrate and products were detected using a refractive index monitor (RI model 504, GL Science, Tokyo, Japan). (Adapted from Honda and Kitaoka, 2004.)

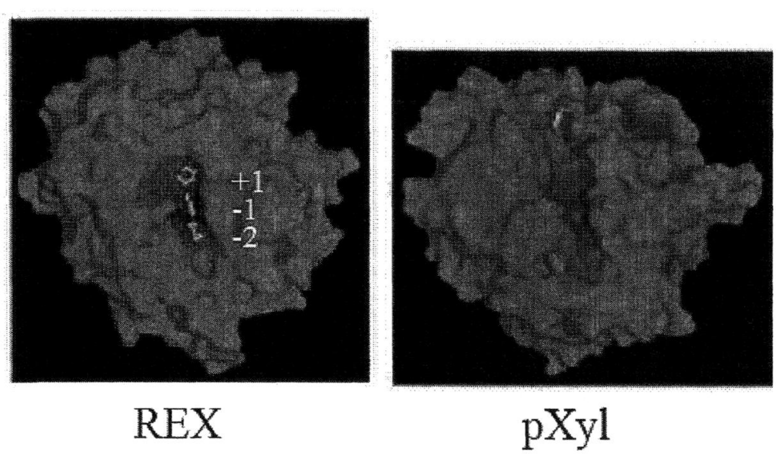

Figure 4 Molecular surfaces of Rex and pXyl showing the substrate binding cleft
(Adapted from Fushinobu et al., 2005.)

Structural analyses of Rex

To understand the mechanism of the reducing-end-specific exo-lytic activity, structural analyses of Rex were carried out. The optimal crystallization conditions were found with solution No. 37 of Crystal Screen Cryo (Hampton Research, CA, USA) at 293 K by the hanging-drop vapor diffusion method for 5 days (Honda et al., 2005). The crystal structures of Rex in unliganded and complex forms at 1.35–2.20-Å resolution (PDB accession Number: 1WU4, 1WU5, and 1WU6) were determined to reveal the structural aspects of its three subsites ranging from –2 to +1 (Fushinobu et al., 2005). The structure of Rex is very similar with that of pXyl (Van Petegem et al., 2003), the GH8 endo-xylanase, composed of the $(\alpha/\alpha)_6$ barrel structure (Fig. 4). The catalytic machinery of Rex is well conserved with pXyl. The most significant difference in the structures of Rex and pXyl was found at the binding cleft. Subsite +2 of Rex is blocked by a barrier formed by a kink in the loop before helix α_{10} and H319 in the loop forms a direct hydrogen bond with the β-hydroxyl of xylose residue bound at subsite +1, contributing to the specific recognition of the reducing-end xylose.

HEHRE RESYNTHESIS-HYDROLYSIS OF Rex

Difficulty for proving the Hehre resynthesis-hydrolysis reaction

The Hehre resynthesis-hydrolysis mechanism is often difficult to detect because many enzymes recognize glycosyl fluoride as not only the donor but also as the acceptor (Konstantinidis et al., 1993; Becker et al., 2000; Williams and Withers, 2000; Hehre, 2000), such as β-maltosyl fluoride for β-amylase and β-glucosyl fluoride for glucoamylase. Reliable experimental evidence for the mechanism was reported for trehalase using β-glucosyl fluoride as the donor and glucose or xylose as the acceptor (Hehre et al., 1982; Kasumi et al., 1986). However, experimental difficulty remains because the donor, β-glycosyl fluoride, is hydrolyzed into the acceptor, glucose, both spontaneously and enzymatically, making the reaction complicated.

We notice that Rex is the suitable enzyme to examine the Hehre resynthesis-hydrolysis mechanism. As Rex hydrolyzes X_3 to release X_1 from the reducing end, it is expected to utilize α-X_2F as the donor and X_1 as the acceptor. It should be mentioned that the hydrolytic product of α-X_2F, X_2, will never act as the acceptor molecule due to the reducing-end exo-specificity. Thus, the donor and acceptor are considered to be completely independent during the reaction.

Activity of the wild-type Rex on α-X_2F

At first, the reactions of wild-type Rex on α-X_2F (20 mM) in both the presence and absence of X_1 (20 mM) were examined. The enzyme was inactive on α-X_2F in the absence of X_1, but produced X_2 from α-X_2F in the presence of X_1 (Fig. 5). The results suggest that the reaction was not simple hydrolysis of α-X_2F but the Hehre resynthesis-hydrolysis. No X_3 was detected during the reaction. Most part of the resulting X_3 was supposed to be hydrolyzed into X_1 and X_2 before it escaped from the active center because X_3 was generated at the very position where it was hydrolyzed.

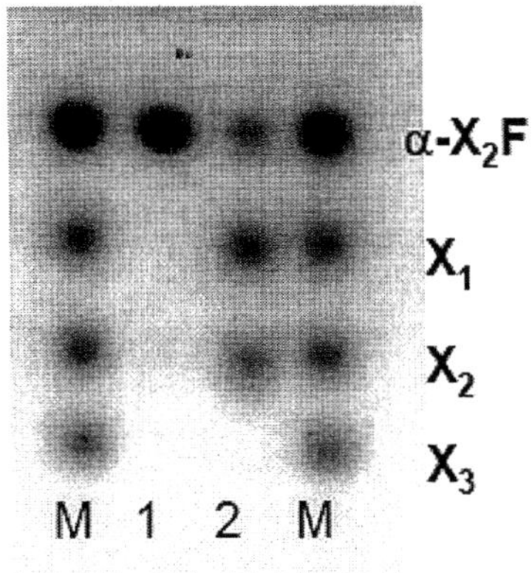

Figure 5 Pattern of action of α-X_2Freaction by wild-type Rex
The α-X_2Freaction by wild-type Rex was carried out in the absence (lane 1) and presence (lane 2) of X_1 for 1 hr with the enzyme concentration of 0.4 μM. M is standard of α-X_2F and X_n (n=1–3). (Adapted from Honda and Kitaoka, 2006.)

CONVERSION OF Rex INTO GLYCOSYNTHASE

Mutation at the base residue (Honda and Kitaoka, 2006)

To convert Rex into glycosynthase, initially the base residue, D263, was mutated as performed for retaining GHs. A saturation mutagenesis library of Rex at D263 was constructed. From the D263X mutant library, 120 colonies were randomly picked and each mutant enzyme was purified by His-tag affinity column chromatography. During the screening of α-X_2F-consuming activity in the presence of X_1 and the accumulation of X_3 on TLC, 71 of the 120 proteins exhibited activity. Other mutants did not consume α-X_2F. Nine types of D263X mutant were found in addition to the wild-type. The numbers of clones obtained were D (wild-type), 15; G, 22; A, 12; V, 8; T, 5; L, 3; N, 2;

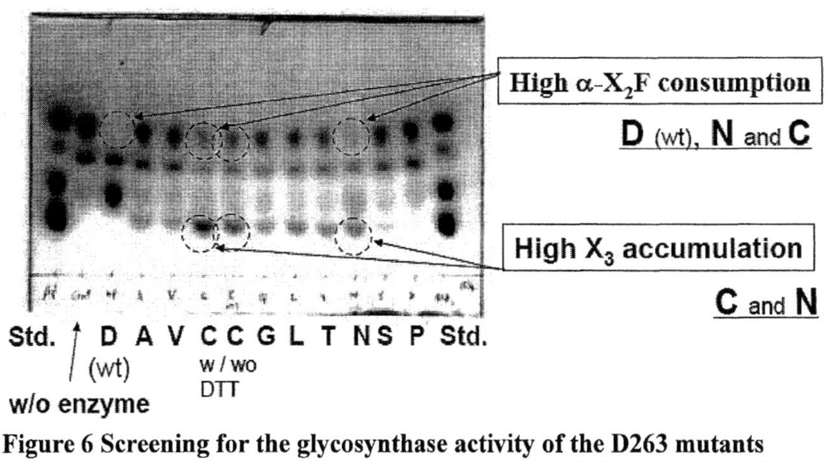

Figure 6 Screening for the glycosynthase activity of the D263 mutants
Each lane name indicates the residue substituting D263. D263C was examined in the presence and absence of 1 mM dithiothreitol.

C, 2; P, 1; and S, 1. Wild-type exhibited the highest α-X_2F consumption without the accumulation of X_3. D263C exhibited the greatest production of X_3, and D263N did the second highest among the mutants (Figure 6). Thus, we selected D263C and D263N as targets for further analysis of glycosynthase properties.

The F^- releasing activity from 10 mM α-X_2F in the presence of 10 mM X_1 and the hydrolytic activities to 2.6 mM X_3 of the wild type and D263 mutants are summarized in Table 1. Both D263C and D263N showed approximately 1/10 F^- releasing activity of the wild type. The mutations caused drastic decreases in the hydrolytic activity (4.5×10^{-4} for D263C, 2.9×10^{-3} for D263N). It should be noted that the remaining hydrolytic activity of D263N was 6.4 times higher than that of D263C, whereas their F^- releasing activities were almost identical, resulting the difference in the ratio of the F^- releasing activity against the hydrolytic activity (F/H ratio). Figure 7 shows the time courses of α-X_2F and X_1 reaction by the enzymes. In the wild-type reaction, F^- and X_2 concentrations increased in parallel but X_3 was not detected. On the other hand, X_3 was detected in the reactions catalyzed by the mutants. In the case of D263N, X_2 concentration was higher than X_3 in the reaction mixture, indicating that the hydrolytic reaction was higher than the transfer reaction. In contrast, X_3 concentration was higher than X_2 in the reaction with D263C, indicating that the transfer activity was higher than the hydrolytic activity. The difference agrees with the remaining hydrolytic activities of the mutants.

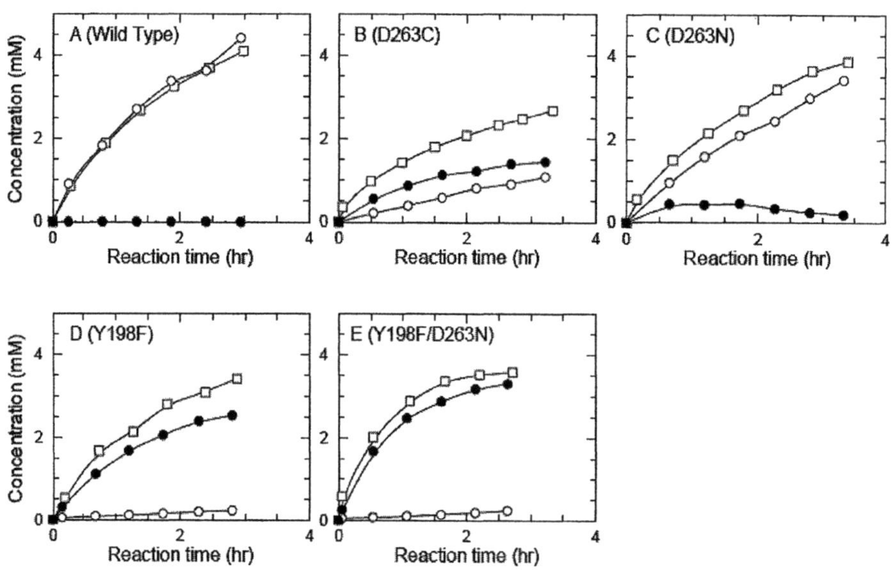

Figure 7 Time courses of α-X_2F and X_1 reaction

The enzymatic reaction was carried out in 0.1 M MOPS buffer (pH 7.0) at 30°C. Both substrate concentrations were 5.0 mM. A: Wild-type (enzyme concentration was 0.5 μM); B: D263C (7.2 μM); C: D263N (7.3 μM), D: Y198F (0.1 μM); E: Y198F/D263N (7.7 μM). Symbols indicate F^- (open diamonds), X_2 (open circles), and X_3 (closed circles). (Adapted from Honda and Kitaoka, 2006, and Honda et al., 2008 with permission from Oxford University Press.)

Table 1 Activities of each mutant of Rex

	F^- release* (F)	X_3 hydrolysis** (H)	F/H
Wild type	3.1	31.2	0.1
D263C	0.29	0.014	21
D263N	0.32	0.09	3.6
Y198F	4.7	0.06	78
Y198F/D263C	0.03	0.0006	50
Y198F/D263N	0.23	0.0016	144

*F- releaseing activitiy from 10 mM α-X_2F and 10 mM X_1. The values are given as s^{-1}.
**Hydrolytic activity on 2.6 mM X_3. The values are given as s^{-1}.
(The data are from Honda et al., 2008.)

Mutation at the residue supporting the nucleophilic water molecule (Honda et al., 2008)

The residue supporting the nucleophilic water activated by the base residue

Conversion of Rex into glycosynthase was successfully performed by mutating the base residue. However, the D263 mutants retained significant hydrolytic activity that decreased the yield of the synthetic product. Moreover, the F^- releasing activities of the mutants were much lower than that of the wild type (Table 1). The mutation at the general base residue of an inverting GH could not remove the hydrolytic activity completely probably because the water molecule retained some activity as a nucleophilic reagent without the aid of the residue (Honda and Kitaoka, 2006). It is therefore still important to reduce the hydrolytic activity of glycosynthase obtained from inverting GHs.

Collins et al. (2002, 2005) found that a mutation at a conserved tyrosine residue (Y203) in the GH8 endo-xylanase, pXyl, which shares 32.6% amino acid sequence identity with Rex, causes a drastic decrease in the hydrolytic activity of the enzyme toward xylan. The residue formed a hydrogen bond with the nucleophilic water molecule that had another hydrogen bond with the general base residue (D281). The pH dependency of the Y203F mutant indicated that the tyrosine residue was not a general base, but was important to locate the nucleophilic water at the proper position (Collins et al., 2005). Thus, we attempted mutating the corresponding residue of Rex, Y198, confirmed by both the alignment of amino-acid sequences and the structural analysis of Rex (Fushinobu et al., 2005).

Activity of Y198F mutants

The F^- releasing activity and the hydrolytic activities of the mutants at Y198 and/or D263 are given in Table 1. The mutation at Y198 drastically decreased the hydrolytic activity (from 31.2 to 0.06 s^{-1}), with a small increase in the F^- releasing activity. The F^- releasing activity of Y198F was more than 10 times that of the mutants at the general base residue (D263). The F/H ratio was 78, approximately 4 times that of D263C, the best glycosynthase mutant at the general base residue. Double mutation was also tested. Y198F/D263N showed 20 times less F^- releasing activity than Y198F, which was comparable to that of the corresponding single mutant, D263N. The F/H ratio (144) of

D263N was two times higher than that of Y198F. The Y198F/D263C mutant showed only faint F^- releasing activity.

Time courses of the reactions with 5 mM α-X_2F and X_1 by Y198F mutants are given in Figure 7. Much greater accumulation of X_3 (glycosynthase product) and much less formation of X_2 (hydrolytic product from X_3) were observed in the reactions catalyzed by Y198F and Y198F/D263N than in that catalyzed by D263C. The ratio of X_3 in the products $\{X_3/(X_2 + X_3)\}$ when half of α-X_2F was consumed was 0 (wild type), 0.59 (D263C), 0.19 (D263N), 0.93 (Y198F), and 0.96 (Y198F/D263N). Although Y198F/D263N showed a slightly larger X_3 ratio than Y198F, the F^- releasing activity of Y198F was much higher than that of Y198F/D263N. It should be noted that the experiments shown in Figure 7D and E were performed with 0.1 μM Y198F and 7.7 μM Y198F/D263N, respectively. Thus, we have judged that Y198F was the best glycosynthase of the mutants.

CONCLUSION

The relationship between the F^--releasing activity and the hydrolytic activity of Rex during acquiring the glycosynthase activity was completely different from those of the retaining enzymes. In the case of retaining enzymes, the parent enzyme does not possess F^--releasing activity (Williams and Withers, 2000; Sugimura et al., 2006), and the mutation at the base residue completely remove the hydrolytic activity because the base residue directly attacks the C1 of the hydrolyzed glycoside. Thus, while converting retaining GHs into glycosynthases, it is important to acquire F^--releasing activity by the mutation.

In contrast, the water molecule activated by the base residue attacks the C1 of the glycoside in the case of the inverting enzymes. Thus, the mutation of the base residue cannot remove the hydrolytic activity completely because the water molecule retains some activity without the aid of the catalytic residue. Furthermore, the parent enzyme possesses F^- releasing activity and the mutations at the base residue cause decreases in the activity. Thus, while converting inverting GHs into glycosynthases, it is important to minimize the decrease in F^--releasing activity, while maximizing the decrease in hydrolytic activity.

We here conclude a concept for creating a glycosynthase from an inverting GH by mutating a residue holding the nucleophilic water molecule with the general base residue while keeping the general base residue intact.

ACKNOWLEDGEMENT

This work was supported in part by a grant from the Program for Promotion of Basic Research Activities for Innovative Biosciences (PROBRAIN).

REFERENCES

Becker D, Johnson K S, Koivula A, Schulein M, and Sinnott M L (2000), 'Hydrolyses of alpha- and beta-cellobiosyl fluorides by Cel6A (cellobiohydrolase II) of *Trichoderma reesei* and *Humicola insolens*', *Biochem J*, 345, 315-319.

Collins T, de Vos D, Hoyoux A, Savvides S N, Gerday C, van Beeumen J, and Feller G (2005), 'Study of the active site residues of a glycoside hydrolase family 8 xylanase', *J Mol Biol*, 354, 425-435.

Collins T, Meuwis M A, Stals I, Claeyssens M, Feller G, and Gerday C (2002), 'A novel family 8 xylanase, functional and physicochemical characterization', *J Biol Chem*, 277, 35133-35139.

Davies G and Henrissat B (1995), 'Structures and mechanisms of glycosyl hydrolases', *Structure*, 3, 853-859.

Drone J, Feng H Y, Tellier C, Hoffmann L, Tran V, Rabiller C, and Dion M (2005), '*Thermus thermophilus* glycosynthases for the efficient synthesis of galactosyl and glucosyl β-(1→3)-glycosides', *Eur J Org Chem*, 2005, 1977-1983.

Ducros V M, Tarling C A, Zechel D L, Brzozowski A M, Frandsen T P, von Ossowski I, Schulein M, Withers S G, and Davies G J (2003), 'Anatomy of glycosynthesis: structure and kinetics of the *Humicola insolens* Cel7B E197A and E197S glycosynthase mutants', *Chem Biol*, 10, 619-628.

Fort S, Boyer V, Greffe L, Davies G, Moroz O, Christiansen L, Schulein M, Cottaz S, and Driguez H (2000), 'Highly efficient synthesis of β(1→4)-oligo- and -polysaccharides using a mutant cellulase', *J Am Chem Soc*, 122, 5429-5437.

Fushinobu S, Hidaka M, Honda Y, Wakagi T, Shoun H, and Kitaoka M (2005), 'Structural basis for the specificity of the reducing end xylose-releasing exo-oligoxylanase from *Bacillus halodurans* C-125', *J Biol Chem,* 280, 17180-17186.

Guerin D M, Lascombe M B, Costabel M, Souchon H, Lamzin V, Beguin P, and Alzari P M (2002), 'Atomic (0.94 A) resolution structure of an inverting glycosidase in complex with substrate', *J Mol Biol*, 316, 1061-1069.

Hehre E J, Brewer C F, and Genghof D S (1979), 'Scope and mechanism of carbohydrase action. Hydrolytic and nonhydrolytic actions of beta-amylase on alpha- and beta-maltosyl fluoride', *J Biol Chem*, 254, 5942-5450.

Hehre E J, Sawai T, Brewer C F, Nakano M, and Kanda T (1982), 'Trehalase: stereocomplementary hydrolytic and glucosyl transfer reactions with alpha- and beta-D-glucosyl fluoride', *Biochemistry,* 21, 3090-3097.

Hehre E J (2000), 'A fresh understanding of the stereochemical behavior of glycosylases: structural distinction of "inverting" (2-MCO-type) versus "retaining" (1-MCO-type) enzymes', In *Adv Carbohydr Chem Biochem,* Vol. 55, pp. 265-310.

Henrissat B (1991), 'A classification of glycosyl hydrolases based on amino acid sequence similarities', *Biochem J*, 280, 309-316.

Henrissat B and Bairoch A (1993), 'New families in the classification of glycosyl hydrolases based on amino acid sequence similarities', *Biochem J*, 293, 781-788.

Henrissat B and Bairoch A (1996), 'Updating the sequence-based classification of glycosyl hydrolases', *Biochem J*, 316, 695-696.

Hommalai G, Withers S G, Chuenchor W, Cairns J R, and Svasti J (2007), 'Enzymatic synthesis of cello-oligosaccharides by rice BGlu1 β-glucosidase glycosynthase mutants', *Glycobiology,* 17, 744-753.

Honda Y and Kitaoka M (2004), 'A family 8 glycoside hydrolase from *Bacillus halodurans* C-125 (BH2105) is a reducing end xylose-releasing exo-oligoxylanase', *J Biol Chem,* 279, 55097-55103.

Honda Y, Fushinobu S, Hidaka M, Wakagi T, Shoun H, and Kitaoka M. (2005) 'Crystallization and preliminary X-ray analysis of reducing-end-xylose releasing exo-oligoxylanase from *Bacillus halodurans* C-125', *Acta Crystallogr, Sect. F: Struct Biol Cryst Commun,* 61, 291-292.

Honda Y and Kitaoka M (2006), 'The first glycosynthase derived from an inverting glycoside hydrolase', *J Biol Chem,* 281, 1426-1431.

Honda Y, Fushinobu S, Hidaka M, Wakagi T, Shoun H, Taniguchi H, and Kitaoka M (2008), 'Alternative strategy for converting an inverting glycoside hydrolase into a glycosynthase', *Glycobiology,* 18, 325-330.

Hrmova M, Imai T, Rutten S J, Fairweather J K, Pelosi L, Bulone V, Driguez H, and Fincher G B (2002), 'Mutated barley (1,3)-beta-D-glucan endohydrolases synthesize crystalline (1,3)-beta-D-glucans', *J Biol Chem,* 277, 30102-30111.

Kasumi T, Brewer C F, Reese E T, and Hehre E J (1986), 'Catalytic versatility of trehalase: synthesis of alpha-D-glucopyranosyl alpha-D-xylopyranoside from beta-D-glucosyl fluoride and alpha-D-xylose', *Carbohydr Res,* 146, 39-49.

Kim Y W, Fox D T, Hekmat O, Kantner T, McIntosh L P, Warren R A, and Withers S G (2006), 'Glycosynthase-based synthesis of xylo-oligosaccharides using an engineered retaining xylanase from *Cellulomonas fimi*', *Org Biomol Chem,* 4, 2025-2032.

Konstantinidis A K, Marsden I, and Sinnott M L (1993), 'Hydrolyses of alpha- and beta-cellobiosyl fluorides by cellobiohydrolases of *Trichoderma reesei*', *Biochem J,* 291, 883-888.

Mackenzie L F, Wang Q, Warren R A J, and Withers S G (1998), 'Glycosynthases: Mutant glycosidases for oligosaccharide synthesis', *J Am Chem Soc,* 120, 5583 -5584.

Malet C and Planas A (1998), 'From beta-glucanase to beta-glucansynthase: glycosyl transfer to alpha-glycosyl fluorides catalyzed by a mutant endoglucanase lacking its catalytic nucleophile', *FEBS Lett,* 440, 208-212.

Mayer C, Zechel D L, Reid S P, Warren R A, and Withers S G (2000), 'The E358S mutant of *Agrobacterium* sp. beta-glucosidase is a greatly improved glycosynthase', *FEBS Lett,* 466, 40-44.

Nashiru O, Zechel D L, Stoll D, Mohammadzadeh T, Warren R A J, and Withers S G (2001), 'β-Mannosynthase: Synthesis of β-mannosides with a mutant β-mannosidase', *Angew Chem Int Ed,* 40, 417-420.

Okuyama M, Mori H, Watanabe K, Kimura A, and Chiba S (2002), 'Alpha-glucosidase mutant catalyzes "alpha-glycosynthase"-type reaction', *Biosci Biotechnol Biochem,* 66, 928-933.

Shintate K, Kitaoka M, Kim Y K, and Hayashi K (2003), 'Enzymatic synthesis of a library of β(1→4) hetero-D-glucose and D-xylose based oligosaccharides employing cellodextrin phosphorylase', *Carbohydr Res,* 338, 1981-1990.

Sinnott M L (1990) 'Catalytic mechanism of enzymic glycosyl transfer', *Chem Rev,* 90, 1171-1202.

Sugimura M, Nishimoto M, and Kitaoka M (2006), 'Characterization of glycosynthase mutants derived from glycoside hydrolase family 10 xylanases', *Biosci Biotechnol Biochem,* 70, 1210-1217.

Takami H, Nakasone K, Takaki Y, Maeno G, Sasaki R, Masui N, Fuji F, Hirama C, Nakamura Y, Ogasawara N, Kuhara S, and Horikoshi K (2000), 'Complete genome sequence of the alkaliphilic bacterium *Bacillus halodurans* and genomic sequence comparison with *Bacillus subtilis*', *Nucleic Acids Res*, 28, 4317-4331.

Trincone A, Perugino G, Rossi M, and Moracci M (2000), 'Highly productive autocondensation and transglycosylation reactions with *Sulfolobus solfataricus* glycosynthase', *Bioorg Med Chem Lett,* 10, 365-368.

van Lieshout J, Faijes M, Nieto J, van der Oost J, and Planas A (2004), 'Hydrolase and glycosynthase activity of endo-1,3-beta-glucanase from the thermophile *Pyrococcus furiosus*', *Archaea*, 1, 285-292.

van Petegem F, Collins T, Meuwis M A, Gerday C, Feller G, and Van Beeumen J (2003), 'The structure of a cold-adapted family 8 xylanase at 1.3 A resolution. Structural adaptations to cold and investgation of the active site', *J Biol Chem*, 278, 7531-7539.

Williams S J and Withers S G (2000), 'Glycosyl fluorides in enzymatic reactions', *Carbohydr Res*, 327, 27-46.

CHARACTERIZATION OF NOVEL GLYCOSIDES USING THE GLUCANSUCRASE

Young-Hwan Moon, Young-Min Kim, Doman Kim

ABSTRACT

Six epigallocatechin gallate glycosides were synthesized by the acceptor reaction of a glucansucrase produced by *Leuconostoc mesenteroides* B-1299CB with EGCG and sucrose. Each of these glycosides was then purified, and the structures were assigned as follows: epigallocatechin gallate 7-*O*-α-D-glucopyranoside (EGCG-G1), epigallocatechin gallate 7,4″-*O*-α-D-glucopyranoside (EGCG-G2), epigallocatechin gallate 7,4′-*O*-α-D-glucopyranoside (EGCG-G3), epigallocatechin gallate 4″-*O*-α-D-glucopyranoside (EGCG-G4), epigallocatechin gallate 4'-*O*-α-D-glucopyranoside (EGCG-G5), and epigallocatechin gallate 4′,4″-*O*-α-D-glucopyranoside (EGCG-G6). Three of these compounds (EGCG-G1, EGCG-G2 and EGCG-G4) were novel compounds. The EGCG glycosides exhibited similar or slower antioxidant effects, depending on their structures (EGCG ≥ EGCG-G1 > EGCG-G4 > EGCG-G5 > EGCG-G2 > EGCG-G3> EGCG-G6), and also manifested a higher degree of browning resistance than was previously noted in EGCG. Also, EGCG-G1, EGCG-G5, EGCG-G4, EGCG-G3, EGCG-G6, and EGCG-G2 were 49, 55, 71, 114, 125, and 130 times as water soluble, respectively, as was EGCG. Two arbutin glucosides were also synthesized via the acceptor reaction. The purified glucosides were elucidated as 4-hydroxyphenyl α-isomaltoside (arbutin-G1), 4-hydroxyphenyl α-isomaltotrioside (arbutin-G2). Arbutin glucoside (4-hydroxyphenyl β-isomaltoside) exhibited slower effects on DPPH radical scavenging and similar effects on tyrosinase inhibition, and increased inhibitory effect on MMP-1 production induced by UVB than arbutin.

Key words: epigallocatechin gallate; glycosides; glucansucrase; arbutin; antioxidant

INTRODUCTION

Glucosyltransferases (GTFs) are enzymes that synthesize either dextrans or glucans, using sucrose as a substrate (Robyt and Martin, 1983). Robyt et al. (1995) initially reported that glucosyltransferases can catalyze the transfer of a sucrose-derived glucose to other carbohydrates, thereby allowing for oligosaccharide synthesis. This reaction is referred to as the *acceptor reaction*, and the added carbohydrates are called *acceptors* (Koepsell et al., 1953). Other than carbohydrates, glucosyltransferases (GTFs) are known to be able to transfer mono-, di-, or higher glucose units to other acceptors, via a variety of glycosidic linkages (Seo et al., 2005). *L. mesenteroides* NRRL B-1299CB is a constitutive mutant, developed via the EMS method (Kim and Robyt, 1995). 1299CB glucansucrase in culture media therefore, does not form a complex with dextran, thereby ensuring higher acceptor reaction efficiency.

Enzymatic transglycosylation, using *L. mesenteroides* glucansucrases, has previously been applied to the modification of a variety of bioactive substances, in efforts to improve their functionality. Acarbose analogues have been synthesized via the reaction of acarbose with sucrose and glucansucrases from *L. mesenteroides* B-512FMC and B-742CB (Yoon and Robyt, 2002). These acarbose analogues exhibited excellent inhibitory effects for a variety of enzymes related to diabetes, including α-glucosidase, as was reported by Kim et al. (1999). Salicin analogues have also been synthesized using glucansucrases, with sucrose as a substrate, by both Seo et al. and Yoon et al. (Seo et al., 2005; Yoon et al., 2004). Both of these research groups determined that the salicin analogues exhibited an inhibitory effect on blood coagulation.

Recently, these enzymatic acceptor reactions have been employed in the modification of natural bioactive compounds, thereby improving their physicochemical properties. Lee et al. (2004) reported that glycosylated ascorbic acids exhibited profound antioxidant effects that prevented lipid oxidation, and also reported that they manifested synergistic effects superior to those associated with ascorbic acid. Gilly et al. (2003) also reported that the glycosylation of resveratrol resulted in a profound inhibition of enzymatic oxidation. With regard to solubility, Lee et al. (1999) reported that glycosylated naringin was 250 times more soluble in water than naringin, and 10 times less bitter. Li et al. (2004) also reported that the solubility of transglycosylated puerarin was 14-168 times higher than that of puerarin. Glycosylated catechin has proven quite stable against UV radiation, although catechin itself is quite readily degraded (Kitao et al., 1993).

Green tea, a popular and commonly consumed beverage in Asia, is an important source of flavonoids called catechins. The green tea catechins, specifically EGCG (epigallocatechin gallate), EGC (epigallocatechin), ECG (epicatechin gallate), and EC (epicatechin), have been extensively studied for their utility as bioactive substances, for use as antioxidant and anticancer agents (Bushman, 1998; Ichihashi et al., 2000). EGCG is the most abundant of the green tea catechins (Cabrera, 2003). It exhibits potent antioxidant (Salah et al., 1995), anticancer (Huh et al., 2004), antitumorigenic (Jung et al., 2001), and antibacterial (Lee et al., 2004) effects, and also appears to play a part in the prevention of dental caries (Otake, 1991), and the regulation of plasma lipid levels (Raederstorff, 2003). However, EGCG is only minimally water-soluble, and is readily degraded in aqueous solutions (Kitao et al., 1995). Due to these disadvantages, the use of EGCG in the food and cosmetic industries remains somewhat limited. In order to overcome this problem, a great deal of research has been done concerning the glycosylations of a variety of polyphenols (Kitao et al., 1993, 1995; Sato et al., 2000). The resultant transglycosylated compounds have sometimes exhibited increased solubility in water, increased stability against light or oxidation, improved taste qualities, and stronger tyrosinase inhibitory effects (Kitao et al., 1993, 1995; Sato et al., 2000; Nakano et al., 2002). When improved in such ways, glycosylated bioactive substances have proven much more useful as food additives and cosmetics.

For glycosylated polyphenols, however, it is quite difficult to determine their bioavailability. Hollman and Katan found that human absorption of the quercetin glycosides (52%) from onions is far better than that of the pure aglycon (24%) (Hollman and Katan, 1993). Researchers proposed that glycosylation might be an important factor in its bioavailability, but critically significant factors were the number and positions of the glucose moieties (Karakaya, 2004). The bioavailability study of glycosylated polyphenols is now going forward.

Arbutin (4-hydroxyphenyl β-D-glucopyranoside) has been used as a cosmetic whitening agent (Maeda and Fukuda, 1996). 4-Hydroxyphenyl α-D-glucopyranoside (α-arbutin), were enzymatically synthesized from hydroquinone and maltopentaose, and 4-hydroxylphenyl β-maltoside and 4-hydroxylphenyl β-maltotrioside were also synthesized using cyclomaltodextrin glucanotransferase (CGTase) with arbutin and starch (Sugimoto et al., 2005). These arbutin-α-D-glucopyranosides have been shown to exhibit competitive type inhibition effects on human tyrosinase, and K_i values of 4-hydroxylphenyl β-maltoside and 4-hydroxylphenyl β-maltotrioside were 0.7 mM and 0.9 mM, respectively. Arbutin-α-glycosides showed stronger inhibitory activity than arbutin, but less activity than α-arbutin (Sugimoto et al, 2003, 2005). Due to these favorable qualities, glycosylated arbutins are expected to prove useful as additives to cosmetics.

MATERIALS AND METHODS

Materials

The EGCG (epigallocatechin gallate), DMSO-d_6 (dimethyl sulfoxide), DPPH (1,1-diphenyl-2-picrylhydrazyl) and DHB (2,5-dihydroxybenzoic acid) were obtained from the Sigma Aldrich (St. Louis, MO, USA). The 4-hydroxyphenyl α-D-glucopyranoside (arbutin or β-arbutin), D_2O (deuterium oxide), DPPH (1,1-diphenyl-2-picrylhydrazyl), DHB (2,5-dihydroxybenzoic acid) and tyrosinase (T7755) were purchased from the Sigma Aldrich. The Sephadex LH-20 gel was acquired from Amersham Biosciences (Uppsala, Sweden). The Bio-gel P-2 was acquired from Bio-Rad (Hercules, CA, USA).

Enzyme preparation

L. mesenteroides B-1299CB mutant was grown at 28°C on LM medium containing 2% (w/v) glucose as a carbon source. The LM medium consisted of 4 g yeast extract, 2 g peptone, 0.2 g $MgSO_4 \cdot 7H_2O$, 0.01 g $FeSO_4 \cdot 7H_2O$, 0.01 g NaCl, 0.01 g $MnSO_4 \cdot H_2O$, 0.015 g $CaCl_2 \cdot 2H_2O$, and 2 g K_2HPO_4 per liter of deionized water. After fermentation, the culture was harvested, centrifuged, and concentrated with hollow fiber (30 K cut-off, Millipore, Japan).

Assay of glucansucrase activity

Enzyme activity was evaluated via the incubation of the enzyme for different reaction periods at 28°C with 100 mM sucrose, using 20 mM sodium acetate (pH 5.2) as a substrate. Standard assay mixtures consisted of 200 μL of 200 mM sucrose and 200 μL of an enzyme solution. Each of the enzyme reaction samples was spotted on a Whatman K5 TLC plate (Whatman Inc, Clifton, NJ, USA). The TLC plate was ascended twice on an acetonitrile-water (85:15, v/v) solvent system. Each of the carbohydrates was visualized via the dipping of the plates into 0.3% (w/v) N-(1-naphthyl)-ethylenediamine and 5% (v/v) H_2SO_4 in methanol, followed by 10 minutes of heating at 121°C. The amount of fructose released from the sucrose was then analyzed using an NIH Image Program, with standard materials. One unit of glucansucrase activity was defined as the amount of enzyme required to generate one μmol of fructose per minute at 28°C and pH 5.2, in 20 mM sodium acetate buffer.

Glucosylation of EGCG and arbutin as acceptors

The reaction mixture (250 mL) consisted of 0.2% EGCG (500 mg), 80 mM sucrose (6.84 g), and B-1299CB glucansucrase (2.4 U/mL) or 12.25 mM arbutin, 100 mM sucrose, and 2 U/mL B-1299CB glucansucrase. The mixture was incubated at 28°C for

6.5 h, after which the sucrose had been depleted. The reaction mixture was then boiled for 5 min in order to halt the enzyme reaction.

Analysis of acceptor reaction products via thin layer chromatography (TLC)

TLC was conducted at room temperature using silica gel 60 F_{254} TLC plates (Merck Co., Whitehouse station, NJ, USA). One μL of the reaction digests was spotted onto the silica gel plate, and the plate was then developed using a solvent mixture of ethyl acetate-acetic acid-water (3:1:1, v/v/v). The developed plate was then dried and visualized as previously described.

Purification of EGCG acceptor reaction products

The reaction digest (250 mL) was then subjected to Sephadex LH-20 column (47×200 mm) chromatography. The transfer products were washed with distilled water (3 L, 1 mL/min) in order to remove sugars (sugar-polymer, fructose and glucose) from the reaction digest, and then successively eluted with 70% (v/v) ethanol (1 L). The eluent (which contained EGCG glycosides) was then concentrated at 47℃ with a rotary evaporator (N-N series, EYELA, Tokyo, Japan). The eluent was subjected to HPLC (LC-10AD, Shimadzu, Koyto, Japan) under the following conditions: reverse column, μ-Bondapak C_{18} (7.8×300 mm, Waters, Milford, MA, U.S.A.); mobile phase, 23% methanol ; flow rate, 3 mL/min; room temperature; detection, UV detector (280 nm, Waters 2487, Waters).

The arbutin reaction digest (60 mL) was concentrated into 3 mL by using evaporator and then the sample (1.5 mL) was applied to Bio-Gel P-2 column (3.5 × 160 cm). Products were eluted with distilled water (1 ml/min). The eluent was fractionated into two groups (which contained each arbutin glucoside) by confirming each fraction using TLC (data not shown) and then concentrated at 47°C with a rotary evaporator (N-N series, EYELA). The groups containing arbutin glucosides were subjected to HPLC (LC-10AD, Shimadzu) under the following conditions: reverse column, Hypersil APS-2 NH_2 (4.6 × 250 mm; Bio-Rad); mobile phase, acetonitrile: methanol = 4:1 or 3:1 (v/v); flow rate, 0.5 ml/min; room temperature; detection, RI detector (RID-10A, Shimadzu).

MALDI-TOF-MS analysis

The purified EGCG or arbutin glycosides (3 mg/mL) were diluted with deionized water, and then mixed with 2,5-dihydroxybenzoic acid (1mg/mL) dissolved in acetonitrile, at a 1:1 ratio (v/v). The mixed solution (1 μL) was then spotted onto a stainless steel plate, and slowly dried at room temperature. The mass spectrum was obtained using a Voyager DE-STR MALDI-TOF mass spectrometer (Applied Biosystems, Foster City, CA, USA). Mass spectra were obtained in positive linear mode with delayed extraction (average of 75 laser shots), with an acceleration voltage of 65 kV.

NMR Analysis

About 2~3 mg of the purified EGCG or arbutin glycosides were dissolved in DMSO-d_6 (250 μL) and placed into 3 mm NMR tubes. NMR spectra were obtained using a Unity Inova 500 spectrometer (Varian Inc., Polo Alto, CA, USA), operating at 500 MHz for

^{1}H and 125 MHz for ^{13}C at 25°C. Linkages between EGCG and glucose were characterized using the spectra obtained via homonuclear correlation spectroscopy (COSY), heteronuclear single quantum coherence (HSQC), and heteronuclear multiple bond correlation (HMBC).

Browning resistance effect of EGCG and its glycosides

Browning resistance after UV irradiation in an aqueous system was examined in water (1.0 mL) containing 0.5% (w/v) of either epigallocatechin gallate or the EGCG glycosides. The sample solutions were exposed to UV irradiation at a distance of 10 cm from the UV source (254 nm, 10 W, G10T8-AN, Germicidal, Sankyo Denki, Hiratsuka, Japan) for 24 hours at room temperature. The increases in absorbance were determined at 460 nm, using a BIO-RAD SmartSpec™ 3000 spectrophotometer.

Antioxidant activity

The antioxidant activity of EGCG or arbutin glycosides was evaluated via DPPH radical scavenging (Abe et al., 2000). Each of the samples (10, 12.5, 25, 50, 100, 200 μM) was dissolved in ethanol (30 μL) and mixed thoroughly with a 100 μM DPPH ethanol solution (270 μL). After 10 minutes of maintenance in darkness at room temperature, the absorbance of the mixture was monitored at a wavelength of 517 nm on a BIO-RAD SmartSpec™ 3000 spectrophotometer. DPPH radical-scavenging activity was then evaluated according to the decrease in the absorbance of the DPPH radical in the samples as compared to that of a blank (ethanol). The SC_{50} value designates the sample concentration at which the levels of DPPH radicals had been reduced by up to 50%.

Water solubility analysis

All excess EGCG and EGCG glycosides were mixed in 200 μL of water in an Eppendorf tube, at room temperature. A 3510R-DTH ultrasonic cleaner (Branson, Danbury, CT, USA) was used to maximize solubility. After 1 h of sonication at room temperature, each of the samples was diluted, and then filtered through a 0.45 μm MFS membrane (Adventec, Pleasanton, CA, USA) for HPLC analysis, in order to determine the concentrations. A model 1525 HPLC system, connected to a 400 × 3.9 mm i.d. μ-Bondapak C_{18} column (Waters) and a model 2487 UV detector (Waters) at 280 nm were utilized in order to quantify the amounts of EGCG and EGCG glycosides. The mobile phase consisted of 23% methanol, and was conducted via the isocratic method, with a flow rate of 0.5 mL/min. The concentrations of the EGCG and EGCG glycosides were calculated as was described previously, by Li et al. (Li et al, 2004).

Tyrosinase inhibition effect

The reaction mixture (90 μl) contained 3.3 mM-3-(3,4-dihydroxyphenyl)-L-alanine (DOPA) in 330 mM phosphate buffer (pH 7) and enzyme in the presence or absence of inhibitors. Fifteen units of tyrosinase were used to determine the K_i value. The reaction mixtures in 96-well plates were incubated at 37°C for 10 min, and the absorbance was measured at 475 nm in a Benchmark microplate reader (BIO-RAD). One unit of enzyme activity was defined as the amount of enzyme which increased the absorbance value by 0.001 at 475 nm per min under the reaction condition described above.

UV irradiation and MMP-1 production test

Human newborn foreskin fibroblast (HS68 cells) was obtained from American Type Culture Collection (ATCC CRL 1635) (Rockville, MD, USA). The HS68 cells were plated in 100 mm tissue culture dishes and cultured in serum-free Dulbecco's modified Eagle's medium DMEM supplemented with 10% fetal bovine serum (FBS) and 1% antibiotic–antimycotic (Ho et al., 2005). The cells were maintained at 37°C under humidified atmosphere of 5% CO_2 condition. When cells reached above 80% confluence, subculture was conducted at a split ratio of 1:5. A UVB lamp was used as UVB source (312 nm, Spectroline Model EB-160C, New York, NY, USA). In brief, serum-starved confluent cells were rinsed twice with phosphate-buffered saline (PBS), and all irradiations were performed under a thin layer of PBS. Immediately after irradiation, fresh serum-free medium was added to the cells. Responses were measured after 24 h incubation. Prior to UVB irradiation (100 mJ/cm^2), the cells were pretreated with arbutin, arbutin glucoside (10 μM/ml). Epigallocatechin gallate (EGCG, 10 μM/ml) was also pretreated as a standard antioxidant. There were negative control without UVB exposure and positive control with UVB irradiation but no addition of antioxidant (Watanabe et al., 2003; Ho et al., 2005).

Enzyme-linked immunosolvent assay

Matrix metalloproteinase-1 (MMP-1) content was determined with an ELISA kit (TPI Inc., Lynnwood, WA, USA) according to the manufacturer's instructions. Each sample was analyzed in triplicate.

RESULTS AND DISCUSSION

Acceptor reaction of EGCG and purification of EGCG glycosides

After the acceptor reaction involving *L. mesenteroides* NRRL B-1299CB glucansucrase (600 U per reaction digest) with EGCG (500 mg) and sucrose (final 80 mM), we were able to identify six reaction products via HPLC purification (Fig. 1). The yields of EGCG-G1, EGCG-G2, EGCG-G3, EGCG-G4, EGCG-G5 and EGCG-G6 were as follows (respectively): 45.4 mg (9.1% of EGCG), 39.2 mg (7.8% of used EGCG), 99.5 mg (19.9%), 31.2 mg (6.2% of used EGCG), 43.5 (8.7%) mg, and 70.2 (14.1% of used EGCG) mg.

Structural determination of EGCG glycosides

The numbers of glucose units attached to the purified each glycoside were confirmed via MALDI-TOF MS analysis. The molecular weights of the glycosides were greater than that of EGCG by exactly one glucose residue addition; EGCG-G1, EGCG-G4 and EGCG-G5 featured one attached glucose, EGCG-G2, EGCG-G3 and EGCG-G6 featured two attached glucoses. The glucosidic linkages were determined via ^{1}H, ^{13}C, ^{1}H-COSY, HSQC and HMBC analyses, and the results are summarized in Table 1 (Moon et al., 2006a, b).

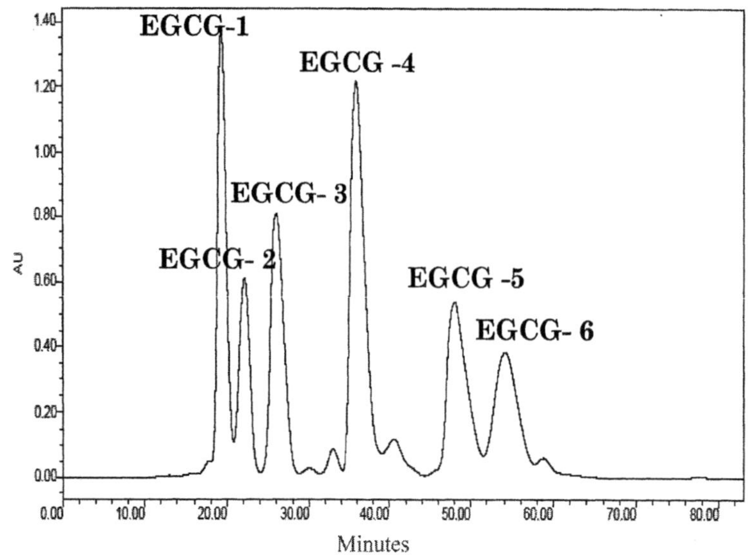

Figure 1 HPLC analysis of EGCG acceptor products after Sephadex LH-20 chromatography

Column, μ-Bondapak™ (C_{18}) (7.8 × 300 mm); Mobile phase, methanol: water (23:77, v/v, pH 2.6 with TFA); Flow rate, 3 mL/min; UV detector, 240 nm.

EGCG-G1

The molecular ions of EGCG-G1 were observed at *m/z* 643 $(M+Na)^+$. In Table 1, a doublet signal at 5.28 ppm (J = 3.5 Hz) was assigned to the anomeric proton, showing that only one glucosyl residue is α-linked to the EGCG. Almost all of the carbon signals assigned to the EGCG moiety were identical to those of EGCG, except for the assignment of signals at 100.9 ppm to C-4a, 96.4 ppm to C-6, and 97.5 ppm to C-8. These signals showed downfield shifts of 3.3, 1.9, and 1.8 ppm, indicating that the transferred glucosyl residue had been attached to C-7 in the EGCG. In our HMBC data, the H-6 was observed at 6.18 ppm, and the couplings were occurred with C-4a, C-5, C-7, and C-8 of A ring and the H-4 was observed at 3.01 and 2.73 ppm, and the couplings were occurred with C-5 and C-8a of A ring (Fig. 2A). These correlations made carbon signals (C-5, C-7, and C-8a) of A ring assigned precisely. After that, the C-1‴of the glucosyl residue was observed at 98.6 ppm, and the coupling appeared to have occurred between proton H-1‴ of the glucosyl residue, and the C-7 of the EGCG. According to these results, we determined that the structure of EGCG-G1 could be most appropriately referred to as epigallocatechin gallate-7-*O*-α-D-glucopyranoside.

EGCG-G2

The molecular ions of EGCG-G2 were observed at *m/z* 805 $(M+Na)^+$ (Fig. 2B). In Table 1, the two doublet signals at 5.29 ppm (J = 3.5 Hz) and 5.09 ppm (J = 3 Hz) were assigned to anomeric protons, thereby indicating that the two glucosyl residues had been connected to EGCG via α-linkages. All of the carbon signals assigned to the EGCG

Table 1 ^{13}C and ^{1}H NMR data of EGCG and EGCG glycosides (units: ppm)

Carbon Position	EGCG (δ)		EGCG-G1 (δ_1)			EGCG-G2 (δ_2)			EGCG-G3 (δ_3)		
	δ_C	δ_H	δ_1c	(δ_1–δ)	δ_H	δ_2c	(δ_2–δ)	δ_H	δ_3c	(δ_3–δ)	δ_H
EGCG											
2	76.6	4.95 (br s)	77.1	(0.5)	5.02 (br s)	76.9	(0.3)	5.05 (br s)	76.9	(0.3)	5.08 (br s)
3	68.2	5.36 (br s)	68.3	(0.1)	5.41 (br s)	68.9	(0.5)	5.43 (br s)	68.1	(-0.1)	5.45 (br s)
4	25.9	2.92 (dd) (*J* = 4.5, 17 Hz) 2.65 (br d) (*J* = 17 Hz)	26.3	(0.4)	3.01 (dd) (*J* = 4.5, 17 Hz) 2.73 (br d) (*J* = 17 Hz)	26.1	(0.2)	3.03 (dd) (*J* = 4.5, 17 Hz) 2.75 (br d) (*J* = 17 Hz)	26.3	(-0.4)	3.01 (dd) (*J* = 4.5, 17 Hz) 2.75 (br d) (*J* = 17 Hz)
4a	97.6		100.9	**(3.3)**		100.8	**(3.2)**		100.9	**(3.3)**	
5	155.8		156.0	(0.2)		155.9	(0.1)		155.8	(0.0)	
6	94.5	5.82 (d) (*J* = 2.5 Hz)	96.4	**(1.9)**	6.18 (d) (*J* = 2 Hz)	96.3	**(1.8)**	6.18 (d) (*J* = 2.5 Hz)	96.5	**(1.0)**	6.19 (d) (*J* = 2 Hz)
7	156.7		157.1	(0.4)		157.2	(0.5)		157.1	(0.4)	
8	95.7	5.93 (d) (*J* = 2.5 Hz)	97.5	**(1.8)**	6.23 (d) (*J* = 2 Hz)	97.6	**(1.9)**	6.23 (d) (*J* = 2.5 Hz)	97.6	**(1.9)**	6.24 (d) (*J* = 2 Hz)
8a	156.7		156.9	(0.2)		156.9	(0.2)		156.9	(0.2)	
1′	128.8		128.9	(0.1)		128.8	(0.0)		135.3	**(6.5)**	
2′/6′	105.7	6.40 (s)	105.9	(0.2)	6.45 (s)	105.8	(0.1)	6.45 (s)	105.9	(0.2)	6.53 (s)
3′/5′	145.6		146.1	(0.5)		146.2	(0.6)		150.7	**(5.1)**	
4′	132.5		132.8	(0.3)		132.8	(0.3)		133.8	(1.3)	
1″	119.5		119.6	(0.1)		125.9	**(6.4)**		119.5	(0.0)	
2″/6″	108.8	6.81 (s)	109.1	(0.3)	6.84 (s)	108.9	(0.1)	6.84 (s)	109.1	(0.1)	6.84 (s)
3″/5″	145.8		145.9	(0.1)		151.2	**(5.4)**		145.9	(0.1)	
4″	138.7		139.0	(0.3)		138.3	(-0.4)		139.1	(0.4)	
COO-	165.4		165.7	(0.3)		165.3	(-0.1)		165.6	(0.2)	
Glucose (-G1/-G5)											
1‴			98.6		5.28 (d) (*J* = 3.5 Hz)	98.6		5.29 (d) (*J* = 3.5 Hz)	98.6		5.29 (d) (*J* = 4 Hz)
2‴			72.1		3.35 (dd) (*J* = 3.5, 9.5 Hz)	72.1		3.35 (dd) (*J* = 3.5, 9 Hz)	72.1		3.36 (dd) (*J* = 3.5, 9.5 Hz)
3‴			73.5		3.62 (m)	73.5		3.61 (m)	73.5		3.62 (m)
4‴			70.3		3.21 (dd) (*J* = 9.5, 9.5 Hz)	70.3		3.23 (dd) (*J* = 9, 9 Hz)	70.3		3.23 (m)
5‴			74.1		3.51 (m)	74.1		3.47 (m)	74.1		3.50 (m)
6‴			61.1		3.60/3.54 (m)	60.1		3.58/3.60 (m)	61.1		3.60/3.54 (m)
Glucose (-G2)											
1⁗						103.4		5.09 (d) (*J* = 3 Hz)	104.3		4.87 (d) (*J* = 3.5 Hz)
2⁗						72.3		3.39 (dd) (*J* = 3, 9.5 Hz)	72.2		3.41 (dd) (*J* = 3.5, 9.5 Hz)
3⁗						73.5		3.64 (m)	73.5		3.62 (m)
4⁗						69.7		3.28 (dd) (*J* = 9.5, 9.5 Hz)	69.7		3.26 (m)
5⁗						74.3		3.49 (m)	74.3		3.53 (m)
6⁗						60.5		3.54/3.61 (m)	60.5		3.60/3.63 (m)

Table 1 Continued

Carbon Position	EGCG (δ)		EGCG-G4 (δ_4)			EGCG-G5 (δ_5)			EGCG-G6 (δ_6)		
	δ_C	δ_H	δ_4c	(δ_4–δ)	δ_H	δ_5c	(δ_5–δ)	δ_H	δ_6c	(δ_6–δ)	δ_H
EGCG											
2	76.6	4.95 (br s)	76.3	(-0.3)	4.97 (br s)	76.3	(-0.3)	5.04 (br s)	76.6	(0.0)	5.04 (br s)
3	68.2	5.36 (br s)	68.7	(0.5)	5.38 (br s)	67.9	(-0.3)	5.42 (br s)	69.1	(0.9)	5.40 (br s)
4	25.9	2.92 (dd) (*J* = 4.5, 17 Hz) 2.65 (br d) (*J* = 17 Hz)	25.6	(-0.3)	2.94 (dd) (*J* = 4, 17 Hz) 2.68 (br d) (*J* = 17 Hz)	25.7	(-0.2)	2.96 (dd) (*J* = 4.5, 17 Hz) 2.69 (br d) (*J* = 17 Hz)	26.1	(0.2)	2.95 (dd) (*J* = 4, 17 Hz) 2.69 (br d) (*J* = 17 Hz)
4a	97.6		97.2	(-0.4)		97.4	(-0.2)		97.6	(0.0)	
5	155.8		155.6	(-0.2)		155.4	(-0.4)		155.8	(0.0)	
6	94.5	5.82 (d) (*J* = 2.5 Hz)	94.3	(-0.2)	5.83 (d) (*J* = 2 Hz)	94.4	(-0.1)	5.87 (d) (*J* = 2 Hz)	94.8	(0.3)	5.86 (d) (*J* = 2 Hz)
7	156.7		156.6	(-0.1)		156.6	(-0.1)		157.1	(0.4)	
8	95.7	5.93 (d) (*J* = 2.5 Hz)	95.6	(-0.1)	5.93 (d) (*J* = 2 Hz)	95.6	(-0.1)	5.97 (d) (*J* = 2 Hz)	96.1	(0.4)	5.96 (d) (*J* = 2 Hz)
8a	156.7		156.5	(-0.2)		156.5	(-0.2)		156.9	(0.2)	
1′	128.8		128.6	(-0.2)		135.1	**(6.3)**		135.4	**(6.6)**	
2′/6′	105.7	6.40 (s)	105.3	(-0.4)	6.40 (s)	105.5	(-0.2)	6.51 (s)	105.8	(0.1)	6.49 (s)
3′/5′	145.6		145.7	(0.1)		150.2	**(4.6)**		150.7	**(5.1)**	
4′	132.5		132.4	(-0.1)		133.3	(0.8)		133.7	(1.2)	
1″	119.5		125.5	**(6.0)**		119.2	(-0.3)		125.8	**(6.3)**	
2″/6″	108.8	6.81 (s)	108.4	(-0.4)	6.81 (s)	108.7	(-0.1)	6.84 (s)	108.9	(0.1)	6.83 (s)
3″/5″	145.8		150.8	**(5.0)**		145.5	(-0.3)		151.3	**(5.5)**	
4″	138.7		137.8	(-0.9)		138.6	(-0.1)		138.3	(-0.4)	
COO-	165.4		164.8	(-0.6)		165.1	(-0.3)		165.3	(-0.1)	
Glucose (-G1/-G5)											
1‴			103.0		5.05 (d) (*J* = 4 Hz)	104.0		4.87 (d) (*J* = 4 Hz)	103.4		5.07 (d) (*J* = 3.5 Hz)
2‴			71.8		3.37 (dd) (*J* = 4, 9.5 Hz)	71.8		3.41 (dd) (*J* = 4, 9.5 Hz)	72.3		3.38 (m)
3‴			73.1		3.63 (m)	73.2		3.62 (m)	73.5		3.61 (m)
4‴			69.2		3.25 (dd) (*J* = 9.5, 9.5 Hz)	69.3		3.21 (dd) (*J* = 9.5, 9.5 Hz)	69.7		3.25 (m)
5‴			73.8		3.97 (m)	73.9		3.98 (m)	74.2		3.96 (m)
6‴			60.1		3.42/3.43 (m)	60.1		3.60/3.54 (m)	60.5		3.56/3.60 (m)
Glucose (-G2)											
1⁗									104.3		4.86 (d) (*J* = 3.5 Hz)
2⁗									72.2		3.38 (m)
3⁗									73.6		3.61 (m)
4⁗									69.6		3.25 (m)
5⁗									74.2		3.96 (m)
6⁗									60.5		3.59/3.63 (m)

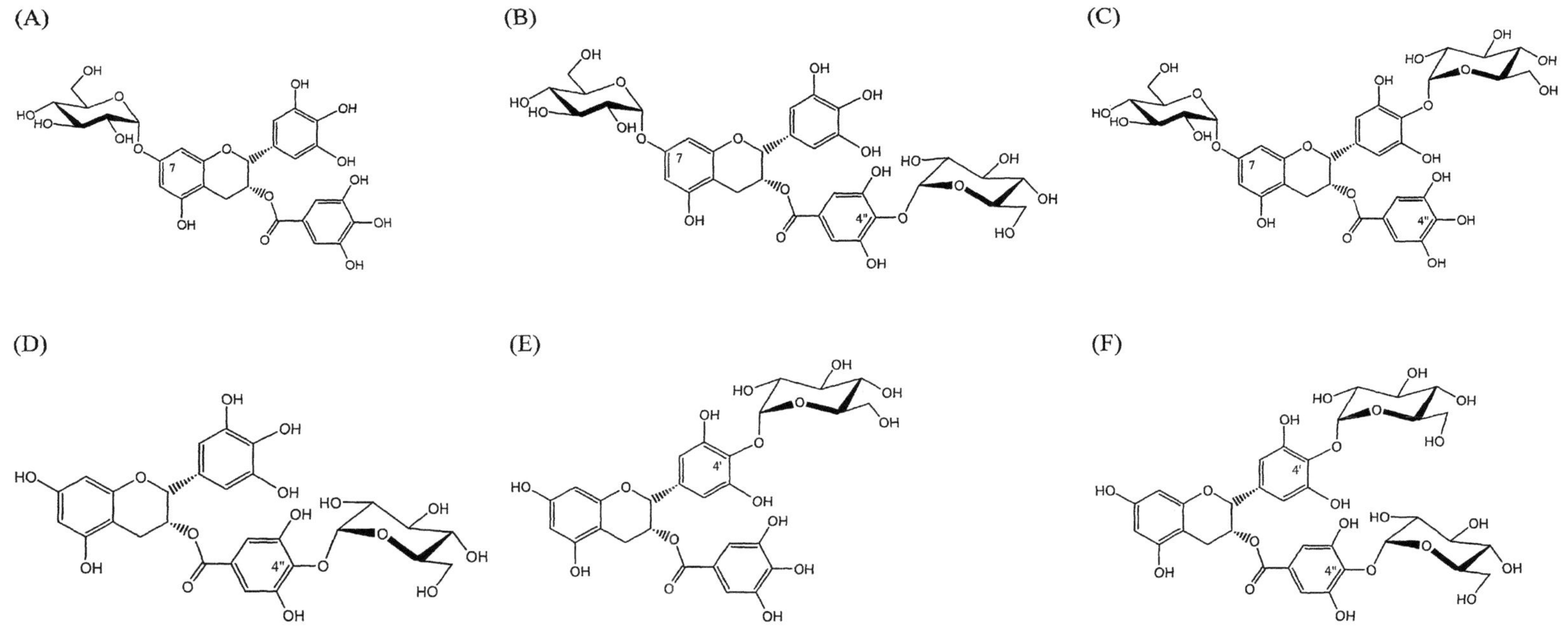

Figure 2 The structures of EGCG glucosides
(A) EGCG-G1, (B) EGCG-G2, (C) EGCG-G3, (D) EGCG-G4, (E) EGCG-G5, and (F) EGCG-G6

moiety were almost identical to those of EGCG, with the exception of the assignment of signals at 100.8 ppm to C-4a, 96.3 ppm to C-6, 97.6 ppm to C-8, 125.9 ppm to C-1", and 151.2 ppm to C-3"/5". These signals evidenced downfield shifts of 3.2, 1.8, 1.9, 6.4 and 5.4 ppm, respectively, and revealed that the two glucoses had been attached to the C-7 and C-4" of EGCG, respectively. According to the HMBC data, these couplings occurred between proton H-1‴ of the glucosyl residue and C-7 of the EGCG, and also between proton H-1⁗ of the remaining glucosyl residue and C-4" of the EGCG residue. These results demonstrate that two α- glucosidic linkages (α-1→7 and α-1→4") were formed during the acceptor reaction, and that EGCG-G2 could appropriately be referred to as epigallocatechin gallate-7,4"-*O*-α-D-glucopyranoside.

EGCG-G3

The molecular ions of EGCG-G2 were observed at *m/z* 805 $(M+Na)^+$. In Table 1, two doublet signals at 4.87 ppm (J = 3.5 Hz) and 5.29 ppm (J = 4 Hz) were assigned to the anomeric protons, thereby indicating that two glucosyl residues were α-linked to EGCG. Almost all of the carbon signals assigned to the EGCG moiety were identical to those of EGCG, except for the assignment of signals at 100.9 ppm to C-4a, 96.5 ppm to C-6, 97.6 ppm to C-8, 135.3 ppm to C-1′, and 150.7 ppm to C-3′/5′. These signals evidenced downfield shifts of 3.3, 1.0, 1.9, 6.5 and 5.1 ppm, respectively, and indicate that the two glucoses had been attached to the C-7 and C-4′ of EGCG, respectively. Comparing the carbon signals of EGCG-G1 and EGCG-G3, the carbon signals of EGCG-G3 were almost same in their assignments. According to our HMBC data, the carbon signals of A ring of EGCG-G3 were shown as EGCG-G1 and EGCG-G3. Furthermore, the couplings appeared to have occurred between the H-1‴ proton of the glucosyl residue and the C-7 of EGCG, and also between the H-1⁗ proton of the other glucosyl residue, and the C-4′ of EGCG (Fig. 2C). These results indicate that two α- glucosidic linkages (α-1→7 and α-1→4′) were formed during the acceptor reaction, and that EGCG-G3 should appropriately be referred to as epigallocatechin gallate-7,4′-*O*-α-D-glucopyranoside.

EGCG-G4

The molecular ions of EGCG-G4 were observed at *m/z* 643 $(M+Na)^+$ (Fig. 2D). In Table 1, a doublet signal at 5.05 ppm (J = 4 Hz) was assigned to the anomeric proton, thereby indicating that only one glucosyl residue was connected to EGCG by an α-linkage. All of the carbon signals assigned to the EGCG moiety were almost identical to those of EGCG, with the exception of the assignment of signals at 125.5 ppm to C-1" and 150.8 ppm to C-3"/5". These signals evidenced downfield shifts of 6.0 and 5.0 ppm, thereby indicating that the transferred glucosyl residue had attached to C-4" in the EGCG residue. According to our HMBC data, the C-4″ of the glucosyl residue was observed at 137.8 ppm, and the coupling occurred between proton H-1‴ of the glucosyl residue and C-4" of the EGCG. According to these results, the structure of EGCG-G4 was designated as epigallocatechin gallate-4"-*O*-α-D-glucopyranoside.

EGCG-G5

The molecular ions of EGCG-G5 were observed at *m/z* 643 $(M+Na)^+$. In Table 1, a doublet signal at 4.87 ppm (J = 4 Hz) was assigned to the anomeric proton, thereby showing that only one glucosyl residue was α-linked to EGCG, similarly to EGCG-G1. Almost all of the carbon signals assigned to the EGCG moiety were identical to those of EGCG, except for the assignment of signals at 135.1 ppm to C-1′, and at 150.2 ppm to C-3′/5′. These signals evidenced downfield shifts of 6.3 and 4.6 ppm, respectively, demonstrating that the transferred glucosyl residue had been attached to C-4′ in the EGCG. In our HMBC data, the carbon signals of A ring of EGCG-G5 were shown as EGCG-G1. Furthermore, the C-1‴ of the glucosyl residue was observed at 104.0 ppm, and the coupling appeared to have occurred between the H-1‴ proton of the glucosyl residue and the C-4′ of the EGCG (Fig. 2E).

EGCG-G6

The molecular ions of EGCG-G6 were observed at *m/z* 805 $(M+Na)^+$ (Fig. 2F). In Table 1, the two doublet signals at 5.07 ppm (J = 3.5 Hz) and 4.86 ppm (J = 3.5 Hz) were assigned to anomeric protons, thereby indicating that the two glucosyl residues had been connected to EGCG via α-linkages. All of the carbon signals assigned to the EGCG moiety were almost identical to those of EGCG, with the exception of the assignment of signals at 135.4 ppm to C-1′, 150.7 ppm to C-3′/5′, 125.8 ppm to C-1", and 151.3 ppm to C-3"/5". These signals evidenced downfield shifts of 6.6, 5.1, 6.3 and 5.5 ppm, respectively, and indicated that the two glucoses had been attached to the C-4′ and C-4" of EGCG, respectively. According to the HMBC data, the couplings occurred between proton H-1‴ of the glucosyl residue and C-4′ of the EGCG, and also between proton H-1⁗ of the remaining glucosyl residue and the C-4" of the EGCG residue. These results demonstrate that two α- glucosidic linkages (α-1→4′ and α-1→4") were formed during the acceptor reaction, and that EGCG-G6 could appropriately be referred to as epigallocatechin gallate-4′,4"-*O*-α-D-glucopyranoside.

Three EGCG glycosides (EGCG-G1, EGCG-G2, EGCG-G4) were identified and reported for the first time in this study, and EGCG-G6 has been previously reported in the study of Kitao *et al* (Kitao et al., 1995), using the sucrose phosphorylase from *L. mesenteroides*. In that study, Kitao et al. synthesized two EGCG glycosides, epigallocatechin gallate-4′-*O*-α-D-glucopyranoside and epigallocatechin gallate-4′,4″-*O*-α-D-glucopyranoside. Both of these glycosides were formed via the attachment of glucose to the B ring or to the galloyl residue. Catechin glycoside has also been previously reported to have been synthesized via the attachment of glucose to the B ring, by an α-linkage (Kitao et al., 1993; Sato et al., 2000). EGCG-G3 was reported previously by Nanjo etc. (Nanjo et al., 1995) who used the cyclodextrin glucanotransferase of *B. stearothermophilus*. EGCG-G5 was also reported by Kitao et al. (Kitao et al., 1995) who used the sucrose phosphorylase from *L. mesenteroides* in their study. That group synthesized two EGCG glycosides, epigallocatechin gallate-4′-*O*-α-D-glucopyranoside and epigallocatechin gallate-4′,4″-*O*-α-D-glucopyranoside. Both of the glycosides were formed via the attachment of glucose to the B ring or to the gallate residue. Catechin glycoside was also formed via the attachment of glucose to the B ring, as an α-linkage (Kitao et al., 1993; Sato et al., 2000). However, the current study is the first to involve the attachment of the glucosyl residue to the A ring or the galloyl residue of EGCG via an α-linkage.

Synthesis and purification of arbutin glucosides

After the acceptor reaction involving *L. mesenteroides* NRRL B-1299CB glucansucrase (final activity of 2 U/ml) with arbutin (final concentration of 12.25 mM, 200 mg) and sucrose (final concentration of 100 mM), we were able to identify two reaction products via TLC analysis. The reaction digest was applied to Bio-Gel P-2 column chromatography. The eluent was fractionated into two groups (Arbutin-glucoside and Arbutin-isomaltoside groups) by confirming each fraction using TLC (data not shown). The two groups were subjected to HPLC, respectively. According to the HPLC chromatogram, arbutin-glucoside could be purified by Bio-Gel P-2 chromatography, but arbutin-isomaltoside was fractionated and purified using repeated HPLC. The yields of arbutin-G1 and arbutin-G2 were as follows, respectively (%, a molar ratio): 80 mg (23.8% of total arbutin in reaction digest), 15 mg (3.3% of total arbutin in reaction digest), using 8.2-fold (sucrose/acceptor ratio) amount of sugar as a donor substrate. The reaction products were obtained as a white powder.

Structural determination of arbutin glucosides

The numbers of glucose units attached to the purified arbutin-glucoside and arbutin-isomaltoside were confirmed via MALDI-TOF MS analysis. The molecular weights of the glycosides confirmed one glucose residues addition for arbutin-glucoside, and two attached glucoses for arbutin-isomaltoside. The glucosidic linkages were determined by ^{1}H, ^{13}C, ^{1}H-^{1}H COSY, HSQC and HMBC analyses, and these results are summarized in Table 2 (Moon et al., 2007).

Arbutin-glucoside

The molecular ions of arbutin-G1 were observed at *m/z* 457 ($C_{18}H_{26}O_{12}Na$). In Table 2, a doublet signal at 4.89 ppm (J = 7.0 Hz, β-linkage) was assigned to the anomeric proton of the glucosyl residue attached to the aromatic ring of arbutin by HMBC (data not shown). The other doublet signal at 4.83 ppm (J = 3.5 Hz, α-linkage) was assigned to the anomeric proton, thereby indicating that one glucosyl residue was α-linked to arbutin. All of the carbon signals assigned to the arbutin moiety were almost negative to those of arbutin comparing the chemical shift (δ_1-δ), with the exception of the assignment of signals at 65.55 ppm to C-6'. This signal evidenced downfield shift of 2.15 ppm, thereby indicating that the transferred glucosyl residue had attached to C-6' in the glucosyl residue of arbutin. According to our HMBC data, the C-6' of the glucosyl residue was observed at 65.55 ppm, and the coupling occurred between proton H-1" of the glucosyl residue and C-6' of the arbutin. According to these results, the structure of arbutin-G1 was designated as 4-hydroxyphenyl α-glucoside (Fig. 3B; Moon et al., 2007).

Arbutin-isomaltoside

The molecular ions of arbutin-G2 were observed at *m/z* 619 $(C_{24}H_{36}O_{17}Na)^+$. In Table 2, a doublet signal at 4.94 ppm (J = 7.0 Hz, β-linkage) was assigned to the anomeric proton of the glucosyl residue of arbutin as like an arbutin-G1. Two doublet signals at 4.86 ppm (J = 3.5 Hz) and 4.81 ppm (J = 3.5 Hz) were assigned to the anomeric protons, thereby indicating that two glucosyl residues were alpha-linked to arbutin. Comparing

Table 2 ^{13}C and ^{1}H NMR data of arbutin and arbutin glucosides (units: ppm)

	Arbutin (δ)	Arbutin-glucoside (δ_1)			Arbutin-isomaltoside (δ_2)		
	δ_C	$\delta_1 c$	$(\delta_1 - \delta)$	δ_H	$\delta_2 c$	$(\delta_2 - \delta)$	δ_H
C-Aromatic							
1	153.12	151.19	(-1.93)		151.21	(-1.91)	
2, 6	121.29	118.27	(-3.02)	6.94 (dt) (J = 4.0, 10 Hz)	118.11	(-0.16)	6.96 (dt) (J = 4.0, 10 Hz)
3, 5	119.08	116.24	(-2.84)	6.76 (dt) (J = 4.0, 10 Hz)	116.28	(+0.04)	6.78 (dt) (J = 4.0, 10 Hz)
4	153.26	150.27	(-2.99)		150.20	(-3.06)	
C-Glu							
1′	104.19	101.09	(-3.10)	4.89 (d) (J = 7.0 Hz)	100.87	(-3.32)	4.94 (d) (J = 7.0 Hz)
2′	75.83	71.40	(-4.43)	3.46 (m)	71.46	(-4.37)	3.48 (m)
3′	78.43	72.91	(-5.52)	3.45 (m)	72.91	(-5.52)	3.47 (m)
4′	72.30	69.30	(-3.00)	3.49 (m)	69.43	(-2.87)	3.48 (m)
5′	78.91	74.33	(-4.58)	3.66 (m)	74.32	(-6.59)	3.71 (m)
6′	63.40	65.55	**(+2.15)**	3.85 (dd) (J = 5, 10 Hz) 3.68 (m)	65.83	**(+2.43)**	3.86 (dd) (J = 5, 10 Hz) 3.73 (m)
1″		97.81		4.83 (d) (J = 3.5 Hz)	97.82		4.86 (d) (J = 3.5 Hz)
2″		71.74		3.55 (m)	70.22		3.76 (m)
3″		73.10		3.63 (m)	71.31		3.47 (m)
4″		69.37		3.31 (dd) (J = 9.5, 9.5 Hz)	69.37		3.43 (dd) (J = 9.5, 9.5 Hz)
5″		75.72		3.50 (m)	73.36		3.63 (m)
6″		60.32		3.61/3.67 (m)	65.29	**(+4.97)**	3.56/3.78 (m)
1‴					97.67		4.81 (d) (J = 3.5 Hz)
2‴					71.73		3.60 (m)
3‴					73.07		3.63 (m)
4‴					69.43		3.33 (dd) (J = 9.5, 9.5 Hz)
5‴					75.73		3.51 (m)
6‴					60.39		3.69/3.75 (m)

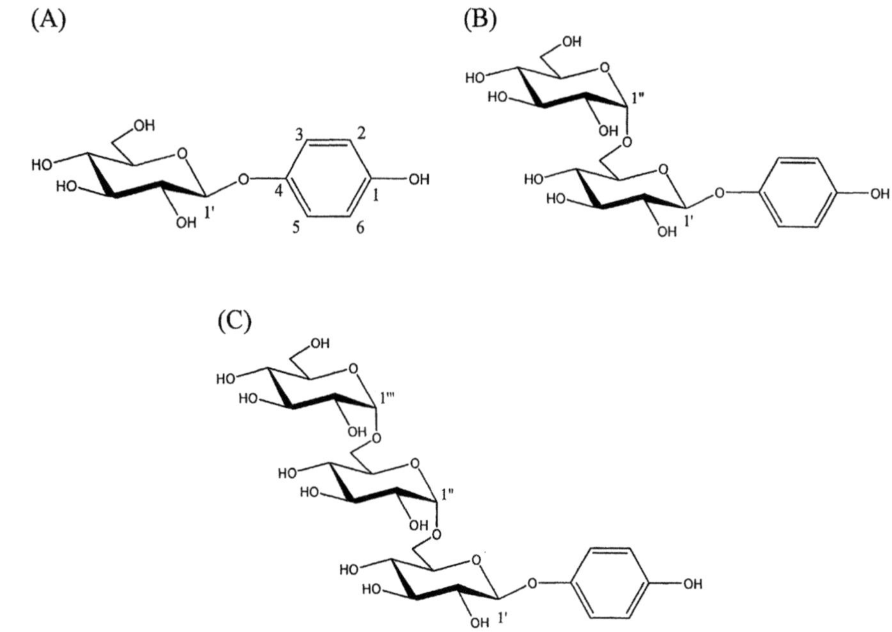

Figure 3 The structures of arbutin (A), arbutin-glucoside (B), and arbutin-isomaltoside (C)

the chemical shift (δ_I-δ) in Table 2, the signal at 65.83 ppm to C-6' and 65.29 ppm to C-6" evidenced downfield shift of 2.43 ppm and 4.97 ppm. These indicate that one glucosyl residue had attached to C-6' of the glucose residue of arbutin and the other glucosyl residue to C-6". According to our HMBC data, the C-6' of the glucosyl residue was observed at 65.83 ppm, and the coupling occurred between proton H-1" of the glucosyl residue and C-6' of the arbutin. Furthermore, the C-6" of the glucosyl residue was observed at 65.29 ppm, and the coupling occurred between proton H-1'" of the glucosyl residue and C-6" of the arbutin. According to these results, the structure of arbutin-isomaltoside was designated as 4-hydroxyphenyl α-isomaltotrioside (Fig. 3C).

The glucosides were the first time reported compounds; 4-hydroxyphenyl α-isomaltoside and 4-hydroxyphenyl α-isomaltotrioside. Previously, Sugimoto et al. synthesized arbutin glucosides (4-hydroxylphenyl β-maltoside and 4-hydroxylphenyl β-maltotrioside) of α-1,4 linkages to arbutin (or β-albutin) using cyclomaltodextrin glucanotransferase (CGTase) from *Bacillus macerans* with soluble starch as a substrate (Sugimoto et al., 2003). The yield of arbutin-glucoside and arbutin-isomaltoside using glucansucrase and sucrose was 27.1% based on arbutin conversion. The yield was lower than synthesis of 4-hydroxylphenyl β-maltoside and 4-hydroxylphenyl β-maltotrioside, approximately 70%, but higher than the yield of glycosylated epigallocatechin gallate (10-20% product yield) using glucansucrase and sucrose.

Properties of EGCG glycosides

Browning resistance effect

Catechin in water is susceptible to degradation and browning as the result of UV irradiation (Sato et al., 2000). Figure 4 shows the browning resistance of EGCG glycosides after UV irradiation. Epigallocatechin gallate (EGCG) resulted in ready and rapid browning. However, the UV-irradiation-associated browning of the EGCG glycosides [EGCG-1 (■), EGCG-2 (▲), EGCG-3 (△), EGCG-4 (×), EGCG-5 (●), EGCG-6 (□)] was quite slow, even after 24 hours [EGCG-G1 (17.5% browning that of EGCG), EGCG-G2 (17.5% of that of EGCG), EGCG-G3 (10.1% browning that of EGCG), EGCG-G4 (19.6% of that of EGCG), EGCG-G5 (15% browning that of EGCG), EGCG-G6 (19.5% of that of EGCG)]. The catechin glycoside (3′-*O*-α-D-glucopyranoside) synthesized via transglycosylation with sucrose phosphorylase also exhibited a browning-resistant quality (Kitao et al., 1993), thus indicating that the glycosylation of the compounds conferred a degree of stability on the EGCG (catechins) with regard to UV irradiation, even though the glycosylation positions and linkages differed between the resultant compounds.

Antioxidant activity

EGCG and its glycosides exhibited differing antioxidant effects, depending on their structural configurations. The SC_{50} of EGCG-G1, according to the results of DPPH radical scavenging measurements, was 1.1 μM, a level of activity comparable with that of purified EGCG (SC_{50} = 0.9 μM). However, the SC_{50} values for EGCG-G2, EGCG-G3, EGCG-G4 and EGCG-G5 were 3.2, 8, 2.1, and 2.8 μM, all lower than that of purified

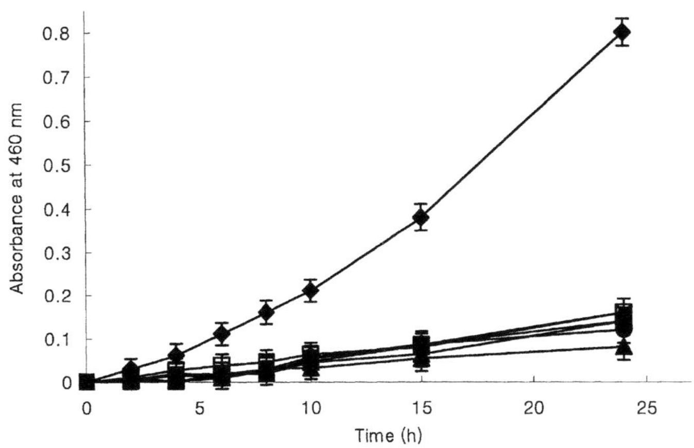

Figure 4 Browning resistance effects of EGCG and EGCG acceptor products
Each sample EGCG (◆), EGCG-1 (■), EGCG-2 (▲), EGCG-3 (△), EGCG-4 (×), EGCG-5 (●), EGCG-6 (□) was exposed to UV irradiation at a distance of 10 for 24 h at room temperature. Increases in absorbance at 460 nm were then determined. Each value is the mean ± standard deviation (*n*=3).

EGCG (Fig. 5). The glucosyl residue attached to the 7-hydroxyl moiety in the A ring exhibited a more profound antioxidant activity than did the EGCG-G5 in which the glucosyl residue was attached to the 4′-hydroxyl moiety in the B ring. Thus, the 4′-hydroxyl moiety in the B ring and the 4"-hydroxyl moiety at the galloyl residue must play an important role in DPPH radical scavenging, whereas the 7-hydroxyl moiety in the A ring has a lesser influence on antioxidant activity, as was previously concluded by Valcic et al. (1999). According to Nanjo et al. (1995), SC_{50} of EGCG, epigallocatechin gallate-4'-*O*-α-D-glucoside, epigallocatechin gallate-3'-*O*-α-D-glucoside, epigallocatechin gallate-7,3'-*O*-α-D-glucoside, and epigallocatechin gallate-4',4"-*O*-α-D-glucoside were 1.2 or 1.8 μM, 1.8 μM, 1.0 or 9.9 μM, 1.2 or 4.7 μM, and 22 μM, respectively. These values showed similar pattern with our results. From these results, we confirmed that *ortho*-trihydroxyl group in the B ring and the galloyl moiety is the most important structural properties for scavenging activity on the DPPH radicals.

For DPPH radical scavenging activity, arbutin and arbutin glucoside were measured at concentrations from 10 to 500 μM as shown in Figure 6. Antioxidant activity of arbutin glucoside was similar with that of arbutin with up to 50 μM. At 100-500 μM, antioxidant activity of arbutin glycoside was 16-20% lower than that of arbutin. This result indicates that attachment of glucose on arbutin decreased its antioxidant activity.

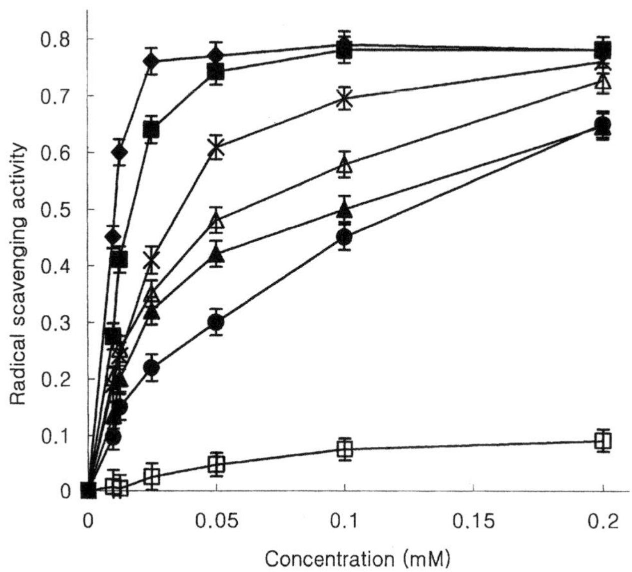

Figure 5 Antioxidant activities of EGCG and EGCG acceptor products
Each sample (30 μL of 10, 12.5, 25, 50, 100, or 200 μM), EGCG (◆), EGCG-1 (■), EGCG-2 (▲), EGCG-3 (△), EGCG-4 (×), EGCG-5 (●), EGCG-6 (□) was mixed with a 100 μM 1,1-diphenyl-2-picrylhydrazyl (270 μL) in darkness at room temperature for 10 minutes and the absorbance was monitored at 517 nm. Each value is the mean ± standard deviation (n=3).

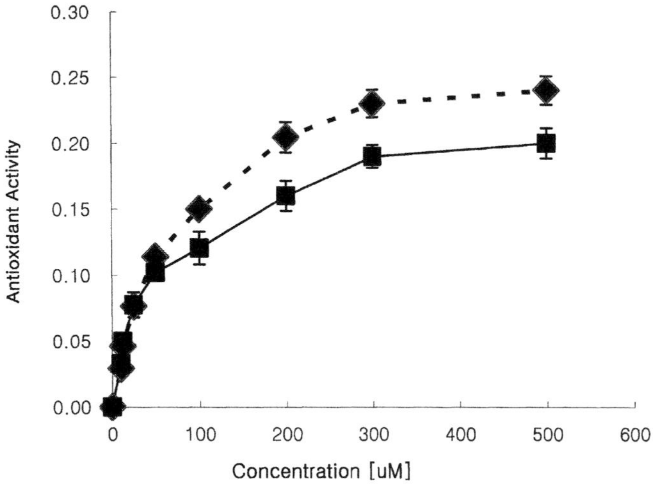

Figure 6 DPPH radical-scavenging activity of arbutin and arbutin glycoside
Samples (30 μL of 10, 12.5, 25, 50, 100, 200, 300, or 500 μM), arbutin (◆), arbutin-G1 (■) were mixed with 100 μM 1,1-diphenyl-2-picrylhydrazyl (270 μL) in darkness at room temperature for 10 min, and the absorbance was monitored at 517 nm. Each value is the mean ± standard deviation (n = 3).

Effects of glycosylation on water solubility

The water solubility of each of the EGCG glycosides was compared to that of EGCG. The solubility of EGCG was 5 mM, whereas the solubility of EGCG-G4, EGCG-G2, EGCG-G6, EGCG-G1, EGCG-G5, and EGCG-G3 were 253, 664.3, 584, 362.8, 281, and 1638.7 mM corresponding to 49, 130, 114, 71, 55, and 125 times the solubility of EGCG, respectively (Table 3). Kitao et al. also reported a 50-fold increase in the solubility of glycosylated catechin (3′-*O*-α-D-glucopyranoside), as compared to purified catechin (Kitao et al., 1993).

Tyrosinase inhibition activity

Tyrosinase inhibitor has important use in the cosmetic industry, as an essential component of skin whitening agent. In this aspect, a great deal of effort has been expended in the search for newly synthesized tyrosinase inhibitors, such as the arbutin derivatives (Sugimoto et al., 2003, 2005). The tyrosine-inhibitory effects exerted by arbutin and arbutin-G1. The type of inhibition for arbutin and arbutin-G1 was identified as competitive type. The K_i values of arbutin and arbutin-G1 was 2.8 mM and 3.7 mM, respectively. In previous studies, arbutin was reported to show a dose-dependent inhibitory effect on the oxidation of L-DOPA catalyzed by mushroom tyrosinase with IC50 of 8.4 mM (Funayama et al., 1995), 24 mM (Hori et al., 2004), 5.4 mM (Jin et al., 1999) and this inhibition was described as competitive (Tomita et al., 1990; Jin et al., 1999) or noncompetitive inhibition (Funayama et al., 1995). Besides, the transglyco-

Table 3 Water solubility of EGCG and EGCG glucosides

Samples	Solubility in water (mM)[a]	Relative solubility
EGCG	5.09 ± 2.54	1
EGCG-1	253.92 ± 6.53	49
EGCG-2	644.42 ± 10.83	126
EGCG-3	584.39 ± 12.65	114
EGCG-4	351.66 ± 7.66	69
EGCG-5	281.85 ± 7.30	55
EGCG-6	623.09 ± 8.33	122

[a]Mean ± standard deviation (n=3); Moon et al., 2006a, b

sylated products of arbutin, 4-hydroxy-phenyl β-maltoside and 4-hydroxy-phenyl β-maltotrioside exhibited stronger inhibitory activities than arbutin against human tyrosinase which origin is different with our study (mushroom). The inhibitory activities of 4-hydroxy-phenyl α-maltoside and 4-hydroxy-phenyl α-maltotrioside were weakened by transglycosylation.

These arbutin-α-D-glucopyranosides have shown to exhibit stronger inhibition effects on human tyrosinase than arbutin. Sugimoto et al. (2003, 2005) also synthesized α-arbutin using α-amylase from *Bacillus* sp. Strain X-23 with hydroquinone and maltopentaose. This α-arbutin showed stronger inhibitory activity than β-arbutin and arbutin glucosides with α-1,4 linkages. Based on those previous results, we tried to synthesize arbutin glucosides of α-1,6 linkages (albutin-G1, 4-hydroxyphenyl β-isomaltoside) using sucrose as a cheap substrate. From current experiment, it is possible that better tyrosinase inhibitor can be α-arbutin glycosides having α-1,6 linked. Thus the synthesis of α-arbutin glucosides with α-1,6 linkages using B-1299CB glucansucrase, hydroquinone and sucrose is in progress. Also, the relationship between the inhibitory effects and physicochemical properties of inhibitors is one of the important subjects to understand catalytic mechanism of tyrosinase, and this study is also in progress.

UV irradiation and MMP-1 production test

Collagen accounts for roughly 90% of the protein in human dermis, and collagen alterations have been considered to be a primary cause of skin aging and wrinkle formation. Furthermore, collagen is crucial material during connective tissue remodeling, e.g., wound healing and fibrosis. The matrix metalloproteinases (MMPs) are a large family of zinc-dependent endo-proteases with a broad range of substrate specificities, and are capable of degrading all extracellular matrix proteins. MMP-1, interstitial collagenases, initiate the degradation of type I and III collagens, and it has been established that single or repeated exposure to UV reduces type I procollagen levels and increases MMP-1 levels in human skin *in vivo* (Watanabe et al., 2004). The cells were pretreated with arbutin, arbutin-G1, or EGCG (10 μM/ml) prior to UVB irradiation (100 mj/cm^2) and harvested 24 h later. MMP-1 production was determined by ELISA. MMP-1 content was normalized to negative control as 100% and to positive control as 200%. Arbutin-G1 (10 μM/mL) showed lower amount of MMP-1 (163%), compared to arbutin

(184%). EGCG showed the highest inhibitory effect with 143% MMP-1 content among three samples. These results indicate that arbutin glucoside showed slightly more inhibitory effect compare to arbutin on MMP-1 production (Ho et al., 2005).

CONCLUSION

In this paper, we report the enzymatic synthesis of a set of novel EGCG and arbutin glycosides using the glucansucrase isolated from *Leuconostoc mesenteroides* B-1299CB. These EGCG glycosides exhibited slower antioxidant activity rates than that of EGCG, but also manifested more profound browning resistance, and far higher water solubility. Arbutin glucoside (4-hydroxyphenyl β-isomaltoside) also exhibited slower effects on DPPH radical scavenging effects than arbutin and almost similar effects on tyrosinase inhibition. However, arbutin glycoside showed better inhibitory effect than arbutin on MMP-1 production induced by UVB. The attachment of a glucosyl residue to EGCG results in an increase in the water solubility of the EGCG glycosides over that of EGCG, and that the number of attached glucosyl residues constitutes an important factor with regard to water solubility. Because EGCG has been the focus of great interest for its bioavailability, the EGCG glycosides should be expected to eventually be useful as materials for use in food additives and cosmetics. For EGCG glycosides, however, the bioavailability such as absorption or antioxidant activity in plasma and tissue has not been carried out. Thus, further study regarding the bioavailability of the EGCG glycosides are in progress as *in vivo*. Arbutin glucoside (albutin-G1) exhibited slower effects on DPPH radical scavenging and similar effects on tyrosinase inhibition, but slightly increased inhibitory effect on MMP-1 production induced by UVB than arbutin.

ACKNOWLEDGEMENTS

This work was supported in part by the Korea Research Foundation Grant funded by the Korean Government (MOEHRD) (KRF-2007-412-J02002). We would also like to express our gratitude to the Korea Basic Science Institute, Gwangju Branch for NMR analysis.

REFERENCES

Abe N, Nemoto A, Tsuchiya Y, Hojo H, and Hirota A (2000), 'Studies of the 1,1-diphenyl-2-picrylhydrazyl radical scavenging mechanism for a 2-pyrone compound', *Biosci Biotech Biochem*, 64, 306-333.

Bushman J L (1998), 'Green tea and cancer in humans: a review of the literature', *Nutr Cancer*, 31, 151-159.

Cabrera C, Gimenez R, and Lopez M C (2003), 'Determination of tea components with antioxidant activity', *J Agric Food Chem*, 51, 4427-4435.

Funayama M, Arakawa H, Yamamoto R, Nishino T, Shin T, and Murao S (1995), 'Effects of alpha and beta arbutin on activity of tyrosinase from mushroom and mouse melanoma', *Biosci Biotechnol Biochem*, 59, 143-147.

Gilly R S, Oded S, Itzhak B, and Zohar K (2003), 'Glycosylation of resveratrol protects it from enzymic oxidation', *Biochem J*, 374, 157-163.

Ho J N, Lee Y H, Park J S, Jun W J, Kim H K, Hong B S, Shin D H, and Cho H Y (2005), 'Protective effects of aucubin isolated from *Eucommia ulmoides* against UVB-induced oxidative stress in human skin fibroblasts', *Biol Pharm Bull*, 28, 1244-1248.

Hollman P C H and Katan M B (1999), 'Dietary flavonoids: intake, health effects and bioavailability', *Food Chem Toxicol*, 37, 937-942.

Hori I, Nihei K N, and Kubo I (2004), 'Structural criteria for depigmenting mechanism of arbutin', *Phytother Res*, 18, 475-479.

Huh S W, Bae S M, Kim Y W, Lee J M, Namkoong S E, Lee I P, Kim S H, Kim C K, and Ahn W S (2004), 'Anticancer effects of (-)-epigallocatechin-3-gallate on ovarian carcinoma cell lines', *Gynecologic Oncology*, 94, 760-768.

Ichihashi M, Ahmed N U, Budiyanto A, Wu A, Bito T, Ueda M, and Osawa T (2000), 'Preventive effect of antioxidant on ultraviolet-induced skin cancer in mice', *J Dermatol Sci*, 23, S45-S50.

Jin Y H, Lee S J, Chung M H, Park J H, Cho T H, and Lee S K (1999), 'Aloesin and arbutin inhibit tyrosinase activity in a synergistic manner via a different action mechanism', *Arch Pharm Res*, 22, 232-236.

Jung Y D, Kim M S, Shin B A, Chay K O, Ahn B W, Liu W, Bucana C D, Gallick G E, and Ellis L M (2001), 'EGCG, a major component of green tea, inhibits tumour growth by inhibiting VEGF induction in human colon carcinoma cells', *Br J Cancer*, 84, 844-850.

Karakaya S (2004), 'Bioavailability of phenolic compounds', *Crit Rev Food Sci Nutr*, 453-464.

Kim D and Robyt J F (1995), 'Dextransucrase constitutive mutants of *Leuconostoc mesenteroides* B-1299', *Enzyme Microb Technol*, 17, 1050-1056.

Kim M J, Lee S B, Lee H S, Lee S Y, Baek J S, Kim D, Moon T W, Robyt J F, and Park K H (1999), 'Comparative study of the inhibition of α-glucosidase, α-amylase, and cyclomaltodextrin glucanosyltransferase by acarbose, isoacarbose, and acarviosine-glucose', *Arch Biochem Biophys*, 371, 277-283.

Kitao S, Matsudo T, Saitoh M, Horiuchi T, and Sekine H (1995), 'Enzymatic syntheses of two Stable (-)-epigallocatechin gallate-glucosides by sucrose phosphorylase', *Biosci Biotech Biochem*, 59, 2167-2169.

Kitao S, Ariga T, Matsudo T, and Sekine H (1993), 'The Syntheses of catechin-glucosides by transglycosylation with *Leuconostoc mesenteroides* sucrose phosphorylase', *Biosci Biotech Biochem*, 57, 2010-2015.

Koepsell H J, Tsuchiya H M, Hellman N N, Kasenko A, Hoffman C A, Shape E S, and Jackson R W (1953), 'Enzymatic synthesis of dextran. Acceptor specificity and chain initiation' *J Biol Chem*, 200, 793-801.

Lee K M, Yeo M, Choue J S, Jin J H, Park S J, Cheong J Y, Lee K J, Kim J H, and Hahm K B (2004), 'Protective mechanism of epigallocatechin-3-Gallate against *Helicobacter pylori*-induced gastric epithelial cytotoxicity via the blockage of TLR-4 signaling', *Helicobacter*, 9, 632-642.

Lee S B, Nam K C, Lee S J, Lee J H, Inouye K, and Park K H (2004), 'Antioxidant effects of glycosyl-ascorbic acids synthesized by maltogenic amylase to reduce lipid oxidation and volatiles production in cooked chicken meat', *Biosci Biotechnol Biochem*, 68, 36-43.

Lee S J, Kim J C, Kim M J, Kitaoka M, Park C S, Lee S Y, Ra M J, Moon T W, Robyt J F, and Park K H (1999), 'Transglycosylation of naringin by *Bacillus stearothermophilus* maltogenic amylase to give glycosylated naringin', *J Agric Food Chem*, 47, 3669-3674.

Li D, Park S H, Shim J H, Lee H S, Tang S Y, Park C S, and Park K H (2004), 'In vitro enzymatic modification of puerarin to puerarin glycosides by maltogenic amylase', *Carbohydr Res*, 339, 2789-2797.

Maeda K and Fukuda M (1996), 'Arbutin: mechanism of its depigmenting action in human melanocyte culture', *J Pharmacol Exp Ther*, 276, 765-769.

Moon Y, Kim G H, Lee J H, Jin W J, Kim D W, and Kim D (2006a), 'Enzymatic synthesis and characterization of novel epigallocatechin gallate glucosides', *Journal of Molecular Catalysis B: Enzyme*, 40, 1-7.

Moon Y, Lee J H, Ahn J S, Nam S H, OH D K, Park D H, Chung H J, Kang S S, Day D F, and Kim D (2006b), 'Synthesis, structure analyses, and characterization of novel epigallocatechin gallate (EGCG) glycosides using the glucansucrase from *Leuconostoc mesenteroides* B-1299CB', *J Agric Food Chem*, 54, 1230-1237.

Moon Y, Nam S, Kang J, Kim Y M, Lee J H, Kang H K, Jun W J, Park K D, Kimura A, and Kim D (2007), 'Enzymatic synthesis and characterization of arbutin glucosides using glucansucrase from *Leuconostoc mesenteroides* B-1299CB', *Appl Microbiol and Biotechnol*, 77, 559-567.

Nakano H, Hamayasu K, Nakagawa K, Tabata A, Fujita K, Hara K, Kiso T, Murakami H, and Kitahata S (2002), 'Transglycosylation of hydroquinone and epicatechin by β-fructofuranosidase from *Arthrobacter* sp.', *J Appl Glycosci*, 49, 115-121.

Nanjo F, Hara M, Bandai T, and Shibuya T (1995), 'Preparation of polyphenol (epicatechin and epigallocatechin) glycosides as phenol oxidase inhibitors', *Japan Kokai Tokyo Koho* 7-179489.

Otake S, Makimura M, Kuroki T, Nishihara Y, and Hirasawa M (1991), 'Anticaries effects of polyphenolic compounds from Japanese green tea', *Caries Res*, 25, 438-443.

Raederstorff D G, Schlachter M F, Elste V, and Weber P (2003), 'Effect of EGCG on lipid absorption and plasma lipid levels in rats', *J Nutr Biochem*, 14, 326-332.

Robyt J F and Martin P J (1983), 'Mechanism of synthesis of D-glucans by D-glucosyltransferase from *Streptococcus mutans* 6715', *Carbohydr Res*, 113, 301-315.

Robyt J F (1995), 'Mechanism in the glucansucrase synthesis of polysaccharides and oligosaccharides from sucrose', *Adv Carbohydr Chem Biochem*, 51, 133-168.

Salah N, Miller N J, Paganga G, Tijburg L, Bolwell G P, and Rice-Evans C (1995), 'Polyphenolic flavanols as scavengers of aqueous phase radicals and as chain-breaking antioxidant', *Arch Biochem Biophys*, 322, 339-346.

Sato T, Nakagawa H, Kurosu J, Yoshida K, Tsugane T, Shimura S, Kirimura K, Kino K, and Usami S (2000), 'α-Anomer-selective glucosylation of (+)-catechin by the crude enzyme, showing glucosyl transfer activity, of *Xanthomonas campestris* WU-9701' *J Biosci Bioeng*, 90, 625-630.

Seo E S, Lee J H, Park J Y, Kim D, Han H J, and Robyt J F (2005), 'Enzymatic synthesis and anti-coagulant effect of salicin analogs by using the *Leuconostoc mesenteroides* glucansucrase acceptor reaction', *J Biotechnol*, 117, 31-38.

Sugimoto K, Nishimura T, Nomura K, Sugimoto K, and Takashi K (2003), 'Syntheses of arbutin-α-glycosides and a comparison of their inhibitory effects with those of α-arbutin and arbutin in human tyrosinae', *Chem Pharm Bull*, 51, 798-801.

Sugimoto K, Nomura K, Nishimura T, Kiso T, Sugimoto K, and Takashi K (2005), 'Syntheses of α-arbutin-α-glycosides and their inhibitory effects on human tyrosinae', *J Biosci Bioeng*, 99, 272-276.

Tomita K, Fukudo M, and Kawasaki K (1990), 'Mechanism of arbutin inhibitory effect on melanogenesis and effect on the humna skin with cosmetic use', *Fragrance J*, 6, 72-77.

Valcic S, Muders A, Jacobsen N E, Liebler D C, and Timmermann B N (1999), 'Antioxidant chemistry of green tea catechins. Identification of products of the reaction of (-)-epigallocatechin gallate with peroxyl radicals', *Chem Res Toxicol*, 12, 382-386.

Watanabe H, Shimizu T, Nishihira J, Abe R, Nakayama T, Taniguchi M, Sabe H, Ishibashi T, and Shimizu H (2004), 'Ultraviolet A-induced production of matrix metalloproteinase-1 is mediated by macrophage migration inhibitory factor (MIF) in human dermal fibroblasts', *J Biol Chem*, 279, 1676-1683.

Yoon S H and Robyt J F (2002), 'Synthesis of acarbose analogues by transglycosylation reactions of *Leuconostoc mesenteroides* B-512FMC and B-742CB dextransucrases', *Carbohydr Res*, 337, 2427-2435.

Yoon S H, Fulton D B, and Robyt J F (2004), 'Enzymatic synthesis of two salicin analogues by reaction of salicyl alcohol with *Bacillus macerans* cyclomaltodextrin glucanyltransferase and *Leuconostoc mesenteroides* B-742CB dextransucrase', *Carbohydr Res*, 339, 1517-1529.

MICROBIAL EXO- AND ENDO-ARABINOSYL HYDROLASES: STRUCTURE, FUNCTION, AND APPLICATION IN L-ARABINOSE PRODUCTION

Tae Jip Kim

ABSTRACT

L-Arabinose is one of main components of hemicellulose widely distributed in plant cell walls, where they are present in significant amounts as arabinan, arabinoxylan, and arabinogalactan. Recently, L-arabinose has been known as a functional sweetener and a food additive for good health to prevent the obesity. Arabinosyl hydrolases are main biocatalysts for the enzymatic production of L-arabinose. Arabinan-degrading enzymes have roughly been classified into the *exo*-type α-L-arabinofuranosidase (AFases; EC 3.2.1.55) and the *endo*-1,5-α-L-arabi-nanases (ABNases; EC 3.2.1.99), on the basis of their modes of action. AFases are typical *exo*-acting enzymes, which hydrolyze terminal non-reducing residues from arabinose-containing polysaccharides, while ABNases are *endo*-type enzymes randomly hydrolyzing internal linkages from mainly linear arabinan, which release a mixture of arabinooligo-saccharides. Both types of enzymes can work in concert with other hemicellulases to completely degrade the backbone of hemicellulosic materials, which makes them the essential enzymes for the industrial production of L-arabinose. In this review, all known primary and three-dimensional structures, enzymatic properties of both AFases and ABNases are comparatively investigated to develop the most cost-effective processes for the production of L-arabinose. It can be applicable to various industrial fields such as food, nutritional, and pharmaceutical technology, as well as reutilization of various plant biomasses for bio-fuels.

Key words: L-arabinose, arabinan, arabinosyl hydrolase, α-L-arabinofuranosidase, *endo*-1,5-α-L-arabinanase

INTRODUCTION

The plant cell wall is composed of complexes of several carbohydrate polymers including cellulose, hemicellulose, lignin, and pectin. However, the degradation of the plant cell wall materials is often inefficient because most polymers of cellulose and hemicellulose are likely to be insoluble or tightly associated with each other. Nevertheless, these wide-spread natural biomasses contain high portion of valuable carbohydrate polymers, which can be utilized via synergistically controlled enzymatic degradation.

Hemicelluloses are the most abundant source of renewable carbon except cellulose (Saha, 2000). Heteroxylans are the most common hemicelluloses and complex poly-saccharides composed of β-(1,4)-linked xylopyranose backbone substituted with neutral (arabinose, galactose) and acidic (methyl-glucuronic acid) sugars, as well as acetyl group and phenolic acids (Paës et al., 2008). The efficient utilization of xylans as sources of bio-fuels and industrial chemicals requires total understanding of the enzyme systems for their conversion. Due to the high complexity and structural variability of heteroxylans, their enzymatic hydrolysis can be achieved via the concerted treatment of various hydrolases that include the main chain-cleaving enzymes, *endo*-β-(1,4)-

xylanases (EC 3.2.1.8) and β-xylosidases (EC 3.2.1.37), and the side chain-active enzymes, acetyl-xylanesterases, α-glucuronidases, ferulic acid esterases, α-L-arabinofuranosidases (EC 3.2.1.55), and *endo*-1,5-α-L-arabinanases (EC 3.2.1.79) (Filho et al., 1996). Some combinations of these enzymes acting against main chains and side chains have been shown to affect synergistically on the complete hydrolysis of appropriate substrates.

Recently, it has been proved that L-arabinose selectively inhibits intestinal sucrase in an uncompetitive manner and reduces the glycemic response after sucrose ingestion in animals. Based on these observations, L-arabinose can be considered as a physiologically functional sugar possessing inhibitory activity against sucrose digestion (Seri et al., 1996). They reported that neither D-arabinose nor L-arabinobiose inhibited sucrase activity and L-arabinose showed no inhibitory effect on the activities of intestinal maltase, isomaltase, trehalase, lactase, and glucoamylase, or pancreatic amylase. It has also been reported that L-arabinose dose-dependently suppressed the blood glucose increase in mice after the ingestion of sucrose. Especially, the simultaneous uptake of sucrose with about 3% (w/w) L-arabinose can reduce 40~50% of sucrose digestion and absorption in blood stream. In this way, Seri et al. (1996) reported that L-arabinose is useful in preventing postprandial hyper-glycemia in diabetic patients when foods containing sucrose are ingested. Therefore, effective production of L-arabinose from arabinoxylan and arabinan is very important in the other emerging fields as well as the food industry.

Functional sugar, L-Arabinose, is present at high concentrations in arabinoxylans, arabinans, and arabinogalactan. The arabinan and arabinoxylan backbone comprise mainly α-(1,5)-linked arabinofuranose units and β-(1,4)-linked xylopyranose moieties, respectively. These polysaccharides commonly contain arabinofuranose molecules decorated as side chains via α-(1,2)- or α-(1,3)-linkage. For more efficient hydrolysis of these polymers, the removal of arabinose side chains by arabinofuranosidases from both arabinoxylan and arabinan is likely to be one of critical steps. At least four types of arabinofuranosidase have been known to remove terminal arabinose side chains from hemicellulose and pectic materials. On the other hand, the debranched arabinan backbone that is free of side-chain constituents is hydrolyzed by *endo*-1,5-α-L-arabinanases, which can act as typical *endo*-acting enzymes generating a variety of arabinooligosaccharides as final products. Especially, the typical *exo*-acting α-L-arabinofuranosidase (AFase; EC 3.2.1.55) exhibits broad substrate specificity and eliminates L-arabinose side chains, allowing *endo*-1,5-α-L-arabinase (ABNase; EC 3.2.1.99) to cleave the internal linkages in main backbone of debranched arabinan. Accordingly, these complementary enzymes can act synergistically in degradation of highly branched arabinan to generate functional sugar, L-arabinose.

Among various arabinosyl hydrolases, AFases have been received increased attention, due to their essential roles in hemicellulose degradation and wide distribution in nature, especially within microbial world. Saha (2000) reported very stimulating review for biochemistry, molecular biology of various AFases and their application in biotechnology. Recently, Numan and Bhosle (2006) have intensively summarized its updated version for the potential applications of AFases in biotechnology. However, these efforts have been limited to mainly those of AFases. Moreover, the additional investigation of novel arabinosyl hydrolases, such as ABNases, can be necessary for the development of more efficient enzymatic hydrolysis process using complex polymers in nature.

In this work, both arabinosyl hydrolases, AFases and ABNases, were considered for the synergistic treatment in L-arabinose production from arabinose-containing polymers, especially for sugar beet arabinan. For the purpose, the following topics have been studied here: (1) classification of various arabinosyl hydrolases (2) enzymatic properties of known microbial arabinosyl hydrolases (3) comparison of primary structures (4) structural aspects of *exo*- and *endo*-arabinosyl hydrolases (5) genome-wide mining of useful arabinosyl hydrolase genes in nature (6) possibility of high-yield enzymatic process for L-arabinose production via simultaneous treatment of arabinosyl hydrolases.

CLASSIFICATION OF ARABINOSYL HYDROLASES

Arabinosyl hydrolases belong to the subset of the glycosyl hydrolase family enzymes. Based on their substrate specificity and mode of action on polysaccharide substrates, these enzymes can be categorized into the following three Enzyme Commission classes: (1) EC 3.2.1.55 (α-L-arabinofuranoside hydrolases or alternatively α-L-arabinosidases), which catalyzes the hydrolysis of terminal non-reducing α-L-arabinose residues from various α-L-arabinosides (2) EC 3.2.1.88 (β-L-arabinosidases), which is specific for β-arabinosides; and (3) EC 3.2.1.99 (arabinan *endo*-1,5-α-L-arabinanases), which catalyzes the random hydrolysis of internal arabinofuranosidic linkages in α-L-(1,5)-arabinans.

Table 1 Glycoside hydrolase families related to arabinosyl hydrolases [1)]

GH	Clan	Mechanism	Structure	Known enzyme activities
GH 3	ND [2)]	retaining	ND	β-glucosidase (EC 3.2.1.21) xylan 1,4-β-xylosidase (EC 3.2.1.37) β-*N*-acetylhexosaminidase (EC 3.2.1.52) glucan 1,3- β-glucosidase (EC 3.2.1.58) glucan 1,4-β-glucosidase (EC 3.2.1.74) *exo*-1,3-/1,4-glucanase (EC 3.2.1.-) α-L-arabinofuranosidase (EC 3.2.1.55)
GH 43	GH-F	inverting	5-fold β-propeller	β-xylosidase (EC 3.2.1.37) β-1,3-xylosidase (EC 3.2.1.-) α-L-arabinofuranosidase (EC 3.2.1.55) arabinanase (EC 3.2.1.99) xylanase (EC 3.2.1.8) galactan 1,3-β-galactosidase (EC 3.2.1.145)
GH 51	GH-A	retaining	$(\beta/\alpha)_8$-barrel	α-L-arabinofuranosidase (EC 3.2.1.55) endoglucanase (EC 3.2.1.4)
GH 54	ND	retaining	ND	α-L-arabinofuranosidase (EC 3.2.1.55) β-xylosidase (EC 3.2.1.37)
GH 62	GH-F	ND	5-fold β-propeller	α-L-arabinofuranosidase (EC 3.2.1.55)
GH 93	ND	retaining	ND	*exo*-1,5-α-L-arabinanase (EC 3.2.1.-)

[1)] Classification with specific enzyme information was proposed from Carbohydrate-Active Enzymes Server at http://www.cazy.org/fam/acc_fam.html.

[2)] ND means that its classification is 'not defined' yet.

Henrissat and Davies (1997) proposed a novel classification strategy for the glycosyl hydrolases, on the basis of amino acid sequences and primary structure similarities. This classification is probably useful to investigate molecular evolutionary factors, action mechanisms, and structure-function relationships of these hydrolases. Most of arabinosyl hydrolases have been found mainly in five GH families 3, 43, 51, 54, and 62. Hydrolases with known arabinoside-hydrolyzing activities were categorized into GH families or Clans listed in Table 1. To date, well-known microbial AFases and ABNases have been assigned mainly to GH51 and GH43, respectively.

In general, the complete hydrolysis of complex heteroxylan polymers needs the combinatorial use of hydrolases in their enzymatic processes. Various specialty enzymes involved in the hydrolysis of heteroxylan can be listed as follows: (1) *endo*-xylanase, mainly hydrolyzing interior β-(1,4)-xylose linkages of xylan backbone (2) *exo*-xylanase, hydrolyzing β-(1,4)-xylose linkages to release xylobiose (3) β-xylosidase, releasing xylose from xylobiose and short chain xylooligosaccharides (4) α-L-arabinofuranosidase, hydrolyzing terminal non-reducing α-arabinoses from arabinoxylans (5) α-glucuronidase, releasing glucuronic acid from glucuronoxylans (6) acetylxylan esterase, hydrolyzing acetylester bonds in acetyl xylans (7) ferulic acid esterase, hydrolyzing feruloylester bonds in xylans (8) *p*-coumaric acid esterase, hydrolyzing *p*-coumaryl ester bonds in xylans (Saha, 2000).

As shown in Table 1, the enzymes with arabinosyl hydrolyzing activities are not well-organized into any specific class, which is caused by the complexity of various substrate polymers in nature. Arabinosyl hydrolases are very important in the degradation of polysaccharides, such as arabinoxylans. Small amount of arabinose residues are linked to O-2 and/or O-3 of the main arabinoxylan backbone, which can be hydrolyzed by *exo*-acting AFases. Accordingly, AFases can contribute the part of microbial xylanolytic systems necessary for complete breakdown of arabinoxylans. They have been isolated from various fungi, bacteria, and plants. These enzymes have been actively studied in recent years due to their industrial applications in food and agricultural processes, such as wine making to increase aroma production, clarification of juices, and improvement of feed digestibility (Saha, 2000).

Beldman et al. (1997) also proposed the following classes of *exo*-AFases, based on their mode of action and substrate specificity: (1) not active towards polymers (2) active towards polymers (3) specific for arabinoxylans (4) not active on the synthetic substrate, *p*-nitrophenyl-α-L-arabinofuranoside (*p*NPA). According to their suggestions, three sub-classes were introduced into the existing arabinoxylan α-L-arabinofuranohydrolases class as follows: (1) AXHB-md 2,3 sub-class includes enzymes that release arabinose from substituted xylose and hydrolyze *p*NPA at a similar rate to that for oligosaccharide substrates (2) AXHB-m 2,3 sub-calss includes enzymes that hydrolyze arabinose residues from α-(1,2)- or α-(1,3)-linked to a single-substituted xylose and do not hydrolyze *p*NPA (3) AXHd3 sub-class includes enzymes that are able to release only α-(1,3)-linked arabinose residues from double-substituted xylose residues but do not hydrolyze *p*NPA (Numan and Bhosle, 2006).

Recently, new types of arabinosyl hydrolases have been isolated with novel properties that have not been reported earlier. Such enzymes could not be classified to any of the existing arabinosyl hydrolase classes. For example, *Cellvibrio japonicus* (formally known as *Pseudomonas cellulosa*) α-L-arabinanase 43A was reported to exhibit both *exo*- and *endo*-hydrolase activities (McKie et al., 1997). On the other side, extremely thermostable AFase from *Thermotoga maritima* MSB8 has the ability to act on both interior α-(1,5)-linked backbone and terminal α-(1,3)-linked side chains of arabinan and debranched arabinan. In addition, it can hydrolyze *p*-nitrophenyl α-L-arabinofuranoside (Yoon et al., 2004; Miyazaki, 2005). Accordingly, these newly found arabinosyl hydrolases should be assigned into a new class due to their unique enzymatic properties. Many works for the finding of new way to categorize these novel arabinosyl hydrolases is still in progress, but it may not be simple because of their muti-substrate specificities or versatile action modes. As a number of biochemical and molecular biological data have been accu-mulated, the better understanding of these types of enzyme can be possible in near future.

ENZYMATIC PROPERTIES OF ARABINOSYL HYDROLASES

α-L-Arabinofuranosidases

As mentioned previously, the degradation of polysaccharides with hemicellulases is often limited by the presence of arabinose residues attached to the main backbones as side chains. α-L-Arabinofuranosidases (AFases; EC 3.2.1.55) specifically catalyze the hydrolysis of terminal non-reducing α-L-(1,2)-, α-L-(1,3)-, and/or α-L-(1,5)-arabinofuranosyl residues from arabinose-containing oligo- and poly-saccharides. Especially, these enzymes can synergistically accelerate the hydrolysis of the glycosidic bonds in polymers in combination with other hemicellulases. Moreover, as shown in Table 2, novel AFases possessing β-xylosidase activity or xylanases with AFase activity also have been reported (Mai et al., 2000; Lee et al., 2003). Based on their mode of action and substrate specificity, AFases can be categorized into three types (Beldman et al., 1997). Type-A AFases preferentially degrade α-(1,5)-L-arabinooligosaccharides to arabinose monomers and are inactive against arabinosyl linkages in polysaccharides. In contrast, type-B AFases show considerable activity on debranching L-arabinose residues from side chains in arabinan or arabinoxylan. Both types of AFases also act on synthetic substrate, *p*-nitrophenyl-α-L-arabinofuranoside. The third type of AFase, called as α-L-arabino-furanohydrolases, is specifically active on arabinosidic linkages in arabinoxylans from oat spelt, wheat, or barley.

Generally, microorganisms interacting with plant materials produce two or more arabinosyl hydrolases as a set of enzymes for the efficient degradation and utilization of them. Sakamoto and Kawasaki (2003a) reported that *Penicillium chrysogenum* 31B secretes at least five distinct arabinan-degrading enzymes including an *exo*-arabinanase, two *endo*-arabinanases, and two AFases in the culture broth. Recently, the enzymatic properties and three-dimensional structure of *Thermobacillus xylanilyticus* AFase (*Tx*-AFase) belonging to family GH51 were determined (Debeche et al., 2000; Paës et al., 2008). This 56 kDa thermostable enzyme displays an optimual activity at 75°C and remains active for several hours at 60°C. Same as the other GH51 hydrolases, *Tx*-AFase is a retaining enzyme and catalyzes both hydrolysis and transglycosylation reactions on pNP-β-D-galactofuranoside and pNP-β-D-xylopyranoside (Rémond et al., 2004 and 2005). However, *Tx*-AFase showed very low catalytic activity on compounds containing L-arabinose. Among natural sugars, it was particularly active on arabino-xylo-oligomers, while much less activity on arabinoxylans was observed. For its bond specificity, *Tx*-AFase displayed a preference for α-(1,2)-bonds and, to a lesser extent, α-(1,3)-bonds. In contrast, α-(1,5)-bonds are very poorly hydrolyzed (Debeche et al., 2000). In case of *Tx*-AFase, its hydrolyzing activity on main arabinose backbone can be much lower than its trimming activity for arabinose residues in side chains.

Table 2 Summarized enzymatic and biochemical characteristics of various microbial AFases

Microbial origins	Oligomeric Structure	M.W. [1] (kDa)	pI	Optimum		Polymer attacked	References
				Temp(°C)	pH		
Aspergillus							
A. awamori IFO4033	-	81	3.3	60	4.0	BA	Kaneko et al., 1998
A. awamori IFO4033	-	62	3.6	60	4.0	BA	Kaneko et al., 1998
A. kawachii IFO4308	-	65		55	4.0	BA, DA, RA, WA	Koseki et al., 2006
A. nidulans	-	65	3.3	65	4.0	-	Ramon et al., 1993
A. niger 5-16 (intracellular)	-	67	3.5	60	4.0	BA	Kaneko et al., 1993
A. niger 5-16 (extracellular)	-	53	3.6	-	4.0	AX, BA	Kaji and Tagawa, 1970
A. oryzae OSI1013	tetramer	55	-	50	5.0	-	Matsumura et al., 2004
A. oryzae HL15	dimer	60	-	60	5.5	AX,AG,,OSX	Hashimoto and Nakata, 2003
A. sogae	-	34	3.9	50	5.0	-	Kimura et al., 2000
Aureobasidium							
A. pullulans NRRL Y-12974	dimer	105	-	75	4.0~4.5	AX, BA, OSX	Saha and Bothast, 1998
A. pullulans NRRL Y-2311-1	-	49	5.0	55	3.5~4.0	AO, MF, OSX, WAX	de Wet et al., 2008
Bacillus							
B. polymyxa CECT153	-	53	9.0	55	6.5	OSX, WA	Morales et al., 1995
B. polymyxa CECT153	-	64	8.7	55	6.5	OSX, WA	Morales et al., 1995
B. pumilus PS213	tetramer	60	5.2	55	7.0	AG, OSX	Degrassi et al., 2003
B. stearothermophilus T-6	hexamer	57	6.5	70	5.5~6.0	BA, OSX	Gilead and Shoham, 1995
B. subtilis 3-6	-	61	-	60	7.0	BA	Kaneko et al., 1994
Bacterium sp. PRI-1686	hexamer	57	-	70	6.0	BA, DA,OSX	Birgisson et al., 2004
Bifidobacterium							
B. adolescents DSM20083	-	100	-	30	6.0	WA	van Laere et al., 1997
B. breve K-110	tetra-/hexamer	60	-	45	4.5	GRb2, GRc	Shin et al., 2003
B. longum B667	tetramer	61	-	45	6.0	AO, AX, BA	Margolles et al., 2003

(continued to next page)

Table 2 Continued

Microbial origins	Oligomeric Structure	M.W. [1] (kDa)	pI	Optimum Temp(°C)	Optimum pH	Polymer attacked	References
Butyrivibrio fibrisolvens GS113	octamer	31	6.0	45	6.0~6.5	AX, BA, OSX	Hespell and O'bryan, 1992
Clostridium							
C. cellulovorans ATCC35296	-	59	-	40~50	6.0	AX, BA	Kosugi et al., 2002
C. stercorarium	-	56	-	70	5.0	-	Schwarz et al., 1995
C. thermocellum ATCC27405	hexamer	-	-	82	-	AO, AXO, WA	Taylor et al., 2006
Cytophaga xylanolytica XM3	tri-/tetramer	56	6.1	45	5.8	AX, BA	Renner and Breznak, 1998
Fusarium							
F. oxysporum F3	-	66	-	60	6.0	AX, BA	Christakopoulos et al., 2000
F. oxysporum	trimer	65	6.0	50~60	6.0	AX, BA	Panagiotou et al., 2003
F. oxysporum	trimer	56	7.3	50~60	6.0	AX, BA	Panagiotou et al., 2003
Geobaillus							
Geobacillus sp. KCTC3012	-	65	-	60	5.0	AO, BA, DA, OSX, RA	Park et al., 2007
G. caldoxylolyticus TK4	tetramer	58	-	75~80	6.0	AO, BA	Canakci et al., 2007
Humicola insolens	-	94	-	40	6.0	WA	Sorensen et al., 2006
Meripilus giganteus	-	69	-	48	3.8	WA	Sorensen et al., 2006
Penicillium							
P. capsulatum	-	65	4.2	60	4.0	AX, AXO, BA	Filho et al., 1996
P. capsulatum	-	63	4.5	55	4.0	AX, AXO, BA	Filho et al., 1996
P. chrysogenum 31B	-	79	-	50	4.0~6.5	AX, AO, BA, DA, SAG	Sakamoto et al., 2003a
P. chrysogenum 31B	-	52	-	50	3.3~5.0	AX, AO, BA, DA, SAG	Sakamoto et al., 2003a
P. purpurogenum	-	58	6.5	50	5.0	AX, BX, IAG, OSX, WB, WS	Carvallo et al., 2003
P. purpurogenum MYA-38	monomer	70	5.3	60	5.0	BA, DA, WA	Fritz et al., 2008
Pichia capsulata X91	tri-/tetramer	72	5.1	50	6.0	AG, BA	Yanai and Sato, 2000
Pseudomonas cellulosa	-	57	-	<55	5.5	AO, BA, DA, WA	Beylot et al., 2001

(continued to next page)

Table 2 Continued

Microbial origins	Oligomeric Structure [1)]	M.W. [2)] (kDa)	pI	Optimum Temp(°C)	Optimum pH	Polymer attacked [3)]	References
Rhodothermus marinus ITI376	-	-	-	85	5.5~7.0	-	Gomes et al., 2000
Rhizomucor pusillus HHT-1	-	88	4.2	65	4.0	AO, BA, DA	Shofiqur Rahman et al., 2001
Ruminococcus albus	tetramer	75	3.8	-	6.9	PGSC, BSC,	Greve et al., 1984
Streptomyces							
S. chartreusis GS901	-	80	6.6	55	5.5	AO, AG, AX, BA	Matsuo et al., 2000
S. chartreusis GS901	-	37	7.5	50	7.0	AO, AG, AX, BA,	Matsuo et al., 2000
S. diastaticus ET	-	38	8.8	-	5.0~6.5	AX	Tajana et al., 1992
S. diastaticus ET	-	60	8.3	-	5.0~6.5	AX	Tajana et al., 1992
S. lividans 66	-	69	4.6	60	6.0	AX, BA	Manin et al., 1994
S. thermoviolaceus OPC-520	-	37	-	60	5.0	AX, OSX	Tsujibo et al., 2002
Trichoderma koningii G-39	-	50	-	2.8~3.2	-	-	Wan et al., 2007
Trichoderma reesei PC-3-7	-	53	7.5	60	4.0	OSX	Nogawa et al., 1999
Thermobacillus xylanilyticus D3	hexamer	56	-	75	5.6~6.2	LX, OSX, WA,	Debeche et al., 2000
Thermomonospora fusca	-	92	7.9	65	9.0	-	Tuncer and Ball, 2003
Thermotoga maritima MSB8	hexamer	55	-	90	7.0	BA, DA	Miyazaki, 2005

[1)] Oligomeric structures are determined by crystal structures or estimated from gel permeation chromatography.

[2)] Each molecular mass of AFase as a monomer is shown here.

[3)] Abbreviations of AX, arabinoxylan; BA, beet arabinan; OSX, oat-spelt xylan; AXO, arabinoxylan oligosaccharides; LX, larchwood xylan; AG, arabinogalactan; DA, debranched arabinan; AO , arabinooligosaccharide; GRc, ginsenoside Rc; GRb2, ginsenoside Rb2; MF, maize fiber; CC, corn cob; CH, corn husk; BX, sugar beet pulp xylan; IAG, larchwood arabinogalactan; WS, wheat straw; WB, wheat bran; WA, wheat arabinoxylan; SAG, soybean arbinogalactan; RA, Rye arabinoxylan; PGSC, polygalacturonase-soluble carbohydrate; BSC, base-soluble carbohydrate (Saha, 2000; Numan and Bhosle, 2006).

On the other hand, a highly thermostable AFase GH51 gene was cloned and characterized from hyperthermophilic bacterium, *Thermotoga maritima* (Yoon et al., 2004; Miyazaki, 2005). *Tm*-AFase has a molecular mass of 55 kDa as a monomer, but 332 kDa on gel permeation chromatography. Therefore, it was supposed to comprise six identical subunits as in case of homologous AFase from *Geobacillus stearothermophilus*. Regarding substrate specificity, *Tm*-AFase can hydrolyze arabinan, debranched arabinan, and arabinoxylan, but not arabinogalactan and carboxymethyl cellulose. *Tm*-AFase is extremely thermophilic, displaying an optimal reaction temperature of 85~90°C, which can be the most thermostable AFase reported thus far.

First of all, various hydrolases with adequate action patterns or substrate specificity can be necessary for development any efficient enzymatic hydrolysis process. Therefore, biochemical properties, optimal reaction conditions, and substrate specificity of known AFases have been summarized and compared in this review (Table 2). Most AFases are able to catalyze various substrates including arabinooligosaccharides, branched arabinan, debranched arabinan, arabinoxylan. Their substrate preferences and action patterns are quite variable depending on microbial origin of each enzyme. In general, typical AFases can cleave *p*NPA synthetic substrates, while they show relatively low activity against hetero-polymeric substrates.

endo-1,5-α-L-Arabinanases

endo-1,5-α-L-Arabinanase (ABNase; EC 3.2.1.99) is an *endo*-acting hydrolase which cleave the α-(1,5)-L-arabinofuranoside linkages of arabinan polymers. Compared with the intensive works on AFases, up to date, only a few experimental results for ABNases have been reported as shown in Table 3.

Table 3 Summarized enzymatic and biochemical characteristics of various microbial ABNases

Type	Microbial origins	M.W.[1)] (kDa)	pI	Optimum Temp (°C)	Optimum pH	Polymer attacked [2)]	Ref.
endo	*Aspergillus nidulans*	40	3.3	68	5.5	-	a
	Bacillus subtilis 168T+	31	9.0	60	6.0	BA, LA	b
	B. subtilis 168T+	46	-	50	7.0	BA, LA, PT	c
	B. thermodenitrificans TS-3	35	4.5	70	6.0	BA, DA	d
	Penicillium Chrysogenum 31B	35	-	30-40	6.0-7.0	BA, DA	e
exo	*Penicillium Chrysogenum* 31B	47	-	40	4.0	AO, BA, DA, SAG	f
exo+endo	*Cellvibrio japonicus* [3)]	34	-	-	-	AO, CMA, LA	g

[1)] Each molecular mass of AFase as a monomer is shown here.

[2)] Abbreviations of BA, beet arabinan; DA, debranched arabinan; SAG, soybean arbinogalactan; PT, Pectin; CMA, carboxymethyl arabinan; LA, linear α-(1,5)-arabinan; AO , arabinooligosaccharides.

[3)] The bacterium was known as *Pseudomonas cellulosa* in the previous reports.

a, Ramon et al., 1993; b, Leal and Sá-Nogueira, 2004; c, Inácio and Sá-Nogueira, 2008; d, Takao et al., 2002; e, Sakamoto et al., 2003b; f, Sakamoto and Thibault, 2001; g, McKie et al., 1997

Microbial mesophilic ABNases have been described in several species of *Bacillus*, *Pseudomonas*, and *Aspergillus*. Interestingly, the α-(1,5)-arabinanse from *Cellvibrio japonicus* was known to hydrolyze linear arabinans almost exclusively into arabinotriose via *exo*-acting catalysis, but this inverting enzyme is also able to cleave in *endo*-action (Beylot et al. 2001). Recently, a thermophilic *Bacillus thermodenitrificans* ABNase GH43 (*Bt*-ABNase) was cloned and characterized. The *Bt*-ABNase showed the highest activity at 70°C and pH 6.0. The most favorable substrates of ABNases can be arabinan polymers, especially debranched or linear α-(1,5)-arabinan. However, the more intensive works on ABNases should be done for the invention of its synergistic effects on the enzymatic L-arabinose production.

STRUCTURES OF ARABINOSYL HYDROLASES

Although common biotechnological applications may favor *endo*-acting enzymes, the complete saccharification of complex and highly branched polysaccharides strongly needs the synergy effect of *endo*- and *exo*-acting hydrolases. The structural basis for *endo*- or *exo*-acting glycoside hydrolases is critical in the rational design of enzymes that are tailored for specific industrial applications. Based on cases of other hydrolases such as cellulases and amylases, the structural bases for both *exo*- and *endo*-activities have seemed to be clear. *endo*-Acting enzymes generally contain an open and extended substrate binding cleft that can interact at random positions along the saccharide chains. On the other hand, the structures of the active site of *exo*-acting (releasing monosaccharides) glycoside hydrolases have the typical pocket that only the terminal sugar residues of polymeric chains can enter the active site. Recently, a more complex and subtle basis for such *exo*- and *endo*-preference has emerged. Thus, selected enzymes that contain open clefts can display typical *exo*-activity, indicating that binding-energies and steric factors within individual subsites can contribute to the modes of action of these biocatalysts in a similar manner to more gross and obvious structural features (Pages et al., 2001). On the basis of the structural aspects, we can consider the possibility in the protein engineering of industrially relevant action modes into highly active carbohydrate modifying enzymes, thus increasing their role in biotechnological processes.

exo-Acting α-L-arabinofuranosidases

To date, three different 3D-structures of α-L-arabinofuranosidases in GH family 51 have been reported from *Geobacillus stearothermophilus* T-6 (Hövel et al., 2003), from *Clostridium thermocellum* ATCC27405 (Taylor et al., 2006), and from *Thermobacillus xylanilyticus* (Paës et al., 2008). All structures revealed that α-L-AFases GH51 possess a typical $(\beta/\alpha)_8$-barrel catalytic domain and a C-terminal domain of unknown function with jelly-roll topology. Their quaternary structures are known to be homo-hexamer, trimer of dimers, composed of catalytically independent six monomers. Family GH51 enzymes perform catalysis with net retention of anomeric configuration through the formation, and subsequent breakdown of a covalent β-L-linked arabinofuranosyl-enzyme inter-mediate. Two essential catalytic residues are required for such a reaction, an acid/base and an enzymatic nucleophile. The structure of the $(\beta/\alpha)_8$-barrel domain places GH51 AFases into the super-family defined as clan GH-A in which both catalytic functions are glutamates that are found on strands β-4 (acid/base) and β-7 (nucleophile), respectively.

Hövel et al. (2003) determined the first novel 3D-structure of retaining α-L-AFase GH51 from *G. stearothermophilus* at 1.75 Å resolution. *Gs*-AFase contains all of the 502 amino acid residues except the first four residues at the N-terminus and its catalytic residues, Glu175 as the acid/base and Glu294 as the nucleophile, were recently identified (Shallom et al., 2002a and 2002b). *Gs*-AFase was found to be a homo-hexamer, a trimer of dimers, and each monomeric subunit is organized into two domains. First domain is a catalytic one of $(\beta/\alpha)_8$-barrel structure and the second is comprised of 12 β-strands with an uncommon jelly-roll structure (Figure 1).

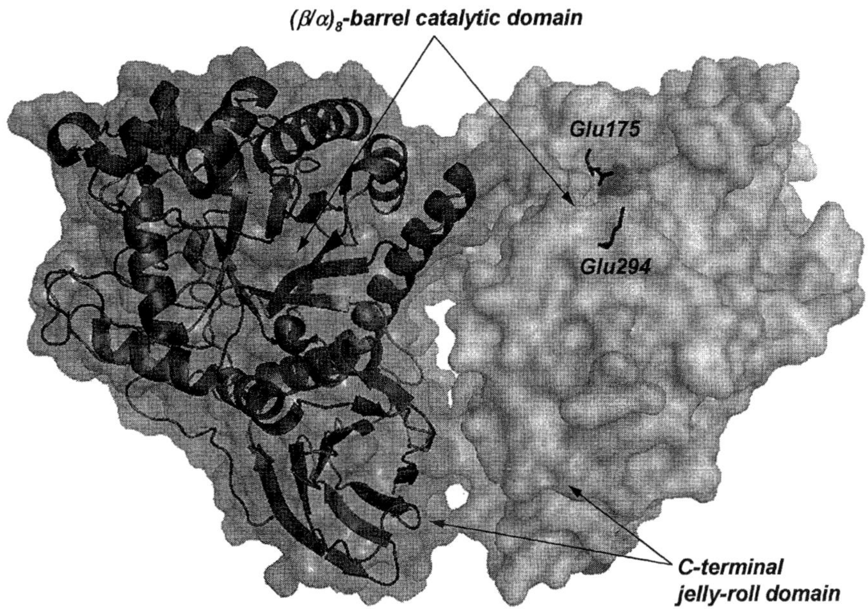

Figure 1 Homo-dimer structure of *Geobacillus stearothermophilus* T-6 AFase
From interior of homo-dimeric structure surrounded by surface, a monomeric *Gs*-AFase (Hövel et al., 2003) is shown as a cartoon diagram. Both of catalytic residues, Glu175 (acid/base) and Glu294 (nucleophile), are apparently emphasized in the surface model on the right side. This image was created from PDB id of 1PZ3 using PyMol software (De Lano Scientific, San Carlos, CA, USA).

In addition, the enzyme complex with the arabinofuranosyl-α-(1,3)-xylose revealed that the arabinofuranose moiety at the -1 subsite is tightly bound by intensive hydrogen bonds, while the xylose at the +1 subsite forms only one hydrogen bond. This report may explain the capability of *Gs*-AFase to bind and hydrolyze substrates with structurally very different leaving groups (Shallom et al., 2002b). This crystal structures proposed nine key residues in *Gs*-AFase responsible for catalysis and substrate binding interactions: Glu29, Arg69, Asn74, Asn174, Glu175, His244, Tyr246, Glu294 and Gln351 (shown in Figure 2). Especially, the residue Asn174 near the acid/base Glu175 is involved in a hydrogen bond to the C-2 sugar hydroxyl group of the arabinofuranosyl moiety. This residue is invariant in the GH-A clan members and was shown to have a critical structural and functional role in maintaining the conformation of the covalent glycosyl-enzyme intermediate and protonation state of the active site residues. Arg69 of *Gs*-AFase is located at the bottom of the active site and seems to keep the catalytic nucleophile Glu294 deprotonated. Hövel et al. (2003) also isolated the covalent glycosyl-enzyme intermediate between *Gs*-AFase Glu175Ala mutant and 2',5'-dinitrophenyl α-L-arabinofuranoside. The catalytic residues of *Gs*-AFase, the acid/base Glu175 and the nucleophile Glu294, are located at the C-terminal ends of β-strands 4 and 7 in the TIM-barrel domain, respectively, which is similar to a GH-A clan enzyme (Zverlov et al., 1998). The structure of intermediate complex revealed that *Gs*-AFase have similar mechanistic features to those found in other glycosidases. The arabinofuranosyl sugars in the active site are tightly bound and distorted by an extensive network of hydrogen bonds. The two catalytic residues are 4.7 Å apart, and together with other conserved residues they contribute to the stabilization of the oxocarbenium ion-like transition state via charge delocalization and specific protein-substrate interactions.

Clostridium thermocellum α-L-Arabinofuranosidase GH51

The second structure of AFase from *Clostridium thermocellum* was crystallized and solved at 3.1 Å resolution by molecular replacement using the related *G. stearothermophilus* AFase by Taylor et al. (2006). *C. thermocellum* is an anaerobic thermophilic soil bacterium that displays a spectrum of plant cell wall-degrading enzymes. *Ct*-AFase shares approximate 65% sequence identity with *Gs*-AFase. The catalytic domain in monomer has a typical $(\beta/\alpha)_8$-barrel structure comprising residues 24~389, which is followed by a 12-stranded β-sandwich domain. As with the *Gs*-AFase, the enzyme is composed of six functional monomers both in crystal and in solution. Both catalytic residues of Glu173 (acid/base) and Glu292 (nucleophile) were found on strands β-4 and β-7 respectively. Glu173Ala mutant of *Ct*-AFase were obtained and co-crystallized with α-(1,5)-linked arabinotriose and arabinosyl-α-(1,3)-xylobiose. Similar interactions of the arabinoside moiety with the -1 subsite were also observed as those of *G. stearothermophilus* AFase (Hövel et al., 2003). An aromatic residue of Trp178 at +1 subsite forms hydrophobic interaction with the aglycone moiety, xylose or arabinose. According to the topology of its active centre, the ligand into the funnel possibly binds the decorated xylans (and arabinans) in which the arabinofuranoside moiety is perpendicular to the main hemicellulose chain. This key specificity determinant for the aglycone sugar in *Ct*-AFase, Trp178, is well-conserved in the *Gs*-AFase as the flexible loop extending from 176 to 179 (refer to Figure 2).

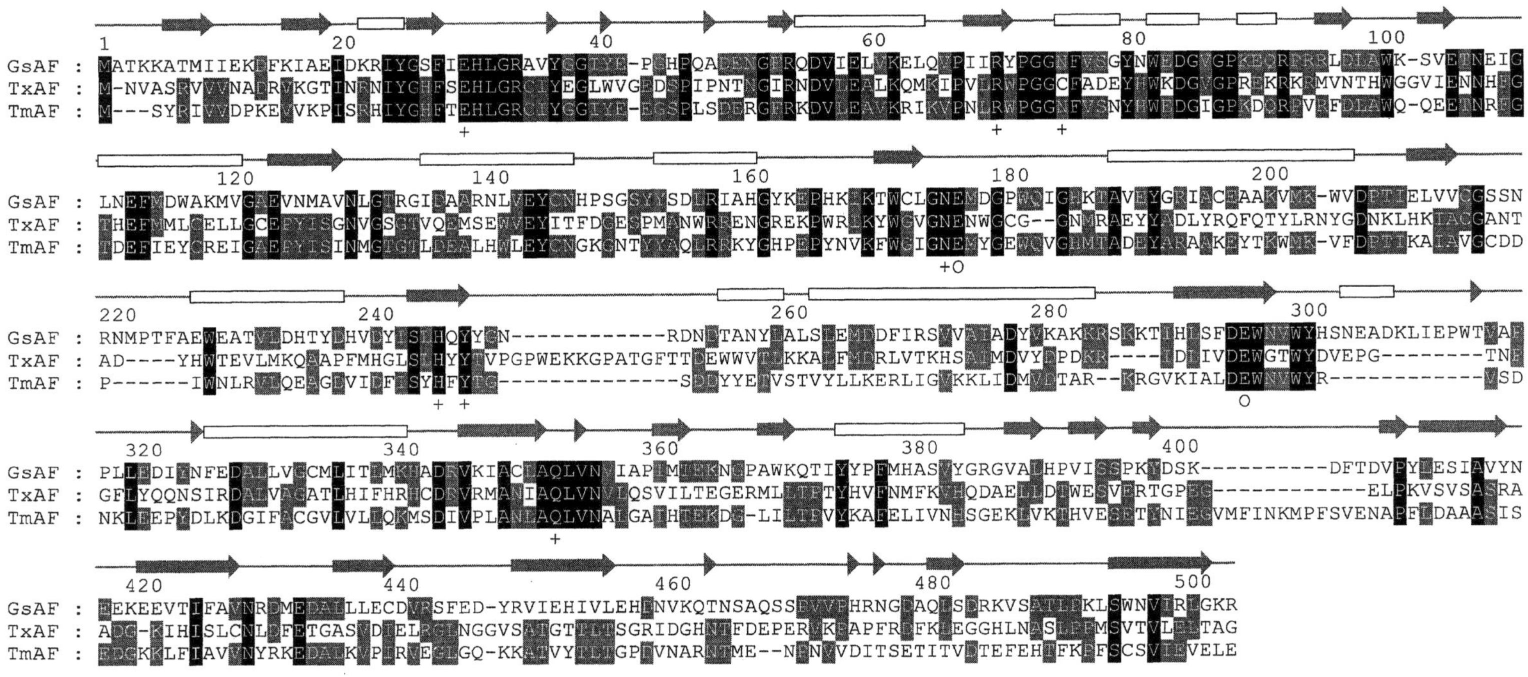

Figure 2 Amino acid sequence alignment among well-known microbial AFases with their secondary structures

Primary structures of three microbial AFases from *Geobacillus stearothermophilus* (*Gs*AF; Hövel et al., 2003), from *Clostridium thermocellum* (*Ct*AF; Taylor et al., 2006), and from *Thermobacillus xylanilyticus* (*Tx*AF; Paës et al., 2008), are compared by amino acid sequence alignment. Above the sequences, secondary structure of *Gs*AF is shown as open boxes (α-helices) and closed arrows (β-seets). Catalytic residues and putative substrate-binding ligands are indicated as open circles and crosses, respectively.

Recently, Paës et al. (2008) reported the complex structure between *Thermobacillus xylanilyticus* AFase Glu176Gln mutant and a branched pentasaccharide. As known structures of other AFase GH51, its overall structure was composed of the catalytic $(\beta/\alpha)_8$-barrel domain and the C-terminal domain with jelly-roll architecture. A branched pentasaccharide, xylanosyl-β-(1,4)-arabinosyl-α-(1,3)-xylotriose, was bound in a groove on the surface of the enzyme, with the mono-arabinosyl branch entering a tight pocket harboring the catalytic dyad (Figure 3). Detailed analyses of both structures and comparisons with the two previously determined structures from *Geobacillus stearothermophilus* and *Clostridium thermocellum* reveal important details unique to the *Thermobacillus xylanilyticus* enzyme. In the absence of substrate, the enzyme adopts an open conformation. In the substrate-bound form, the long loop connecting β-strand 2 to α-helix 2 closes the active site and interacts with the substrate through residues His98 and Trp99. The results of kinetic and fluorescence titration studies using mutants showed the importance of this loop, and support the notion of an interaction between Trp99 and the bound substrate. It was suggested that the changes in loop conformation are an integral part of the *T. xylanilyticus* α-L-arabinofuranosidase reaction mechanism, and ensure efficient binding and release of substrate.

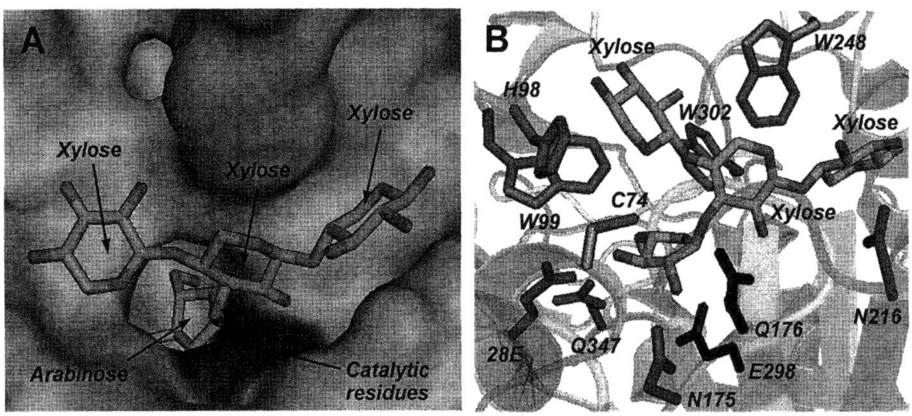

Figure 3 Closed-up view of enzyme-substrate complex structure between *Thermobacillus xylanilyticus* AFase and a branched substrate analogue
A branched substrate analogue, xylanosyl-β-(1,4)-arabinosyl-α-(1,3)-xylotriose, was co-crystllized with *Tx*-AFase mutant (Paës et al., 2008). Closed-up view near the interface between enzyme and substrate in a surface model is shown in panel (A). Catalytic amino acid residues (in black) and substrate-binding ligands (in dark grey) are located near substrate analogue (B). The images were created from PDB id of 2VRQ using PyMOL.

Aspergillus kawachii α-L-Arabinofuranosidase GH54

An *α*-L-arabinofuranosidase belonging to GH family 54 (AkAbf54) was firstly reported from an industrially important fungus, *Aspergillus kawachii* IFO4308 (Miyanaga et al., 2004), which consists of two domains, a catalytic and an arabinose-binding domain

(AkCBM42; Miyanaga et al., 2006). The catalytic domain has a β-sandwich fold similar to those of Clan-B glycoside hydrolases and its AkCBM42 with β-trefoil fold is classified into carbohydrate-binding module (CBM) family 42 (Figure 4). The nucleophile and acid/base residues of AkAbf54 were determined to be Glu221 and Asp297, respectively. In the arabinose-complex structure, one of three arabinofuranose molecules is bound to the catalytic domain and the other two molecules are bound to AkCBM42. In the complex between AkAbf54 and an arabinofuranosyl-α-(1,2)-xylobiose, the arabinose moiety occupies the binding pocket of AkCBM42, whereas the xylobiose moiety is exposed to the solvent. Based on isothermal titration calorimetry and frontal affinity chromatography, they proposed that AkCBM42 binds the non-reducing end arabinofuranosidic moiety of hemicellulose. It means that a CBM can specifically recognize the side-chain mono-saccharides of branched hemicelluloses.

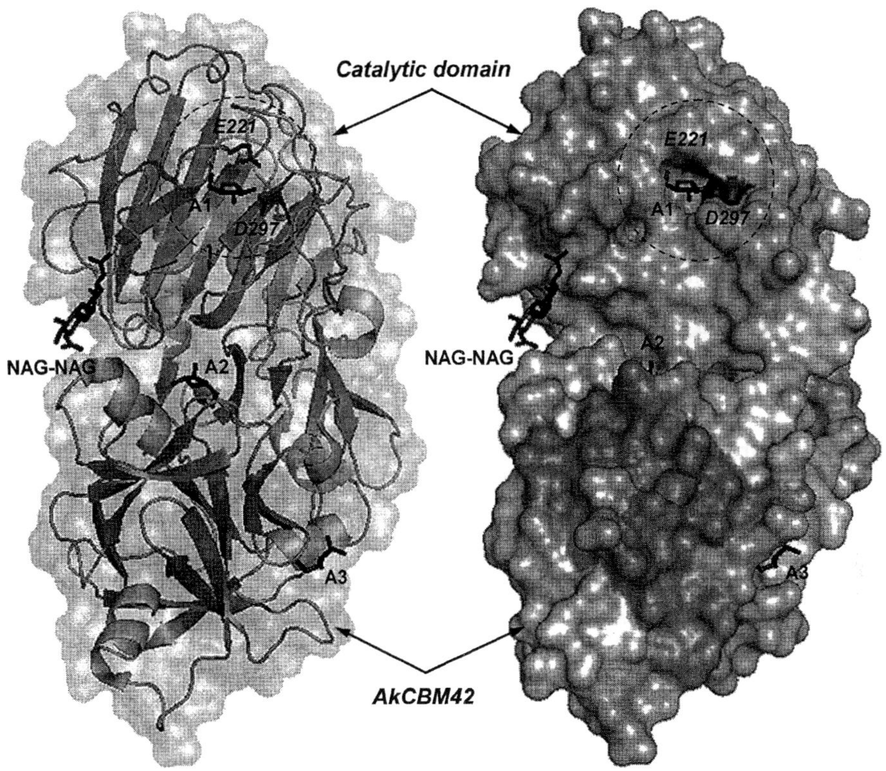

Figure 4 Three-dimensional structure of *Aspergillus kawachii* AFase GH54 with carbohydrate-binding module (CBM) family 42

Structures of the fungal AFase GH54 with CBM42 are shown as cartoon diagram surrounded by (A) transparent and (B) non-transparent surface models, respectively. Three arabinoses (A1 to A3) and two *N*-acetyl glucosamine-linked molecules (NAG-NAG) are bound to different parts of the AFase molecule. Catalytic cleft is indicated by a dashed circle and the key residues are shown in black. These images was created from PDB id of 1WD4 (Miyanaga et al., 2004) using PyMOL program.

endo-Acting α-L-arabinanases

Linear or debranched arabinan comprising α-(1,5)-linked L-arabinofuranosides can readily be hydrolyzed mainly by *endo*-1,5-α-L-arabinanases. ABNases are members of GH family 43, which hydrolyze α-(1,5)-L-arabinofuranoside linkages in arabinan. Recent crystallographic studies (Nurizzo et al., 2002; Yamaguchi et al., 2005) showed that known ABNases have monomeric structure with a typical β-propeller topology (Figure 5). Combined with the structural insights of AFases, a variety of protein engineering between *exo*- and *endo*-hydrolases will be one of the emerging topics in applied and industrial enzymology.

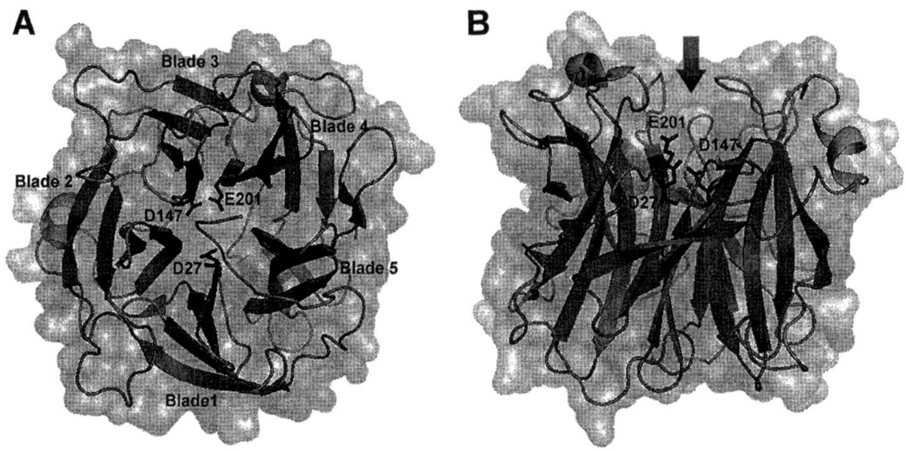

Figure 5 Structure of *Bacillus thermodenitrificans* ABNase GH43
Three-dimensional structure of *Bacillus endo*-1,5-α-L-arabinanase is shown as cartoon diagram of β-propeller surrounded by transparent surface. About its catalytic pocket, top view (A) and side view are compared and its catalytic amino acid residues, D27, D147, and E201, are drawn in black. These images were created from PDB id of 1W7L (Nurizzo et al., 2002) using PyMOL program.

Cellvibrio japonicus endo-1,5-α-L-Arabinanase GH43

Nurizzo et al. (2002) determined the first three-dimensional structure of novel ABNase (*Cj*Arb43A) from *Cellvibrio japonicus*. The structure determined at 1.9Å resolution revealed a novel five-bladed β-propeller fold topology. A long V-shaped surface groove, partially enclosed at one end, forms a single extended substrate-binding surface across the face of the propeller. Three carboxylates, Asp38, Asp158, and Glu221, deep in the active site cleft provide the general acid/base residues for the hydrolysis of glycosidic linkages as an inverting mechanism. They proposed that Arb43A possesses six sugar-binding subsites, with cleavage of arabinohexaose occurring between arabinofuranose residues 3 and 4. Glu221 residue was supposed to be the nucleophile on the basis of its structural location (Figure 6).

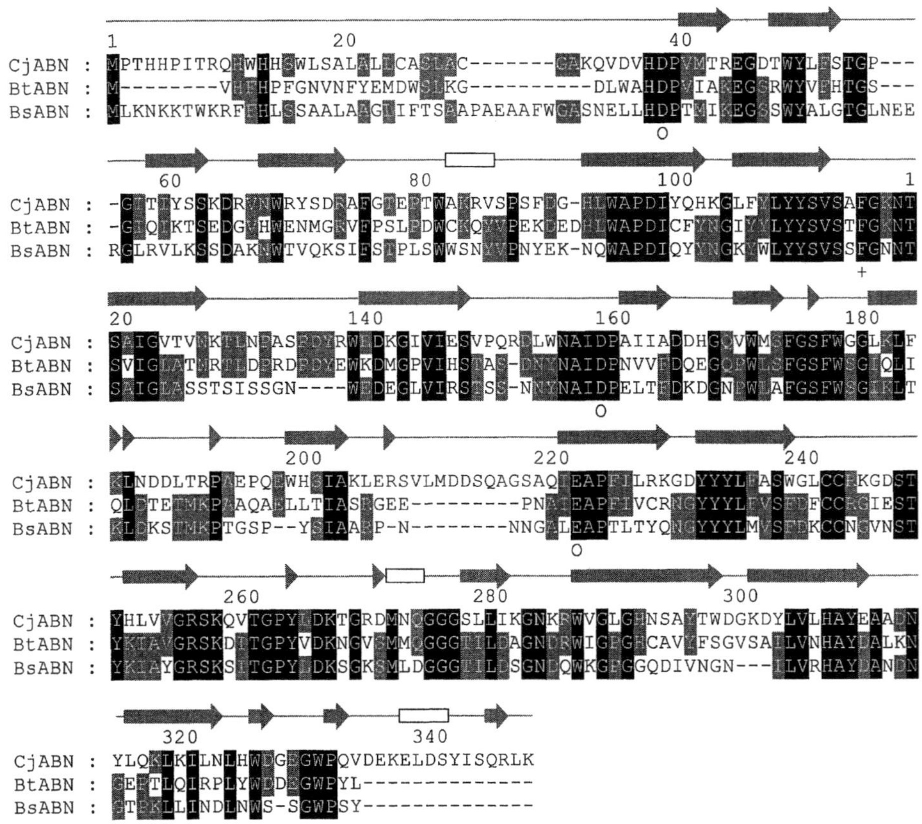

Figure 6 Amino acid sequence alignment among well-known three microbial ABNases with their secondary structures

Primary structures of three microbial ABNases from *Cellvibrio japonicus* (*Cj*ABN; Nurizzo et al. 2002), from *Bacillus thermodenitrificans* (*Bt*ABN; Yamaguchi et al., 2005), and from *Bacillus subtilis* (*Bs*ABN; Proctor et al., 2005), are compared by amino acid sequence alignment. Above the sequences, secondary structure of *Cj*ABN is shown as open boxes (α-helices) and closed arrows (β-seets). Putative catalytic residues are indicated as open circles below the sequences.

According to the substrate-enzyme complex structure, its substrate-binding cleft is positioned at its surface and the catalytic carboxylates are located at its center (Figure 7). *Cj*Arb43A was known as a novel arabinanase with both *exo*- and *endo*-activities to generate mainly arabinotriose (McKie et al., 1997). They also found that the substrate-binding groove is not obviously enclosed but is partially blocked at one end by a large loop between third and fourth propeller blades (Nurizzo et al., 2002). Therefore, Proctor et al. (2005) have tried the rational design approach that led to the conversion of the *Cj*Arb43A enzyme from an *exo*- to an *endo*-mode of action. They reported that a double mutant of Asp35Leu/Gln316Ala displays similar activity to wild-type enzyme and the removal of the steric block mediated by the side chains of Gln316 and Asp53 at the -3 subsite confers its capacity to attack the internal glycoside bonds.

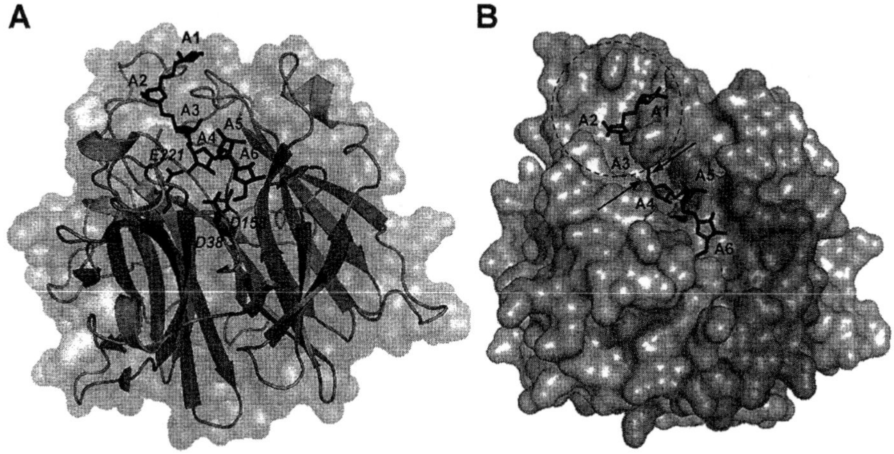

Figure 7 Structure of *endo*-acting ABNase GH43 from *Cellvibrio japonicus*
Three-dimensional structure of *Cellvibrio japonicus endo*-1,5-α-L-arabinanase is shown as cartoon diagram of β-propeller surrounded by (A) transparent and (B) non-transparent surface models. Arabinohexaose substrate is bound at the V-shaped long active groove region. The unique blocking-wall structure at one end in catalytic pocket of ABNase is marked by a dashed circle. These images were created from PDB id of 1GYE (Nurizzo et al., 2002) using PyMOL program.

Bacillus thermodenitrificans endo-1,5-α-L-Arabinanase GH43

Recently, a thermostable and typical *endo*-acting ABNase (ABN-TS) from *Bacillus thermodenitrificans* TS-3 was cloned and its three-dimensional structure was determined at 1.9 Å resolution (Yamaguchi et al., 2005). The enzyme showed a five-bladed β-propeller fold, which is identical to *Cellvibrio* ABNase (Nurizzo et al., 2002). The substrate-binding cleft formed across one face of the propeller is open on both sides to allow random binding of several sugar units in arabinan polymer. This wide-open active cleft structure enables *Bt*-ABN to hydrolyze polymers with *endo*-action.

On the contrary, one side of active groove in *Cellvibrio* ABNase is blocked (as shown in Figure 7B), which causes its *exo*-favored hydrolysis pattern. From the crystal structure, Asp27, Asp147, and Glu201 were supposed to be the catalytic residues of ABN-TS. Yamaguchi et al. (2005) also proposed that the β-propeller fold is stabilized through a ring closure. In ABN-TS, the fifth blade is composed of the C-terminal residues. Thus, instead of the classical "velcro"-like structure (Figure 8), ABN-TS exhibits a new closure-mode in the β-propeller, which is composed of the N-terminal residues, from Phe7 to Gly21, located between the first and last blades. In addition, they reported that deletion of 16 amino acid residues (Val2 to Trp17) from N-terminus decreased remarkably the thermostability of ABN-TS. To date, any approaches in protein engineering such as mutagenesis have rarely been done. However, it can be one of good trials for making more powerful arabinosyl hydrolases.

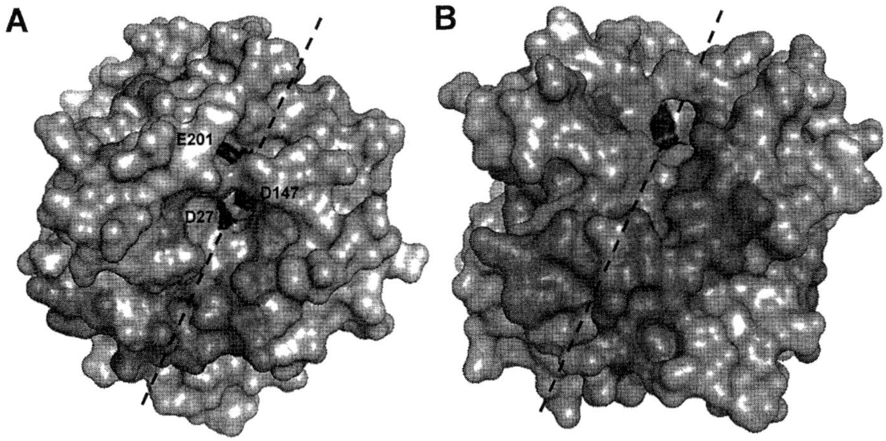

Figure 8 Unique closed active site structure of *B. thermodenitrificans* ABNase
Three-dimensional structure of *B. thermodenitrificans endo*-1,5-α-L-arabinanase is shown as surface diagrams with different angles of view (A) and (B). Each dashed line shows the interface between any long-chain substrate and *endo*-acting enzyme. Black-colored regions correspond each catalytic residues. These images were created from PDB id of 1W7L using PyMOL.

ARABINOSYL HYDROLASE GENES IN NATURE

To date, many AFase-type enzymes have been isolated and characterized from various microbial sources. Recently, genome-wide BLAST search showed that various bacterial strains possess the putative AFase genes with well-conserved motif sequences at the nucleotide and amino acid sequence levels. According to the genome database, various microorganisms have AFase genes in their genome and share a broad range of similarity in the primary structure with each other. Park et al. (2007) have proposed two sets of degenerate PCR primers and successfully PCR-detected various putative AFase genes, based on their three highly conserved amino acid blocks (PGGNFV, GNEMDG, and DEWNVW). If possible, more efficient and powerful techniques to find a variety of novel arabinosyl hydrolase genes from nature should be developed and they can make the versatile biocatalysts be widely applied to various industrial saccharifying processes.

In this study, the amino acid sequence of *Geobacillus* AFase was aligned and compared with various other sequences available from the NCBI database using the BLAST searching program (Table 4). Interestingly, a number of putative AFase genes have been found from database, which share variable amino acid sequence identity (26~96%) with each other. Nevertheless, there has still been some uncertainty or confusion among putative AFases, due to their complexity in substrate specificities. As shown in Table 4, *Thermobacillus xylanilyticus* AFase shares 26% of extremely low identity with *Geobacillus* AFase, but it was already known as one of typical AFases on the basis of its enzymatic properties and three-dimensional structure. Therefore, all putative AFases found here can be the promising candidates for the development of arabinosyl hydrolases.

Table 4 Comparison of deduced amino acid identity among known or putative AFase genes found in microbial genome databases

	Microbial origins	Protein ID[1)]	ASI[2)] (%)	References[3)]
1	*Geobacillus stearothermophilus* T-6	ACE73682	100	Gilead (1995)
2	*Geobacillus thermoleovorans*	ABD48560	96	-
3	*Geobacillus thermodenitrificans* NG80-2	ABO67151	95	-
4	*Geobacillus stearothermophilus* KCTC 3012	ABM68633	91	Park (2007)
5	*Geobacillus caldoxylosilyticus* TK4	ABI34800	90	Canakci (2007)
6	*Anoxybacillus kestanbolensis* AC26Sari	ACB54691	90	-
7	*Bacillus halodurans* C-125	BAB05580	78	-
8	*Bacillus licheniformis* DSM 13	AAU41887	76	-
9	*Paenibacillus* sp. HanTHS1	ABZ10760	72	-
10	*Bacillus subtilis*	CAB14832	71	Kaneko (1994)
11	*Bacillus amyloliquefaciens* FZB42	ABS74936	70	-
12	*Clostridium beijerinckii* NCIMB 8052	ABR34520	70	-
13	*Bacillus clausii* KSM-K16	BAD62939	69	-
14	*Clostridium thermocellum* ATCC 27405	ABN53749	68	Taylor (2006)
15	*Caldicellulosiruptor saccharolyticus* DSM8903	ABP67153	66	-
16	*Clostridium phytofermentans* ISDg	ABX41547	64	-
17	*Lactobacillus brevis* ATCC 367	ABJ64817	59	-
18	*Rhizobium etli* CIAT 652	ACE93495	58	-
19	*Rhizobium leguminosarum* bv. *viciae*	CAK03231	57	-
20	*Rhizobium etli* CFN 42	ABC93630	57	-
21	*Marinomonas* sp. MWYL1	ABR71625	55	-
22	*Agrobacterium tumefaciens* str. C58	AAK90279	54	-
23	*Mesorhizobium loti* MAFF303099	BAB50453	54	-
24	*Arthrobacter aurescens* TC1	ABM06849	54	-
25	*Verminephrobacter eiseniae* EF01-2	ABM59178	54	-
26	*Streptomyces coelicolor* A3(2)	CAB86096	54	-
27	*Streptomyces lividans*	AAA61708	54	Manin (1994)
28	*Streptomyces avermitilis* MA-4680	BAC73455	54	-
29	*Sinorhizobium meliloti* 1021	CAC49446	53	-
30	*Sinorhizobium medicae* WSM419	ABR64329	52	-
31	*Clavibacter michiganensis* subsp. *Sepedonicus*	CAQ00998	52	-
32	*Clavibacter michiganensis*	CAN00332	51	-
33	*Kineococcus radiotolerans* SRS30216	ABS02525	49	-
34	*Reinekea* sp.	EAR07724	49	-
35	*Bifidobacterium longum* DJO10A	ACD99184	48	-
36	*Arthrobacter* sp. FB24	ABK01627	48	-

(continued to next page)

Table 4 Continued

	Microbial origins	Protein ID [1)]	ASI [2)] (%)	References [3)]
37	*Bifidobacterium longum* NCC2705	AAN24971	45	-
38	*Bifidobacterium longum* B667	AAO84266	45	Margolles (2003)
39	*Roseiflexus castenholzii* DSM 13941	ABU56727	45	-
40	*Opitutus terrae* PB90-1	ACB75101	44	-
41	*Roseiflexus* sp. RS-1	ABQ91872	44	-
42	*Solibacter usitatus* Ellin6076	ABJ84616	40	-
43	*Thermotoga maritima* MSB8	AAD35369	37	Miyazaki (2005)
44	*Aspergillus fumigatus* Af293	XP_752357	37	-
45	*Thermotoga* sp. RQ2	ACB09009	36	-
46	*Emericella nidulans*	ABF50847	36	-
47	*Pyrenophora triticrepentis* Pt-1C-BFP	XP_001935209	36	-
48	*Thermotoga petrophila* RKU-1	ABQ46651	35	-
49	*Aspergillus clavatus* NRRL 1	XP_001269270	35	-
50	*Neosartorya fischeri*	XP_001264777	35	-
51	*Thermotoga lettingae* TMO	ABV33713	33	-
52	*Cryptococcus neoformans* var. *neoformans*	XP_572197	33	-
53	*Caulobacter* sp. K31	ABZ72739	30	-
54	*Cytophaga xylanolytica* XM3	AAC38456	30	Renner (1998)
55	*Cellvibrio japonicus* Ueda107	ACE86344	30	-
56	*Clostridium stercorarium*	AAC28125	29	Schwarz (1995)
57	*Flavobacterium johnsoniae* UW101	ABQ04151	29	-
58	*Clostridium cellulovorans*	AAN05450	28	Kosugi (2002)
59	*Caulobacter crescentus* CB15	AAK23403	28	-
60	*Gramella forsetii* KT0803	CAL65667	27	-
61	*Xanthomonas oryzae* pv. *oryzae* PXO99A	ACD60479	27	-
62	*Bacteroides thetaiotaomicron* VPI-5482	AAO75455	27	-
63	*Thermobacillus xylanilyticus*	CAA76421	26	Debeche (2000)

[1)] Each gene product can be identified and obtained from NCBI database by using its own number.
[2)] ASI (%) means relative amino acid sequence identity determined by sequence alignment on the basis of *G. stearothermophilus* T-6 AFase as 100 percent.
[3)] References for molecular cloning or enzymatic characterization were shown here.

At the same way, various putative ABNases were also found from NCBI database and the results were shown in Table 5. It has been known that various microbes possess putative ABNase genes in their own genomes. Compared with AFases, the distribution spectrum of ABNases in nature seems to be narrow and their application may be restricted to arabian homopolymer. However, ABNase can synergistically accelerate *exo*-acting hydrolysis of AFase in commercial processes, which is one of the reasons why they have to be focused in future.

Table 5 Comparison of deduced amino acid identity among known or putative ABNase genes found in microbial genome databases

	Microbial origins	Protein ID[1)]	ASI[2)] (%)	References[3)]
1	*Bacillus thermodenitrificans* TS-3	BAB64339	100	Takao (2002)
2	*Bacillus licheniformis* DSM 13	AAU40201	56	-
3	*Bacillus subtilis* 168T+	AAV87172	53	Leal (2004)
4	*Bacillus amyloliquefaciens* FZB42	ABS74945	52	-
5	*Bacillus subtilis* subsp. subtilis str. 168	CAB14841	52	-
6	*Opitutus terrae*	ACB74675	50	-
7	*Shewanella* sp. MR-7	ABI42969	50	-
8	*Shewanella* sp. ANA-3	ABK48312	50	-
9	*Saccharophagus degradans* 2-40	ABD80276	50	-
10	*Shewanella* sp. MR-4	ABI39068	49	-
11	*Cellvibrio japonicus* [4)]	CAA71485	48	McKie (1997)
12	*Shewanella putrefaciens*	ABP75772	47	-
13	*Shewanella* sp. W3-18-1	ABM24796	47	-
14	*Gramella forsetii* KT0803	CAL65671	45	-
15	*Kineococcus radiotolerans* SRS30216	ABS01569	40	-
16	*Aspergillus clavatus* NRRL 1	XP_001271414	37	-
17	*Neosartorya fischeri* NRRL 181	XP_001263002	36	-
18	*Aspergillus niger*	AAA32682	36	-
19	*Emericella nidulans*	ABF50890	35	-
20	*Aspergillus fumigatus* Af293	XP_754164	35	-
21	*Penicillium chrysogenum*	BAD89094	34	Sakamoto (2003b)
22	*Bacteroides fragilis* NCTC 9343	CAH06067	34	-
23	*Aspergillus niger* CBS 513.88	XP_001393437	33	-
24	*Rhodococcus* sp. RHA1	ABG97171	32	-
25	*Bacteroides thetaiotaomicron*	AAO75474	31	-
26	*Saccharopolyspora erythraea* NRRL2338	CAM03699	30	-
27	*Bifidobacterium longum* NCC2705	AAN24036	26	-

[1)] Each gene product can be identified and obtained from NCBI database by using its own number.
[2)] ASI (%) means relative amino acid sequence identity determined by sequence alignment on the basis of *B. thermodenitrificans* TS-3 ABNase as 100 percent.
[3)] References for molecular cloning or enzymatic characterization were shown here.
[4)] The bacterium was previously known as *Pseudomonas cellulosa*.

APPLICATION IN L-ARABINOSE PRODUCTION

In a previous review for AFases, Numan and Bhosle (2006) mentioned that AFases are accessory enzymes that cleave α-L-arabinofuranosidic linkages and act synergistically with other hemicellulases and pectic enzymes for the complete hydrolysis of hemicelluloses and pectins. These enzymes warrant substantial research efforts because they represent potential rate-limiting enzymes in the degradation of lignocelluloses from agricultural residues (Saha, 2000). The action of α-L-AFase alone or in combination with other lignocellulose-degrading enzymes represents a promising biotechnological tool as the alternatives to some of the existing chemical technologies such as in pulp and paper industry, synthesis of oligosaccharides and pretreatment of lignocelluloses for bio-ethanol production. For such reasons, researches on AFase have been continuously increased for over a decade. On the contrary, there has been relatively little attention to ABNases, due to their narrow distribution in nature and limitations in substrate-preference. However, together with *exo*-acting AFases, ABNases can be one of essential enzymes especially for the production of L-arabinose from arabinan polymer. Arabinan is an α-(1,5)-linked polymer of L-arabinofuranose units partially branched with α-(1,2)- or α-(1,3)-linked arabinose moieties (Figure 9).

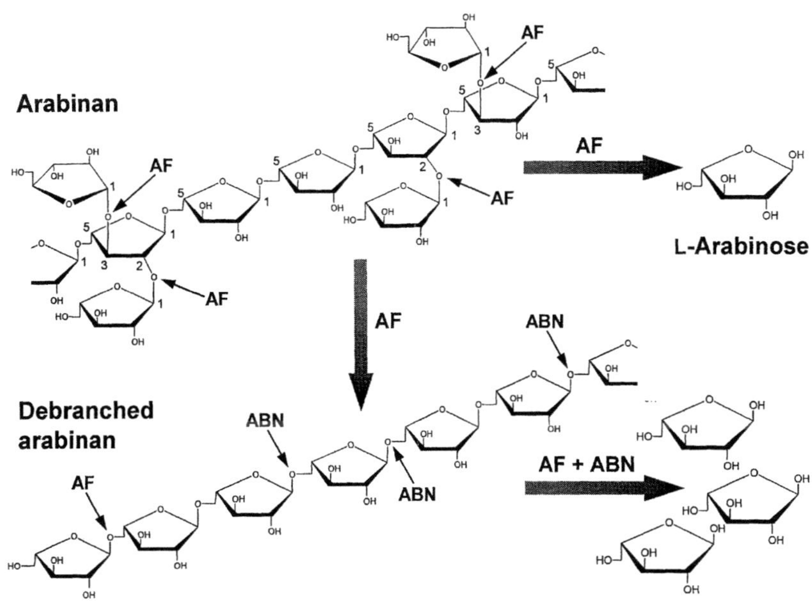

Figure 9 Scheme of arabinan hydrolysis using both AFase and ABNase
Arabinan is an α-L-(1,5)-linked arabinose polymer with α-L-(1,2)- or α-L-(1,3)-linked arabinose branches. At the first stage, AFase can remove terminal branched arabinoses to generate debranched or linear arabinan. Then ABNase and AFase can simultaneously be treated for the more efficient L-arabinose recovery via their synergistic actions.

Due to the steric hindrance in branched structure, typical ABNases are likely to possess very poor accessibility to branched structured arabinan polymer. However, the pre-treatment of AFase can trim various branched arabinoses from whole arabinan polymer, which can provide the high possibility in *endo*-hydrolysis of debranched arabinan by ABNases. Moreover, ABNase can supply AFases with more shortened arabinooligosaccharides as better substrates. Undoubtedly, common AFase is supposed to hydrolyze α-(1,5)-linked debranched arabinan backbone to complete saccharification to arabinose monomers. However, the preference of various AFases to individual α-(1,2)-, α-(1,3)-, or α-(1,5)-arabinofuranosyl linkages can be varied and the efficacy in hydrolysis reaction may be differ from each other. Accordingly, the development of highly synergistic processes using both *exo*- and *endo*-acting arabinosyl hydrolases will stimulate the cost-effective industrial production of functional sugar, L-arabinose. For these purposes, novel hydrolases with versatile activity and specificity should be picked out from the probable candidates studied here and also the existing enzymes should be improved to more useful ones via a variety of emerging protein engineering technologies in future.

ACKNOWLEDGEMENTS

This work was supported by the Quality of Life project from the Ministry of Knowledge Economy (MKE).

REFERENCES

Beldman G, Schols H A, Piston S M, Searl-van Leeuwen M J F and Voragen A G J (1997), 'Arabinan and arabinan degrading enzymes', *Adv Macromol Carbohydr Res*, 1, 1-64.

Beylot M H, McKie V A, Voragen A G J, Doeswijk-Voragen C H L and Gilbert H J (2001), 'The *Pseudomonas cellulosa* glycoside hydrolase family 51 arabinofuranosidase exhibits wide substrate specificity', *Biochem J*, 358 (Pt 3), 607-614.

Birgisson H, Fridjonsson O, Bahrani-Mougeot F K, Hreggvidsson G O, Kristjansson J K and Mattiasson B (2004), 'A new thermostable α-L-arabinofuranosidase from a novel thermophilic bacterium', *Biotechnol Lett*, 26 (17), 1347-1351.

Canakci S, Belduz A O, Saha B C, Yasar A, Ayaz F A and Yayli N (2007), 'Purification and characterization of a highly thermostable α-L-arabinofuranosidase from *Geobacillus caldoxylolyticus* TK4', *Appl Microbiol Biotechnol*, 75 (4), 813-820.

Carvallo M, de Ioannes P, Navarro C, Chavez R, Peirano A, Bull P and Eyzaguirre J (2003), 'Characterization of an α-L-arabinofuranosidase gene (abfl) from *Peni-cillium purpurogenum* and its expression', *Mycol Res*, 107 (Pt 4), 388-394.

Christakopoulos P, Katapodis P, Hatzinikolaou D G, Kekos D and Macris B J (2000), 'Purification and characterization of an extracellular α-L-arabinofuranosidase from *Fusarium oxysporum*', *Appl Biochem Biotechnol*, 87 (2), 127-133.

Debeche T, Cummings N, Connerton I, Debeire P and O'Donohue M J (2000), 'Genetic and biochemical characterization of a highly thermostable α-L-arabinofuranosidase from *Thermobacillus xylanilyticus*', *Appl Environ Microbiol*, 66 (4), 1734-1736.

Degrassi G, Vindigni A and Venturi V (2003), 'A thermostable α-arabinofurano-sidase from xylanolytic *Bacillus pumilus*: purification and characterisation', *J Biotechnol*, 101 (1), 69-79.

de Wet B J M, Matthew M K A, Storbeck K H, van Zyl W H and Prior B A (2008), 'Characterization of a family 54 α-L-arabinofuranosidase from *Aureobasidium pullulans*', *Appl Microbiol Biotechnol*, 77 (5), 975-983.

Filho E X, Puls J and Coughlan M P (1996), 'Purification and characterization of two arabinofuranosidases from solid-state cultures of the fungus *Penicillium capsulatum*', *Appl Environ Microbiol*, 62 (1), 168-173.

Fritz M, Ravanal M C, Braet C and Eyzaguirre J (2008), 'A family 51 α-L-arabinofuranosidase from *Penicillium purpurogenum*: purification, properties and amino acid sequence', *Mycol Res*, 112 (Pt 8), 933-942.

Gilead S and Shoham Y (1995), 'Purification and characterization of α-L-arabinofuranosidase from *Bacillus stearothermophilus* T-6', *Appl Environ Microbiol*, 61 (1), 170-174.

Gomes J, Gomes I I, Terler K, Gubala N, Ditzelmuller G and Steiner W (2000), 'Optimisation of culture medium and conditions for α-L-arabinofuranosidase production by the extreme thermophilic eubacterium *Rhodothermus marinus*', *Enzyme Microb Technol*, 27 (6), 414-422.

Greve L C, Labavitch J M and Hungate R E (1984), 'α-L-Arabinofuranosidase from *Ruminococcus albus* 8: purification and possible role in hydrolysis of alfalfa cell wall', *Appl Environ Microbiol*, 47 (5), 1135-1140.

Hashimoto T and Nakata Y (2003), 'Synergistic degradation of arabinoxylan with α-L-arabinofuranosidase, xylanase and β-xylosidase from soy sauce koji mold, *Aspergillus oryzae*, in high salt condition', *J Biosci Bioeng*, 95 (2), 164-169.

Henrissat B and Davies G (1997), 'Structural and sequence-based classification of glycoside hydrolases', *Curr Opin Struct Biol*, 7 (5), 637-644.

Hespell R B and O'bryan P J (1992), 'Purification and characterization of an α-L-arabinofuranosidase from *Butyrivibrio fibrisolvens* GS113', *Appl Environ Microbiol*, 58 (4), 1082-1088.

Hövel K, Shallom D, Niefind K, Belakhov V, Shoham G, Baasov T, Shoham Y and Schomburg D (2003), 'Crystal structure and snapshots along the reaction pathway of a family 51 α-L-arabinofuranosidase', *EMBO J*, 22 (19), 4922-4932.

Inácio J M and Sá-Nogueira I (2008), 'Characterization of abn2 (yxiA), encoding a *Bacillus subtilis* GH43 arabinanase, Abn2, and its role in arabino-polysaccharide degradation', *J Bacteriol*, 190 (12), 4272-4280.

Kaji A and Tagawa K (1970), 'Purification, crystallization and amino acid composition of α-L-arabinofuranosidase *from Aspergillus niger*', *Biochim Biophys Acta*, 207 (3), 456-464.

Kamitori S, Kondo S, Okuyama K, Yokota T, Shimura Y, Tonozuka T and Sakano Y (1999), 'Crystal structure of *Thermoactinomyces vulgaris* R-47 α-amylase II (TVAII) hydrolyzing cyclodextrins and pullulan at 2.6 Å resolution', *J Mol Biol*, 287 (5), 907-921.

Kaneko S, Shimasaki T and Kusakabe I (1993), 'Purification and some properties of intracellular α-L-arabinofuranosidase from *Aspergillus niger* 5-16', *Biosci Bio-technol Biochem*, 57 (7), 1161-1165.

Kaneko S, Sano M and Kusakabe I (1994), 'Purification and some properties of α-L-arabinofuranosidase from *Bacillus subtilis* 3-6', *Appl Environ Microbiol*, 60 (9), 3425-3428.

Kaneko S, Arimoto M, Ohba M, Kobayashi H, Ishii T and Kusakabe I (1998), 'Purification and substrate specificities of two α-L-arabinofuranosidases from *Aspergillus awamori* IFO 4033', *Appl Environ Microbiol*, 64 (10), 4021-4027.

Kimura I, Yoshioka N, Kimura Y and Tajima S (2000), 'Cloning, sequencing and expression of an α-L-arabinofuranosidase from *Aspergillus sojae*', *J Biosci Bioeng*, 89 (3), 262-266.

Koseki T, Miwa Y, Mese Y, Miyanaga A, Fushinobu S, Wakagi T, Shoun H, Matsuzawa H and Hashizume K (2006), 'Mutational analysis of *N*-glycosylation recognition sites on the biochemical properties of *Aspergillus kawachii* α-L-arabinofuranosidase 54', *Biochim Biophys Acta*, 1760 (9), 1458-1464.

Kosugi A, Murashima K and Doi R H (2002), 'Characterization of two noncellulosomal subunits, ArfA and BgaA, from *Clostridium cellulovorans* that cooperate with the cellulosome in plant cell wall degradation', *J Bacteriol*, 184 (24), 6859-6865.

Leal T F and Sá-Nogueira I (2004), 'Purification, characterization and functional analysis of an endo-arabinanase (AbnA) from *Bacillus subtilis*', *FEMS Microbiol Lett*, 241 (1), 41-48.

Lee R C, Hrmova M, Burton R C, Lahnstein J and Fincher R C (2003), 'Bifunctional family 3 glycoside hydrolases from barley with α-L-arabinofuranosidase and β-D-xylosidase activity. Characterization, primary structures, and COOH-terminal processing', *J Biol Chem*, 278 (7), 5377-5387.

Mai V, Wiegel J and Lorenz W W (2000), 'Cloning, sequencing, and characterization of the bifunctional xylosidase-arabinosidase from the anaerobic thermophile *thermoanaero-bacter ethanolicus*', *Gene*, 247 (1-2), 137-143.

Manin C, Shareek F, Morosoli R and Kluepfel D (1994), 'Purification and characterization of an α-L-arabinofuranosidase from *Streptomyces lividans* 66 and DNA sequence of the gene (abfA)', *Biochem J*, 302 (Pt 2), 443-449.

Margolles A and de los Reyes-Gavilan C G (2003), 'Purification and functional characterization of a novel α-L-arabinofuranosidase from *Bifidobacterium longum* B667', *Appl Environ Microbiol*, 69 (9), 5096-5103.

Matsumura K, Obata H, Hata Y, Kawato A, Abe Y and Akita O (2004), 'Isolation and characterization of a novel gene encoding α-L-arabinofuranosidase from *Aspergillus oryzae*', *J Biosci Bioeng*, 98 (2), 77-84.

Matsuo N, Kaneko S, Kuno A, Kobayashi H and Kusakabe I (2000), 'Purification, characterization and gene cloning of two α-L-arabinofuranosidases from *strepto-myces chartreusis* GS901', *Biochem J*, 346 (Pt 1), 9-15.

McKie V A, Black G W, Millward-Sadler S J, Hazlewood G P, Laurie J I and Gilbert H J (1997), 'Arabinanase A from *Pseudomonas fluorescens* subsp. *cellulosa* exhibits both an endo- and an exo- mode of action', *Biochem J*, 323 (Pt 2), 547-555.

Miyanaga A, Koseki T, Matsuzawa H, Wakagi T, Shoun H and Fushinobu S (2004), 'Crystal structure of a family 54 α-L-arabinofuranosidase reveals a novel carbo-hydrate-binding module that can bind arabinose', *J Biol Chem*, 279 (43), 44907-44914.

Miyanaga A, Koseki T, Miwa Y, Mese Y, Nakamura S, Kuno A, Hirabayashi J, Matsuzawa H, Wakagi T, Shoun H and Fushinobu S (2006), 'The family 42 carbohydrate-binding module of family 54 α-L-arabinofuranosidase specifically binds the arabinofuranose side chain of hemicellulose', *Biochem J*, 399 (3), 503-511.

Miyazaki K (2005), 'Hyperthermophilic alpha-L-arabinofuranosidase from *Thermotoga maritima* MSB8: molecular cloning, gene expression, and characterization of the recombinant protein', *Extremophiles*, 9 (5), 399-406.

Morales P, Sendra J M and Perez-Gonzalez J A (1995), 'Purification and characterization of an arabinofuranosidase from *Bacillus polymyxa* expressed in *Bacillus subtilis*', *Appl Microbiol Biotechnol*, 44 (1-2), 112-117.

Nogawa M, Yatsui K, Tomioka A, Okada H and Morikawa Y (1999), 'An α-L-arabinofuranosidase from *Trichoderma reesei* containing a noncatalytic xylan-binding domain', *Appl Environ Microbiol*, 65 (9), 3964-3968.

Numan M T and Bhosle N B (2006), 'α-L-Arabinofuranosidases: the potential applications in biotechnology', *J Ind Microbiol Biotechnol*, 33 (4), 247-260.

Nurizzo D, Turkenburg J P, Charnock S J, Roberts S M, Dodson E J, McKie V A, Taylor E J P, Gilbert H J and Davies G J (2002), '*Cellvibrio japonicus* α-L-arabinanase 43A has a novel five-blade beta-propeller fold', *Nat Struct Biol*, 9 (9), 665-668.

Paës G, Skov L K, O'Donohue L K, Remond C, Kastrup L K, Gajhede M and Mirza O (2008), 'The structure of the complex between a branched pentasaccharide and *Thermobacillus xylanilyticus* GH-51 arabinofuranosidase reveals xylan-binding determinants and induced fit', *Biochemistry*, 47 (28), 7441-7451.

Pages S, Kester H C, Visser J and Benen H C (2001), 'Changing a single amino acid residue switches processive and non-processive behavior of *Aspergillus niger* endopolygalacturonase I and II', *J Biol Chem*, 276 (36), 33652-33656.

Panagiotou G, Topakas E, Economou L, Kekos D, Macris B J and Christakopoulos P (2003), 'Induction, purification, and characterization of two extracellular α-L-arabinofuranosidases from *Fusarium oxysporum*', *Can J Microbiol*, 49 (10), 639-644.

Park J M, Han N S and Kim T J (2007), 'Rapid detection and isolation of known and putative α-L-arabinofuranosidase genes using degenerate PCR primers', *J Microbiol Biotechnol*, 17 (3), 481-489.

Proctor M R, Taylor E J, Nurizzo D, Turkenburg J P, Lloyd R M, Vardakou M, Davies G J and Gilbert H J (2005), 'Tailored catalysts for plant cell-wall degradation: redesigning the exo/endo preference of *Cellvibrio japonicus* arabinanase 43A', *Proc Natl Acad Sci U S A*, 102 (8), 2697-702.

Ramon D, vd Veen P and Visser J (1993), 'Arabinan degrading enzymes from *Aspergillus nidulans*: induction and purification', *FEMS Microbiol Lett*, 113 (1), 15-22.

Rémond C, Plantier-Royon R, Aubry N and O'Donohue M J (2005), 'An original chemoenzymatic route for the synthesis of β-D-galactofuranosides using an alpha-L-arabinofuranosidase', *Carbohydr Res*, 340 (4), 637-644.

Rémond C, Plantier-Royon R, Aubry N, Maes E, Bliard C and O'Donohue M J (2004), 'Synthesis of pentose-containing disaccharides using a thermostable α-L-arabinofuranosidase', *Carbohydr Res*, 339 (11), 2019-2025.

Renner M J and Breznak J A (1998), 'Purification and properties of ArfI, an α-L-arabinofuranosidase from *Cytophaga xylanolytica*', *Appl Environ Microbiol*, 64 (1), 43-52.

Saha B C (2000), 'α-L-Arabinofuranosidases: biochemistry, molecular biology and application in biotechnology', *Biotechnol Adv*, 18 (5), 403-423.

Saha B C and Bothast R J (1998), 'Effect of carbon source on production of α-L-arabinofuranosidase by *Aureobasidium pullulans*', *Curr Microbiol*, 37 (5), 337-340.

Sakamoto T and Thibault J F (2001), 'Exo-arabinanase of *Penicillium chrysogenum* able to release arabinobiose from α-1,5-L-arabinan', *Appl Environ Microbiol*, 67 (7), 3319-3321.

Sakamoto T and Kawasaki H (2003a), 'Purification and properties of two type-B α-L-arabinofuranosidases produced by *Penicillium chrysogenum*', *Biochim Biophys Acta*, 1621 (2), 204-210.

Sakamoto T, Ihara H, Kozaki S and Kawasaki H (2003b), 'A cold-adapted endo-arabinanase from *Penicillium chrysogenum*', *Biochim Biophys Acta*, 1624 (1-3), 70-75.

Schwarz W H, Bronnenmeier K, Krause B, Lottspeich F and Staudenbauer W L (1995), 'Debranching of arabinoxylan: properties of the thermoactive recombinant α-L-arabinofuranosidase from *Clostridium stercorarium* (ArfB)', *Appl Microbiol Biotechnol*, 43 (5), 856-860.

Seri K, Sanai K, Matsuo N, Kawakubo K, Xue C and Inoue S (1996), 'L-Arabinose selectively inhibits intestinal sucrase in an uncompetitive manner and suppresses glycemic response after sucrose ingestion in animals', *Metabolism*, 45 (11), 1368-1374.

Shallom D, Belakhov V, Solomon D, Shoham G, Baasov T and Shoham Y (2002), 'Detailed kinetic analysis and identification of the nucleophile in α-L-arabinofuranosidase from *Geobacillus stearothermophilus* T-6, a family 51 glycoside hydrolase', *J Biol Chem*, 277 (46), 43667-43673.

Shallom D, Belakhov V, Solomon D, Gilead-Gropper S, Baasov T, Shoham G and Shoham Y (2002), 'The identification of the acid-base catalyst of α-arabinofuranosidase from *Geobacillus stearothermophilus* T-6, a family 51 glycoside hydrolase', *FEBS Lett*, 514 (2-3), 163-167.

Shin H Y, Park S Y, Sung H J and Kim D H (2003), 'Purification and characterization of α-L-arabinopyranosidase and α-L-arabinofuranosidase from *Bifidobacterium breve* K-110, a human intestinal anaerobic bacterium metabolizing ginsenoside Rb2 and Rc', *Appl Environ Microbiol*, 69 (12), 7116-7123.

Shofiqur R A, Kawamura S, Hatsu M, Hoq M K and Takamizawa K (2001), 'Physico-chemical properties of a novel alpha-L-arabinofuranosidase from *Rhizomucor pusillus* HHT-1', *Can J Microbiol*, 47 (8), 767-772.

Sorensen H R, Jorgensen C T, Hansen C H, Jorgensen C I, Pedersen S and Meyer A S (2006), 'A novel GH43 alpha-L-arabinofuranosidase from *Humicola insolens*: mode of action and synergy with GH51 α-L-arabinofuranosidases on wheat arabinoxylan', *Appl Microbiol Biotechnol*, 73 (4), 850-861.

Tajana E, Fiechter A and Zimmermann W (1992), 'Purification and characterization of two α-L-arabinofuranosidases from *Streptomyces diastaticus*', *Appl Environ Micro-biol*,

58 (5), 1447-1450.

Takao M, Akiyama K and Sakai T (2002), 'Purification and characterization of thermostable endo-1,5-α-L-arabinase from a strain of *Bacillus thermodenitrificans*', *Appl Environ Microbiol*, 68 (4), 1639-1646.

Taylor E J, Smith N L, Turkenburg J P, D'Souza S, Gilbert H J and Davies G J (2006), 'Structural insight into the ligand specificity of a thermostable family 51 arabinofuranosidase, Araf51, from *Clostridium thermocellum*', *Biochem J*, 395 (1), 31-37.

Tsujibo H, Takada C, Wakamatsu Y, Kosaka M, Tsuji A, Miyamoto K and Inamori Y (2002), 'Cloning and expression of an α-L-arabinofuranosidase gene (stxIV) from *Streptomyces thermoviolaceus* OPC-520, and characterization of the enzyme', *Biosci Biotechnol Biochem*, 66 (2), 434-438.

Tuncer M and Ball A S (2003), 'Purification and partial characterization of alpha-L-arabinofuranosidase produced by *Thermomonospora fusca*', *Folia Microbiol (Praha)*, 48 (2), 168-172.

van Laere K, Beldman G and Voragen A G (1997), 'A new arabinofuranohydrolase from *Bifidobacterium adolescentis* able to remove arabinosyl residues from double-substituted xylose units in arabinoxylan', *Appl Microbiol Biotechnol*, 47 (3), 231-235.

Wan C F, Chen W H, Chen C T, Chang M D, Lo L C and Li Y K (2007), 'Mutagenesis and mechanistic study of a glycoside hydrolase family 54 α-L-arabinofuranosidase from *Trichoderma koningii*', *Biochem J*, 401 (2), 551-558.

Yanai T and Sato M (2000), 'Purification and characterization of a novel α-L-arabinofuranosidase from *Pichia capsulata* X91', *Biosci Biotechnol Biochem*, 64 (6), 1181-1188.

Yoon H S, Keum I, Han N S, and Kim J H (2004), 'Molecular cloning and characterization of a gene encoding alpha-L-arabinofuranosidase from *Thermotoga maritima*', *Food Sci Biotechnol*, 13 (2), 244-247.

Yamaguchi A, Tada T, Wada K, Nakaniwa T, Kitatani T, Sogabe Y, Takao M, Sakai T and Nishimura K (2005), 'Structural basis for thermostability of endo-1,5-α-L-arabinanase from *Bacillus thermodenitrificans* TS-3', *J Biochem*, 137 (5), 587-592.

Zverlov V V, Liebl W, Bachleitner M and Schwarz W H (1998), 'Nucleotide sequence of arfB of *Clostridium stercorarium*, and prediction of catalytic residues of α-L-arabinofuranosidases based on local similarity with several families of glycosyl hydrolases', *FEMS Microbiol Lett*, 164 (2), 337-343.

ENZYMATIC SYNTHESIS AND PROPERTIES OF TREHALOSE ANALOGUES AS DISACCHARIDE AND TRISACCHARIDE

Soo Bok Lee, Soo-In Ryu, Hye-Min Kim, and Bong-Gwan Kim

ABSTRACT

Trehalose analogue, non-reducing dissacharide of 1-α-D-glucopyranosyl α-D-galactopyranoside, was synthesized by *Pyrococcus horikoshii* glycosyltransferase transglycosylation reaction with sugar nucleotides and galactose. This disaccharide analogue was effective inhibitor for several disaccharidases including rat intestinal trehalase and sucrase. Trehalose was also modified by *Escherichia coli* β-galactosidase transglycosylation reaction with lactose to give trehalose trisaccharide analogues. These trisaccharide analogues have been supposed to be indigestible oligosaccharides exhibiting enhanced hygroscopic, cryoprotective, anti-cariogenic, and prebiotic effects. The enzymatic techniques using glycosyltransferase and glycosidase might lead to create more trehalose-based analogues with a wide variety of acceptor and donor sugars.

Key words: trehalose; trehalose analogue; transglycosylation; *Pyrococcus horikoshii* glycosyltransferase; *Escherichia coli* β-galactosidase

INTRODUCTION

Enzymes are nature's biocatalysts that are particularly suitable to support synthesis of natural products. Generally, biocatalysts exhibit a high degree of substrate specificity, regioselectivity, and stereospecificity, therefore satisfying the increased demand for pure biochemical compounds. They have critical difference from ordinary catalysts by only catalyzing one, or at the most, only a few specific reactions with particular substrates at mild reaction conditions (Robyt et al., 2003). They provide specific products that have common features. Enzymes are highly promising tool used industrially to produce new and specific products that cannot be made by conventional chemical methods. It is true that the enzymatic processes exhibit their potential in the carbohydrate field, too.

In the present study, we have used two types of enzymes, *Pyrococcus horikoshii* trehalose-synthesizing glycosyltransferase (PhTG) and *Escherichia coli* β-galactosidase (EbG), to modify trehalose. Trehalose is a non-reducing dissacharide in which two glucose molecules are bonded in an α,α-(1↔1)-glycosidic linkage (Elbein, 1974). This naturally occurring sugar is widely distributed in various organisms to serve as a source of energy and a protectant of proteins and cellular membranes from a variety of environmental stress conditions, including desiccation, dehydration, heat, freezing, and oxidation (Benaroudj et al., 2001). Trehalose is not very hygroscopic, mildly sweet, stable to wide ranges of pH and heat, and a low or anti-cariogenic compound (Richards et al., 2002). However, trehalose ingested is generally hydrolyzed to glucose by intestinal trehalase (EC 3.2.1.28) and absorbed in the small intestine. Enzymatic transformations by glycosyltransferase and glycosidase in this study have suggested a way of creating indigestible analogues of disaccharide and trisaccharide having a structural feature of trehalose in common.

REPLACEMENT OF ONE GLUCOSYL END OF TREHALOSE WITH GALACTOSE BY *P. HORIKOSHII* TREHALOSE-SYNTHESIZING GLYCOSYLTRANSFERASE

P. horikoshii trehalose-synthesizing glycosyltransferase (PhTG) reacted with nucleoside diphosphate (NDP)-glucose as a donor and D-glucose, rather than glucose-6-phosphate (G-6-P), as an acceptor, directly and efficiently giving rise to free trehalose (see Fig. 1A) (Ryu et

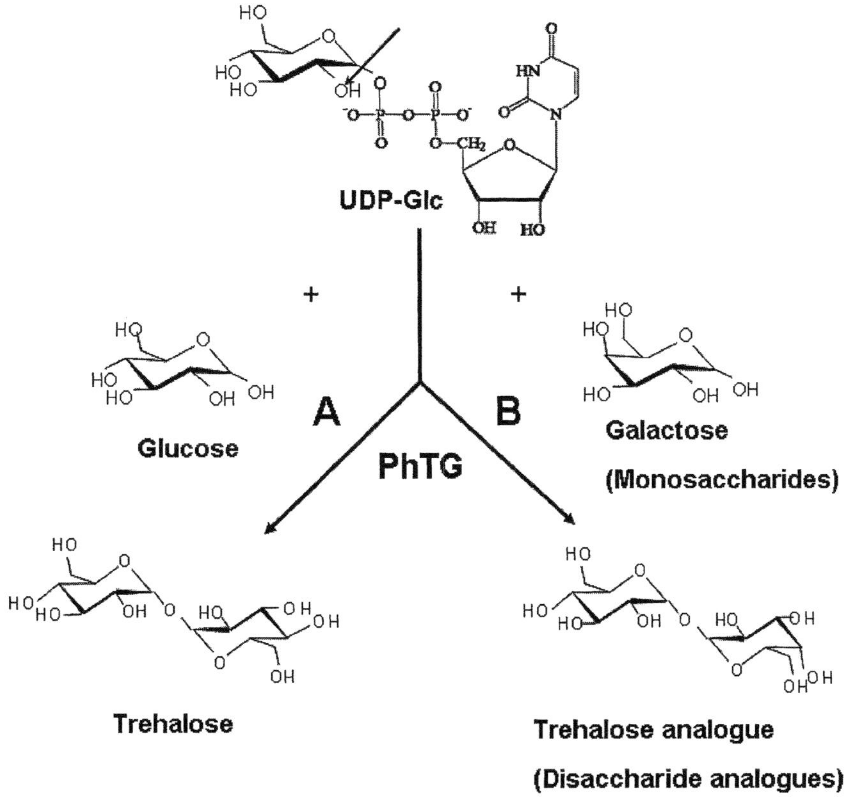

Figure 1 A reaction scheme for the transglycosylation reactions of PhTG with NDP-glucose as donor and D-glucose (A) or D-galactose (B) as acceptor

al., 2005; Qu et al., 2004). In contrast, the most widely distributed pathway synthesizing trehalose in numerous organisms involves co-contribution of a trehalose-6-phosphate (P) synthase (EC 2.4.1.15) that catalyzes the transfer of glucose from sugar nucleotides, typically UDP-glucose to G-6-P to produce trehalose-6-phosphate (T-6-P), and a trehalose-phosphate phosphatase that hydrolyzes T-6-P to free trehalose (Valenzuela-Soto et al., 2004). For overall sequence identity, PhTG did not have significant homology with trehalose 6-P synthases. However, a search of conserved domain database (CDD) including Smart and Pfam (http://www.ncbi.nlm.nih.gov/BLAST) exhibited that PhTG had considerably high homology in nucleotide-sugar binding regions of trehalose 6-P synthases. Three dimensional protein modeling using 3D-PSSM (http://sbg.bio.ic.ac.uk/ ~ 3dpssm/) proposed that PhTG consisted of two non-similar domains with three layers (α/β/α) each. The C-terminal region of PhTG was very homologous with those of the glycosyl transferases group 1 (Pfam00534), which were characterized to catalyze the transfer of sugar moiety from NDP-glucose to specific acceptor molecules. The N-terminal domain of PhTG was supposed to be unique and served as acceptor sugar binding region. Presently, the three dimentional structure determined by X-ray crystallography is under investigation. Interestingly, PhTG exhibited broad substrate specificity with acceptor molecule in the transfer reaction with NDP-glucose donor. During the course of the reaction with galactose as the acceptor, a new trehalose analogue of disaccharide was observed, which was a transfer product in which a glucosyl moiety of NDP-glucose was transferred to C-1 of D-galactose to give an α,α-(1↔1)-glycosidic linked trehalose analogue (1-α-D-glucopyranosyl α-D-galactopyranoside) (see Fig. 1B; Kim et al., 2007).

The analogue transfer product was prepared by using 18 mM solution of UDP-glucose and 56 mM solution of D-galactose in 50 mM sodium acetate buffer (pH 6.0). The mixture was incubated 10 min at 37 °C and the reaction was initiated by the addition of PhTG. The enzyme reaction was carried out for 24-36 h at the same temperature. The products of the reaction were analyzed by thin-layer chromatography (TLC) on Whatman K5F silica gel plates, using an ascent of 3:1:1 volume proportions of isopropyl alcohol/ethyl acetate/water or 7:3:1 volume proportions of *n*-butanol/pyridine/water (Fig. 2). Carbohydrates on the plate were visualized by dipping into 0.3% (w/v) *N*-(1-naphthyl)-ethylenediamine and 5% (v/v) H_2SO_4 in methanol, followed by heating at 110 °C for 10 min. The reaction products were also quantitatively analyzed by high performance anion exchange chromatography (HPAEC), using a Dionex CarboPac PA100 column. Actually, the enzyme was able to react with several donors such as UDP-, ADP-, and GDP-glucoses, yielding the same transfer product. Kinetically, D-galactose was less preferred to D-glucose as the acceptor in the transfer reaction, possessing approximately 3 times higher K_m and 8 times lower V_{max}.

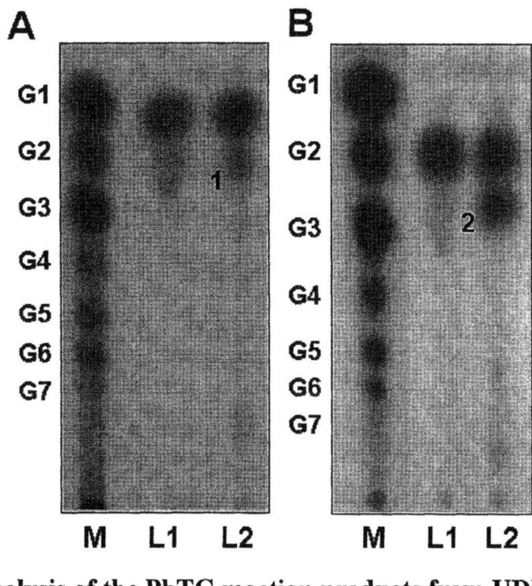

Figure 2 TLC analysis of the PhTG reaction products from UDP-glucose as donor and D-glucose (A) or D-galactose (B) as acceptor

Lane M: maltodextrin standards; lane 1: UDP-glucose and D-glucose (A) or D-galactose (B); lane 2: PhTG reaction products. 1 and 2 indicate 1-α-D-glucopyranosyl α-D-glucopyranoside and 1-α-D-glucopyranosyl α-D-galactopyranoside, respectively. (Adapted from references 5 and 8)

The transfer product was purified by descending paper chromatography and preparative liquid chromatography on a polymeric gel filtration JAI W-251 column. The molecular mass was determined to be 342 Da by Matrix-assisted laser desorption/ionization time-of-flight (MALDI-TOF) mass spectrometer, indicating a hexose disaccharide. The structure of the product was determined by ^{13}C-NMR. The glycosidic linkage that was formed in the transglycosylation reaction from UDP-glucose to the D-galactose acceptor was identified solely to be the α-(1, 1)-linkage. The production yield was approximately 31% based on the molar concentration of UDP-glucose.

INHIBITION STUDIES OF TREHALOSE DISACCHARIDE ANALOGUE

Trehalose is generally hydrolyzed by an intestinal trehalase to yield two glucose molecules, which are absorbed into the small intestine (Benaroudj et al., 2001). On the contrary, the trehalose disaccharide analogue product containing a galactose (1-α-D-glucopyranosyl α-D-galactopyranoside) in this study was neither hydrolyzed by porcine kidney and rat intestine trehalases, nor by α-glucosidase and β-galactosidase. Furthermore, the analogue was not hydrolyzed by rat intestinal enzymes including maltase, sucrase, and isomaltase. Unexpectedly, the analogue was effective at inhibiting trehalase and other disaccharide-hydrolyzing enzymes (Kim et al., 2007).

The disaccharide analogue was competitive inhibitor for porcine kidney and rat intestine trehalases, with K_i values of 0.68 mM and 3.72 mM, respectively. The analogue was also effective competitive inhibitor for baker's yeast α-glucosidase, with K_i value of 0.29 mM, and for other rat intestinal disaccharidases such as sucrase, maltase, and isomaltase, with K_i values of 0.66, 3.04, and 2.14 mM, respectively (see Table 1). Reportedly, L-arabinose was recommended as moderately potent inhibitor sugar only for intestinal sucrase, with K_i value of 2 mM in vitro (Seri et al., 1996). Actually, it was reported that this monosaccharide had no inhibitory action on intestinal maltase, isomaltase, and trehalase. Compared to L-arabinose, the trehalose disaccharide analogue was three times more potent inhibitor for intestinal sucrase, concomitantly inhibiting other intestinal disaccharidases including maltase in vitro. Recently, valienamine (an amino sugar analogue of D-glucose) and 3,4,5-trihydroxybenzoic acid (gallic acid) were also suggested to be effective competitive inhibitors for brush border sucrase, with respective K_i values of 0.8 and 1 mM, which were quite similar to the trehalose analogue in this study (Zheng et al., 2005, Gupta et al., 2007). In fact, disaccharidases such as trehalase, sucrase, and α-glucosidase (maltase) in the brush border membranes of the small intestine are considered to be important targets for inhibition. This inhibition may delay the intestinal digestion of dietary carbohydrates, thus controlling diabetes and obesity (Gupta et al., 2007).

In addition, we also tested a number of monosaccharides such as D-fructose, D-mannose, D-xylose, and D-tagatose, as the acceptor for the PhTG-catalyzed glycosyl-transfer reaction. As a result, we identified some of monosaccharides tested were effectively used to produce the trehalose analogues of disaccharide. The properties of the analogue products were under investigation.

Table 1 Inhibitions of disaccharidase activities by trehalose analogue (1-α-D-glucopyranosyl α-D-galactopyranoside) and some compounds

Enzymes	Inhibitors	Inhibition type	K_i (mM)	Potency[a]
Porcine kidney trehalase	Trehalose analogue	Competitive	0.68	
Baker's yeast α-glucosidase	Trehalose analogue	Competitive	0.29	
Rat intestine trehalase	Trehalose analogue	Competitive	3.72	
Rat intestine sucrase	Trehalose analogue	Competitive	0.66	3.0
	L-Arabinose	Uncompetitive	2.0	1.0
	Valienamine	Competitive	0.8b	2.5
	Gallic acid	Competitive	1.0	2.0
Rat intestine maltase	Trehalose analogue	Competitive	3.04	
Rat intestine isomaltase	Trehalose analogue	Competitive	2.14	

ADDITION OF GALACTOSE TO ONE GLUCOSYL END OF TREHALOSE BY THE REACTION OF LACTOSE AND β-GALACTOSIDASE

Trehalose was modified at one glucosyl end by the reaction of trehalose with lactose and *E. coli* β-galactosidase (EbG), giving rise to trehalose analogues of trisaccharide and D-glucose (Kim et al., 2007). Generally, oligosaccharides have been widely used as a food ingredient due to their favorable properties, such as high water holding, low calories with indigestibility, low sweetness, growth factors for Bifidus, and no dental caries (Ohta et al., 2002). Reportedly, there were attempts to make glycosyl-trehalose trisaccharides by the transglycosylation of cyclomaltodextrin glucanyltransferase or α-glucosidase, yielding trehalose trisaccharides by attaching only glucose residue in the α-(1→4)- or (1→6)-linkage (Kurimoto et al., 1997). However, the properties of these trehalose analogues have yet to be more investigated.

EbG reacted with lactose donor to cleave the galactosyl moiety and transfer it to the one glucosyl end of trehalose acceptor. During the course of the transgalactosylation reaction, two trehalose trisaccharide analogues were observed, which were the transfer products in which D-galactose was transferred to C-6 of one glucose end of trehalose to give an β-(1→6)-linked trehalose analogue (6^{II}-β-D-galactopyranosyl trehalose), and to C-4 of one glucose end of trehalose to give an β-(1→4)-linked trehalose analogue (4^{II}-β-D-galactopyranosyl trehalose), respectively (Fig. 3A).

These transfer products were prepared by using a 15% (w/v) solution of lactose and 30% (w/v) solution of trehalose in 50 mM Tris-HCl buffer/pH 7.5 (Kim et al., 2007). The enzyme reaction was carried out at 45 °C for 48 h. The transfer products were also analyzed by TLC, and their relative amounts were determined by TLC densitometry. The transfer products were purified by Bio-Gel P2 gel permeation chromatography and preparative HPLC equipped with a polymeric JAI W-251 column. The molecular masses of the purified transfer products were determined to be identically 504 Da by liquid chromatography-mass spectrometry (LC-MS), which matched the expected molecular mass of hexose trisaccharide. The detailed glycosidic structures of the products were determined by ^{13}C-NMR, in which it was found that there were large changes in chemical shifts of C-1 of the galactose with 6.9 ppm and 6.5 ppm, and C-6 and C-4 of the one glucose moiety of trehalose with 7.6 ppm and 7.8 ppm, respectively. These results confirmed that C-1 of the transferred galactosyl group was alternatively attached to C-6 and C-4 of one glucose moiety of trehalose. The relative ratio of two transfer products, 6^{II}-β-D-galactopyranosyl trehalose and 4^{II}-β-D-galactopyranosyl trehalose, in amount was about 9:1, indicating that the β-galactosyl transfer by EbG in this study was preferentially performed onto C-6 of the glucose molecule of trehalose (Fig. 3B). The production yields of the two products were approximately 26% and 2.8%, based on the concentration of trehalose.

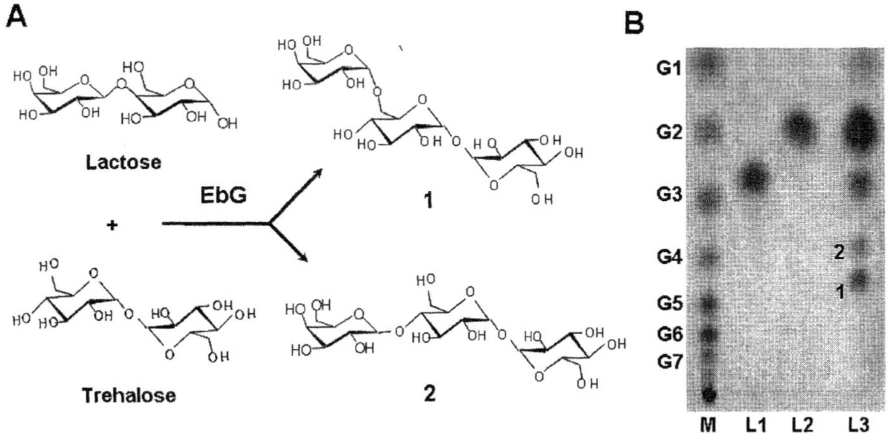

Figure 3 Transgalactosylation reaction of EbG with lactose as donor and trehalose as acceptor

(A) reaction scheme; (B) TLC of the EbG reaction products. Lane M: maltodextrin standards; lane 1: lactose; lane 2: trehalose; lane 3: EbG reaction products. **1** and **2** indicate 6-β-D-galactosyl-trehalose and 4-β-D-galactosyl-trehalose, respectively. (Adapted from Kim et al., 2007)

PHYSICOCHEMICAL AND PHYSIOLOGICAL PROPERTIES OF TREHALOSE TRISACCHARIDE ANALOGUE

The two trehalose trisaccharide analogues were resistant to the hydrolysis by trehalase. This indicated that the analogues might be effective at inhibiting the action of trehalase. To date, these analogues have not been tested as inhibitors for digesting enzymes. The trehalose trisaccharide analogues (a mixture of 9:1 ratio) exhibited significantly enhanced properties in terms of hygroscopicity, anti-cariogenicity, bifidogenicity, and cryoprotective effect with comparison to trehalose (Ryu et al., 2007).

Trehalose is usually found in the dihydrate form. The lyophilized trehalose dihydate has a low hygroscopic property, the water content of which remains stable at 9.1% up to a relative humidity of approximately 90% (Lammert et al., 1998). In this study, the lyophilized mixture of the analogue products had an enormously enhanced ability to absorb moisture compared to the trehalose. The moisture absorption of the analogue mixture was almost saturated at 3 days of equilibration, where the water content gained was approximately 5.0 times higher than that of trehalose and 1.9 times higher than sucrose, respectively (Ryu et al., 2007). This high ability of water absorption might make it applicable as a good humectant. The analogue products were poorly utilized by *Streptococcus sobrinus* as was xylitol. The products exhibited approximately 10 times lower cell-proliferation of *S. sobrinus* than that of sucrose, 2 times for trehalose, and 3 times for sorbitol. The analogue products also showed the significant promotion for the growth of *Bifidobacteria* such as *B. bifidum, B. longum, and B. longum* BORI, and no growth-stimulating effect for the harmful bacteria (*Escherichia coli, Clostridium botulinum,* and *Staphylococcus aureus*) used (Ryu et al., 2007). The results suggested that the trehalose trisaccharide analogues might be effective sugar substitutes as humectant, with anti-cariogenic and prebiotic effects for promoting a gut health. In addition, the analogue product functioned as a cryoprotectant in fish protein. Trehalose is also well known to protect proteins from the damage caused by freezing. Generally, sugar treatment with sorbitol or trehalose retards the denaturation of fish muscle protein caused by repeated freeze-thawing process (Sych et al., 1990). Based on molar concentrations of the sugars employed, the trisaccharide analogue product (0.16 M) was more effective than trehalose (0.23 M) and sorbitol (0.44 M) tested in the cryoprotection of the fish protein against 2 cycles of freeze-thawing (Kim et al., 2008). Accordingly, the trehalose trisaccharide analogues might be alternative non-digestible oligosaccharides with enhanced properties of hygroscopicity, cryoprotectivity, non-cariogenicity, and bifidogenicity.

SUMMARY

We have shown the enzymatic modifications of trehalose by substitution or addition of component monosaccharide. The one glucosyl end of trehalose was replaced by *P. horikoshii* trehalose-synthesizing enzyme catalyzed transglycosylation reaction in which the glucose of NDP-glucose donor was transferred to galactose acceptor, potentially including a variety of monosaccharides as acceptors, giving trehalose disaccharide analogues. To the one glucosyl end of trehalose was added galactose by *E. coli* β-galactosidase catalyzed reaction between lactose and trehalose to give two trehalose trisaccharide analogues. These two enzymatic techniques using glycosyltransferase and glycosidase might be used independently or sequentially with a wide variety of acceptor sugars to produce various trehalose-based analogues.

ACKNOWLEDGEMENT

We thank for financial support from the Marine and Extreme Genome Research Center Program, Ministry of Maritime Affairs and Fisheries, Republic of Korea.

REFERENCES

Benaroudj N, DoHee L, and Goldberg A L (2001) 'Trehalose accumulation during cellular stress cells and cellular proteins from damage by oxygen radicals', *J Biol Chem*, 276, 24261-24267.

Elbein A D (1974) 'The metabolism of α,α-trehalose', *Adv Carbohyd Chem Biochem*, 30, 227-256.

Gupta N, Gupta S, and Mahmood A, (2007) 'Gallic acid inhibits brush border disaccharidases in mammalian intestine', *Nutr Res*, 27, 230-235.

Kim B G, Lee K J, Han N S, Park K H, and Lee S B (2007) 'Enzymatic synthesis and characterization of galactosyl trehalose trisaccharides', *Food Sci Biotechnol*, 16, 127-132.

Kim B G, Ryu S I, and Lee S B (2008) 'Cryo- and thermo-protective effects of enzymatically synthesized β-galactosyl-trehalose trisaccharide', *Food Sci Biotechnol*, 17, 199-202.

Kim H M, Chang Y K, Ryu S I, Moon S G, and Lee S B (2007) 'Enzymatic synthesis of a galactose-containing trehalose analogue disaccharide by *Pyrococcus horikoshii* trehalose-synthesizing glycosyltransferase: Inhibitory effects on several disaccharidase activities', *J Mol Catal B-Enzym*, 49, 98-103.

Kurimoto M, Tabushi A, Mandai T, Shibuya T, Chaen H, Fukuda S, Sugimoto T, and Tsujisaka Y (1997) 'Synthesis of glycosyl-trehaloses by cyclomaltodextrin glucanotransferase through the transglycosylation reaction', *Biosci Biotech Biochem*, 61, 1146-1149.

Lammert A M, Schmidt S J, and Day G A (1998) 'Water activity and solubility of trehalose', *Food Chem*, 61, 139-144.

Ohta M, Pan Y T, Laine R A, and Elbein A D (2002) 'Trehalose-based oligosaccharides isolated from the cytoplasm of *Mycobacterium smegmatis*, *Eur J Biochem*, 269, 3142-3149.

Qu Q, Lee S J, and Boos W (2004) 'TreT, a novel trehalose glycosyltransferring synthase of the hyperthermophilic archaeon *Thermococcus litoralis*', *J Biol Chem*, 279, 47890-47897.

Richards A B, Krakowka S, Dexter L B, Schmid H, Wolterbeek A P M, Waalkens-Berendsen D H, Shigoyuki A, and Kurimoto M (2002) 'Trehalose: a review of properties, history of use and human tolerance, and results of multiple safety studies', *Food Chem Toxicol*, 40, 871-898.

Robyt J F, Park K H, Lee S B, and Yoon S H (2003) 'Enzymatic synthesis of acarbose oligosaccharide analogues as new enzyme inhibitors', *ACS Sym Ser*, 849, 168-181.

Ryu S I, Kim B G, Park M S, Lee Y B, and Lee S B (2007) 'Evaluation of enhanced hygroscopicity, bifidogenicity, and anti-cariogenicity of enzymatically synthesized β-galactosyl-trehalose oligosaccharides', *J Agri Food Chem*, 55, 4184-4188.

Ryu S I, Park C S, Cha J, Woo E J, and Lee S B (2005) 'A novel trehalose-synthesizing glycosyltransferase from *Pyrococcus horikoshii*: Molecular cloning and characterization', *Biochem. Biophys. Res. Commun.*, 329, 429-436.

Seri K, Sanai K, Matsuo N, Kawakubo K, Xue C, and Inoue S (1996) 'L-arabinose selectively inhibits intestinal sucrase in an uncompetitive manner and suppresses glycemic response after sucrose ingestion in animals', *Metabolism*, 45, 1368-1374.

Sych J, Lacroix C, Adambounou L T, and Castaigne F, (1990) 'Cryoprotective effect of lactitol, palatinit, and polydextrose on cod surimi proteins during frozen storage', *J Food Sci*, 55, 356-360.

Valenzuela-Soto E M, Márquez-Escalante J A, Iturriaga G, and Figueroa-Soto C G (2004) 'Trehalose 6-phosphate synthase from *Selaginella lepidophylla*: purification and properties', *Biochem Biophys Res Commun*, 313, 314-319.

Zheng Y G, Shentu X P, and Shen Y C (2005) 'Inhibition of porcine small intestinal sucrase by valienamine', *J Enzym Inhib Med Chem*, 20, 49-53.

GLYCOSIDASES AND THEIR MUTANTS AS USEFUL TOOLS FOR GLYCOSIDE SYNTHESIS

Young-Wan Kim

ABSTRACT

The roles of carbohydrates in human health and diseases are raising topics to develop functional foods and therapeutics. However, their complex structures and complicate synthetic chemical routes are drawbacks to understand the relationships between functions and structures of carbohydrates. Enzymatic syntheses using glycosidases and glycosyltransferases have been proposed as alternatives of the chemical synthesis. The enzymatic synthesis of glycosides is one of the oldest scientific topics, but still challenging. This review describes the recent achievements in the field of enzymatic carbohydrate synthesis using wild type and engineered glycosidases. Here we focus the methodology using retaining glycosidases which produce the same stereochemistry outcomes as that of the original substrates. Several retaining glycosidases are very useful not only to add sugar moiety to non-glycosylated natural products, such as flavonoids and ascorbic acid so on, but also to remodeling the structures of sugar parts in glycosylated compounds. Through enzyme engineering the valuable transglycosylation properties of retaining glycosidases have been improved to enhance selective transglycosylation and to increase yields by modulating hydrolysis and transglycosylation activities. The powerful enzymatic tools for the preparation of oligosaccharides are very promising for understanding the functions of carbohydrates, leading to promote human health and prevent diseases.

Key words: directed evolution; glycosidases; glycosynthases; transglycosylation; rational design

INTRODUCTION

Carbohydrates in living systems involve in not only energy generation/storage but also in numerous biological events such as cancers, inflammations, pathogen infections, cell-to-cell communications so on (Varki, 1993). For these reasons, oligosaccharides have considerable potential as therapeutics, and recent topics of biotechnology have been focused on the roles of oligosaccharides found in important cellular events (Dwek, 1996; Jacob, 1995; Varki, 1993). In addition, the components and the structures of oligosaccharides seriously affect the profiles of their biological activities, solubility, toxicity, bioavailability, and shelf-lives in the circuits of cells so on. Therefore, the optimization of the structure of oligosaccharides found in various therapeutics, including glycoproteins, vaccines, and phytochemicals so on, could lead to enhancement of activity and stability, reduction of dosage, and minimum side effects so on (Shriver et al., 2004). To achieve these aims, efficient and selective protocols for the synthesis of oligosaccharides must be developed.

Many scientists have exploited many chemical synthetic routes based on Organic Chemistry, but the classical chemical synthesis for the preparation of oligosaccharides is often impractical for the synthesis of complex oligosaccharides found in important cellular events because of the need for selective and labor-intensive protection-deprotection steps and difficulties in directing product stereochemistry. To circumvent these limitations, enzymatic syntheses using glycosidases or glycosyltransferases have rapidly gained prominence (Crout and Vic, 1998; Koeller and Wong, 2000; Withers, 2001). However, the later have restrictions such as problems in overproduction and limited availability of nucleotide phosphosugars (NP sugars) as donor substrates so on. In contrast, glycosidases have the attractive against such problems of glycosyltransferases; easy overproduction, cheaper and simpler substrates than NP sugars so on. However, glycosidases also have undesirable properties for the synthesis of glycosidic linkages; re-hydrolysis of transfer products, low yields, and production of regioisomers as by-products so on. To solve these problems many scientists are surveying novel glycosidases and exploit new strategies based on mutagenesis.

MECHANISMS OF GLYCOSIDASES

Glycosidases (E.C. 3.2.1.X) are a widespread enzymes which hydrolyze the glycosidic linkages. Glycosidases can be classified into over 100 glycoside hydrolase families (GHs) based on the amino acid sequence homology (Henrissat, 1991), but the hydrolysis mechanism is either net retention or inversion of anomeric configuration of the substrates (Zechel and Withers, 2000).

The class of glycosidases known as retaining glycosidases carry out the hydrolysis of glycosidic bonds via a two step, double-displacement mechanism in which, in the first step, a covalent glycosyl-enzyme is formed on an active site nucleophile (a glutamic or aspartic acid residue) with acid catalytic assistance from another active site glutamic or aspartic acid residue. Once formed, this intermediate is then hydrolysed by the attack of water on the anomeric center of the glycosyl-enzyme intermediate, with general base assistance from the same residue that originally functioned as the acid (Fig. 1A). In the same fashion, a hydroxyl group of sugar acceptors can function as that of water in the second step of the reactions catalyzed by retaining glycosidases. The result is transferring glycosyl residue from the glycosyl enzyme intermediate to the sugar acceptors, leading to the formation of a new glycosidic linkage instead of hydrolysis. The transglycosylation catalyzed by retaining glycosidases have been applied to synthesis of novel oligosaccharides and the modification of the structure of sugar chain in various glycosides. Although the power of the retaining glycosidases for preparation of oligosaccharides and novel glycosides, the yields of the transfer products are generally moderate due to subsequent hydrolysis by wild type glycosidases. In addition, production of regioisomers needs a series of purification steps, resulting in elevating production cost. To overcome these restrictions of wild type glycosidases several strategies have been attempted (See **ENGINEERING GLYCOSIDASES** section).

The other kingdom of glycosidases is inverting glycosidases which cleave the glycosidic linkages through direct nucleophilc attack by a water molecule which is activated by a general base residue of the inverting enzymes. Another catalytic residue which functions as a general acid catalyst assists the hydrolysis. The hydrolysis result is the inversion of the stereochemistry at the anomeric center from α to β or from β to α. Inverting glycosidases also take hydroxyl group of the sugar acceptors instead of water molecule, producing glycosylated products. However, these reactions are reverse reactions of the hydrolysis, not transglycosylation reaction catalyzed by retaining glycosidases. There were attempts to use for the synthesis of new glycosidic linkages through the reverse reactions of inverting glycosidase, but the yields were not good enough to apply to industrial scale-production.

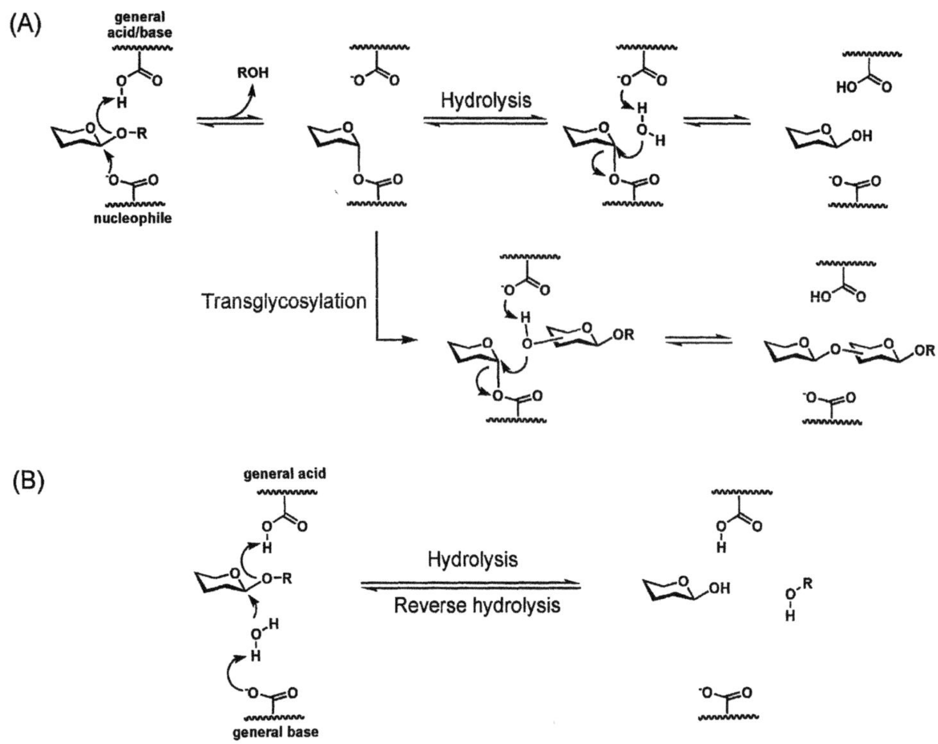

Figure 1 Scheme of mechanisms catalyzed by glycosidases

EXAMPLES OF TRANSGLYCOSYLATION BY GLYCOSIDASES

Cyclodextrin glycosyltransferases

Cyclodextrin glycosyltransferases (CGTases, E.C. 2.4.1.19) are one of the most famous groups of glycosidases with high transglycosylation activity. CGTases are multifunctional enzymes which catalyze the formation of cyclodextrins (CDs), starch hydrolysis, disproportionation between linear oligosaccharides, and coupling reactions (reverse reactions of cyclization reactions) of between CDs and sugar acceptors, such as sugar alcohols, polyols, flavonoids, ascorbic acid, and glycosides so on (Qi and Zimmermann, 2005). The intermolecular glycosylation reactions by CGTases yield novel compounds with different physicochemical and biological properties.

A commercial application of this method is found in glycosylation of the intense sweetener, stevioside. This bitter compound is isolated from the leaves of the plant Stevia rebaudiana and has a low solubility. Glycosylation decreases bitterness and increases solubility (Pedersen et al., 1995). Another famous example is the production of glycosylated ascorbic acid by CGTases (Tanaka et al., 1991). Ascorbic acid (AA) is one of the strong antioxidants, and it is widely used in pharmacy, food, and cosmetics. Its instability in the presence of molecular oxygen and cupper ion or at high temperature was greatly improved by the regioselective glycosylation toward 2-OH position of AA through the coupling reaction by

CGTase from *Bacillus stearothermophilus* (Tanaka et al., 1991). Ascorbic acid-2-*O*-α-D-glucoside produced through a sequential reaction by glucoamylase serves as a stable vitamin C supplement with no cytotoxicity (Wakamiya et al., 1995).

4-α-Glucanotransferases

4-α-Glucanotransferases (α-GTases, EC 2.4.1.25) catalyze not only the formation of a series of maltooligosaccharides with exclusive regioselectivity toward α-(1→4)-glycosidic linkages but also the production of cyclic maltooligosaccharides (cycloamylose) with larger than γ-CD (Takaha and Smith, 1999). To date, α-GTases have been classified as GH13, GH57, or GH77 (Henrissat, 1991). Those enzymes belonging to GH57 or GH77 use maltotriose and longer maltooligosaccharides as sugar donors. In GH13 there are only two α-GTases which were isolated from *Thermotoga maritima* and display catalytic properties that distinguish them from the other α-GTases in GH57 and GH77 (Liebl et al., 1992; Meissner and Liebl, 1998). Particularly, *T. maritima* maltosyltransferase will only transfer maltosyl residues (Meissner and Liebl, 1998), and *T. maritima* α-GTase (TMαGT) has certain substrate restrictions; maltotetraose is the smallest possible donor and glucose cannot serve as an acceptor (Liebl et al., 1992).

The powerful transglycosylation activities of α-GTases are potential to incorporate a new glucosyl moiety to carbohydrates and other glycosylated natural products. Such glycosylation dramatically increase the water-solubility of the compounds such as isoflavones, curcumin, and taxol so on. Most of such glycosylations for glycosides have been carried out CGTases (Qi and Zimmermann, 2005). Recently the glycosylation of genistin using α-GTase has been successfully accomplished by Park and his collegues (Li et al. 2005). As we can expected, α-GTase from *Thermus scotoductus* (TSαGT) produced a series of glycosylated genistins ranging in length from glucose to maltooligosaccharide with degree of polymerization (DP) 18. Maltosyl α-(1→4)-geneistin showed a dramatically increased water-solubility by four orders of magnitude relative to genistin. Interestingly, the authors found inclusion complexes between cycloamylose produced by TSαGT and genisin derivatives after removing the genisitn derivatives with a linear maltooligosaccharides. The profile of mass spectrums suggested that relatively less translgycosylated genistin derivatives (DP1~DP8) formed inclusion complexes with cycloamylose ranging in length from DP6 to DP13. In fact, encapsulation of hydrophobic compounds in CDs has a problem in size-limit of the encapsulated compounds. Therefore, the results by the Park's group are very promising as a solution to encapsulate large molecules. In addition the size of cycloamyloses seems to be adapted to the size of the target compounds.

Maltogenic amylases

Maltogenic amylases (MAases, EC 3.2.1.133), which have been studied by Park and his colleagues, are one of the most useful enzymes in the field of glycoside synthesis. MAases belong to a subclass of glycoside hydrolase family 13 (GH13), also known as cyclodextrin/pullulan-(CD/PUL-)hydrolyzing enzymes, including cyclodextrinases (CDases, EC 3.2.1.54), neopullulanases (NPLases, EC 3.2.1.135), and *Thermoactinomyces vulgaris* α-amylase II (TVAII) (Park et al., 2000; Kamitori et al., 2002; Hondoh et al., 2003). They show unique characteristics differing from typical α-amylases. They have an extra N-terminal domain which mediates their domain-swapping dimeric structure, resulting in narrow and deep active sites (Fig. 2A; Kim et al., 1999; Lee et al., 2002b). Due to the shape of their active sites, only small substrates such as CDs and maltooligosaccharides can easily access the active site, resulting in a preference for CDs over larger substrates such as starch and pullulan (Park et al., 2000). Maltotriose is the smallest substrate for MAases. Therefore, maltose and glucose are end products in hydrolysis of CDs and starch by MAases, and maltose is a main product. Interestingly, panose is the hydrolysis product from pullulan which is a polysaccharides consisting of maltotriose units linked through α-(1→6)-glycosidic linkages (Fig. 2B).

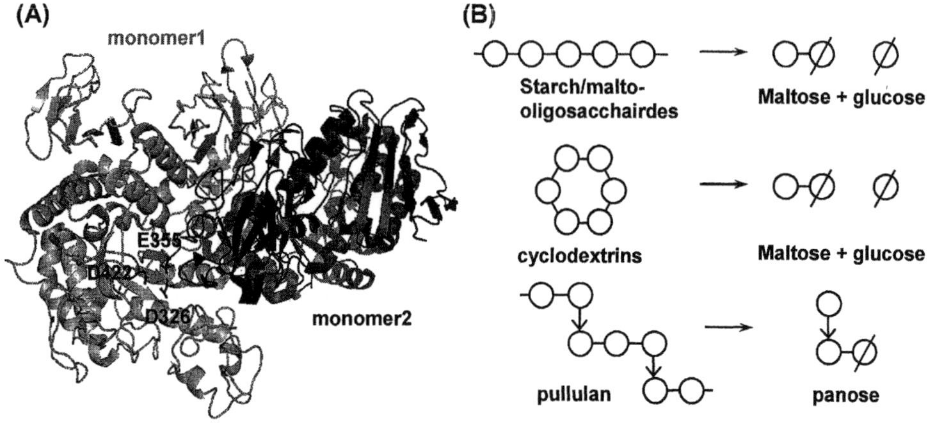

Figure 2 Structure and hydrolysis patterns of maltogenic amylases
(A) Structure of *Thermus* maltogenic amylase (1sma.pdb, Kim et al, 1999). The catalytic residues (D326, E355, and D422, ThMA numbering) are represented as solid lines on the figure. (B) Open circles represent a glucose residue and the open circles with a slash line are the glucose at the reducing end.

MAases display strong transglycosylation activities forming α-(1→6)-glycosidic linkages with less production of α-(1→3)- and α-(1→4)-linked transfer products using acarbose or maltotriose as donors with various sugar acceptors (Park et al., 2000). The dimeric structure of MAases yields an extra sugar bind site at the active site, which plays important roles in transglycosylation catalyzed by MAases (Kim et al., 1999; Kim et al., 2000). In addition, the susceptibility to hydrolysis of α-(1→4)-glycosidic linkages yields accumulation of α-(1→6)-linked transfer products in the reactions catalyzed by MAases (Kim et al., 2000). Their synthetic properties have been applied to production of isomaltooligosaccharides, and a cooperative action mode of MAase from *Bacillus stearothermophilus* (BSMA) and TMαGT has been reported, increasing the yield from 57.5% to 67.4% (Lee et al., 2002a). Furthermore, MAases have been used for glycosylation of natural products including ascorbic acid, neohesperidine, naringin, puerarin so on.

Recently, MAases have been used for modification of the structure of polysaccharides as resistance starch. MAases can add a glucose residue to the non-reducing ends of starch via a α-(1→6)-glycosidic linkage. Lee et al. (2008a) developed a process to produce a novel modified starch, termed highly branched amylopectin cluster through the reaction catalyzed by BSMA with amylopectin cluster which were produced by TSαGT. Determination of digestibility of HBAC using kinetic analysis with fungal glucoamylase and porcine pancreatic α-amylase let us know that HBAC shows less digestibility than unmodified amylopectin cluster; about 4~5 times less k_{cat}/K_M values for HBAC than that of amylopectin cluster (Lee et al., 2008a).

Of particular interests, they can hydrolyze acarbose, a strong inhibitor of α-amylases, to acarviosine-glucose (Acv-G) and glucose, followed by consequential synthesis of acarbose derivatives which have substitution of other sugars acceptors for the glucose at reducing end of acarbose with an α-1,6-glycosidic linkage (Fig. 3A; Park et al., 1998). More promisingly, *Thermus* maltogenic amylase (ThMA) can hydrolyze Acv-G to acarviosin (Acv) and glucose and produce acarviosyl glycosides (Lee et al., 2008b). One of them, α-acarviosinyl-(1→9)-3-α-D-glucopyranosylpropen (Acv-GP2, Figure 3B), is a α-glucosidase-selective inhibitor with 27-fold enhanced selectivity toward α-glucosidase (K_i = 0.1 μM) over α-amylase (K_i = 6.7 μM) relative to acarbose (K_i for α-amylase = 2.3 μM and K_i for α-glucosidase = 0.3 μM). Inhibitors with selective inhibition toward α-glucosidase over α-amylase may help reducing side effects such as flatulence by microbial fermentation of undigested starch. Therefore, Acv-GP2 could be a more potent hypoglycemic agent than acarbose.

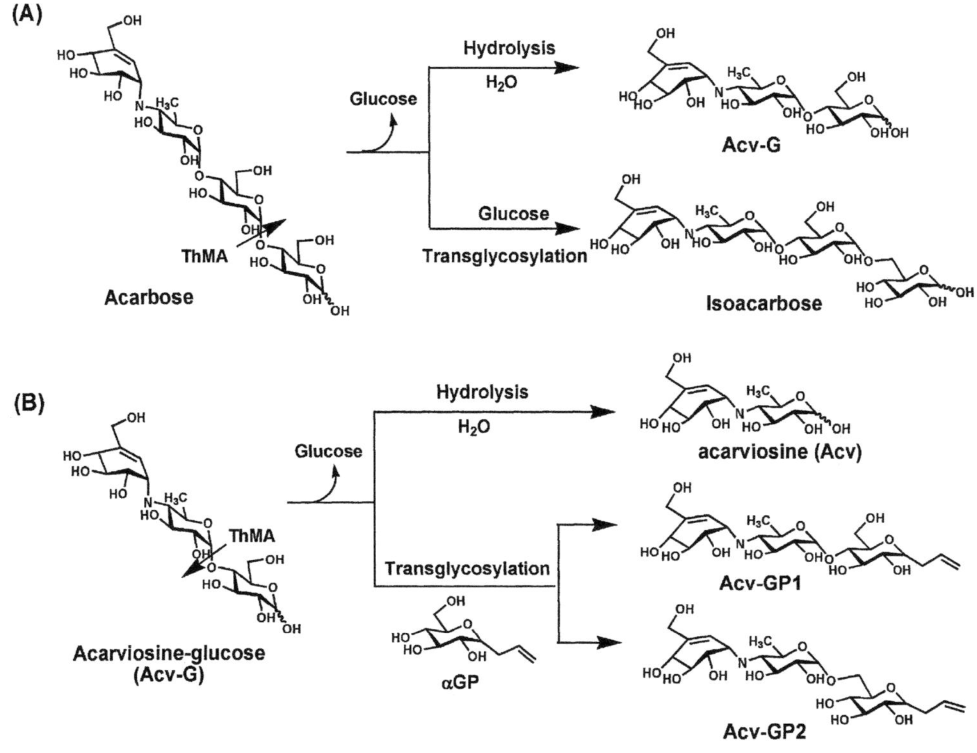

Figure 3 Enzymatic synthesis of acarbose derivatives using ThMA
(A) Transglycosylation reaction with acarbose and glucose as a sugar donor and an acceptor, respectively (B) Transglycosylation reaction with acarviosine-glucose (Acv-G) and 3-α-D-glucopyranosylpropen (αGP) as a sugar donor and an acceptor, respectively.

ENGINEERING GLYCOSIDASES

Rational design of glycosidases

The first trial to engineering catalytic properties of glycosidase is rational design based on the relationship between function and structural information. To succeed this strategy the identification of key residues of glycosidases in transglycosylation is of particular importance. Many scientists have attempted to find the key residues based on multiple sequence alignment and confirm the functions of the candidates through mutagenesis and kinetic analyses.

Park and his coworkers have reported an interesting result on key residues for transglycosylation by GH-13 enzymes (Yang et al., 2007b). *Pyrococcus furiosus* thermostable amylase (PFTA) is one of CD/PUL-hydrolyzing enzymes (Yang et al., 2004). Unlike typical CD/PUL-hydrolyzing enzymes, PFTA has very weak transglycosylation activity. Of PFTA mutants, a mutant with both H414N and G415E in the second conserved region of GH13 enzymes displayed α-GTases-like disproportion activity without intramolecular transglycosylation via cycloamylose formation, which is a general feature of α-GTases (Takaha and Smith, 1999). These mutation sites are substrate binding sites (+1 and +2 subsite). H414N and G415E mutations in PFTA may alter the binding affinity of sugar acceptors (maltooligosaccharides in this case), leading to enhancing transglycosylation activity. The new α-(1→4)-glycosidic linkages synthesized by the PFTA mutant resistant against re-hydrolysis due to the significantly reduced hydrolytic activity of the enzyme caused by the mutations, resulting in the accumulation of long maltooligosaccharides (Yang et al., 2007b).

There are many reports on the mutant glycosidases obtained a higher ratio of transglycosylation to hydrolysis activities by a couple of mutations near the active sites. Rivera, et al. (2003) found that a mutation, V286F in an α-amylase from *Bacillus licheniformis*, differing only by one hydroxyl group, was 3-fold less hydrolytic than the wild-type enzyme and apparently had a higher transglycosylation/hydrolysis ratio. In the case of an α-amylase from Saccharomycopsis its transglycosylation activity was increased by the substitution of W84L (Matsui et al., 1991). In addition, mutations M424K and F426Y in *Pyrococcus furiosus* β-glucosidase significantly improved (18–40%) the synthesis of galactooligosaccharides (Hansson et al., 2000). In most cases, the mutations might affect the affinity of substrate, hydrophobicity, and electrostatic environment of the active site, leading to not only lowering hydrolysis activity but also maintained or relatively less reduced transglycosylation activity.

On the other hand, the case of β-galactosidase from *Bifidobacterium* (BIF3) is very interesting because the enhancing factor for the transglycosylation activity did not come from the active site. BIF3 has a galactose-binding motif at C-terminal, and the deletion of the C-terminal affected both the hydrolysis and transglycosylation activities. Of the truncated mutants, termed BIF3-d3, with deleted C-terminal by 580 amino acid residues displayed a higher transgalactosylation/hydrolysis ratio than the WT enzyme (Jorgensen et al., 2001). Likewise other cases, hydrolysis activity of BIF3-d3 dramatically decreased (only 6 % of that of wild type enzyme), but the transglycosylation activity was 9-fold higher than the hydrolysis activity.

Directed evolution of glycosidases

Even though there are enormous biochemical and structural information on glycosidases, it is still challenging to find the important residues for transglycosylation. Generally, the mutant enzymes yielded by attempts using site-directed mutagenesis obtain relatively higher transglycosylation activity over hydrolysis activity with cost of their overall catalytic power.

To overcome the problem Dion and his coworkers have attempted directed evolution approaches to modulate hydrolysis and transglycosylation activity of glycosidases (Feng et al., 2005; Osanjo et al., 2007). Their strategy is based on screening mutants with high transglycosylation activity from a primary screened library which showed low hydrolysis activity. The mutants derived from a β-galactosidase produced transglcosylation products with high yields (60-75%) with their original substrates. The mutant has two mutations at the -1 subsite (N282T and F401S), and the authors suggest that the reposition of the the glycone of the glycosyl enzyme-intermediate with a better fit of the acceptor in the +1 subsite.

The same strategy has been applied to *Thermoanaerobacterium thermosulfurigenes* CGTase to reduce hydrolysis activity with retaining cyclization activity (Kelly et al., 2008). Screening system for mutants showing low hydrolysis activity and high CD-producing activity using a 96-well plate reader yielded two mutants which has a mutation with either S77P or W239R. of these residues the most effective mutant, S77P, were located on the outer region of the active site, and the three dimensional crystal structure of the mutant showed no alterations to the peptide backbone. However, the conformational changes of the side chains of the active site residues were found, and such structural shifts might cause the increased cyclization/hydrolysis ratio by 15-fold. Detail kinetic analysis revealed that S77P also decreased coupling side reaction activity as well. The authors only reported the improved yield of CD production, but it would be interesting to use the mutant CGTase to glycosylate natural products as sugar acceptors.

Mechanism-based engineering of glycosidases: Glycosynthases

Mechanism of glycosynthases

In 1998, Withers and his colleagues have reported an innovative methodology for preparation of oligosaccharides using an engineered β-glycosidase, termed glycosynthase (Mackenzie et al., 1998). *Agrobacterium* β-glycosidase (Abg) is one of retaining glycosidases, and the engineered Abg mutant has a mutation at its catalytic nucleophile (E358) through site-directed mutagenesis, having a non-nucleophilic residue, alanine. The mutant (AbgE358A) do not catalyze the hydrolysis of glycosidic bonds, but still possesses the power of the formation of glycosidic bonds upon employing α-glycosyl fluorides whose anomeric configuration opposite to that of the original substrate is mimicking the glycosyl enzyme intermediate. Consequently, AbgE358A catalyzes the formation of new glycosidic linkages with the same stereochemistry as the normal substrate through the nucleophilic displacement of the fluoride by attack of a hydroxyl group on glycosyl acceptors (Fig. 4A). AbgE358A was capable of transferring glucosyl and galactosyl moieties from α-glucosyl fluoride and α-galactosyl fluoride, respectively, to various sugar acceptors in very high yields (70~95%) because its hydrolysis activity was removed completely (Mackenzie et al., 1998). With the exception of a unique α-glycosynthase derived from *Schizosaccharomyces pombe* α-glucosidase (Fig. 4B, Okuyama et al., 2002) all glycosynthases have originated from β-glycosidases. The anomeric configuration of the glycosyl enzyme intermediate of α-glycosidases is β-configuration.

Consequently, the unique α-glycosynthase employed β-glucosyl fluorides as a donor.

Given the mechanism of glycosynthase all candidates for glycosynthases were retaining glycosidases. Recently, Honda and Kitaoka (2006) have attempted to expand the glycol-synthase strategy to another glycosidase kingdom, inverting glycosidases which hydrolyze the glycosidic linkage via a single displacement mechanism involving acid/base catalysis without the covalent glycosyl enzyme intermediate. In this instance, the mutation of the general base residue of an inverting xylanase from *Bacillus halodurans* C-125 (Rex) not only suppressed hydrolytic activity by four orders of magnitude but also allowed trans-glycosylation via the Hehre resynthesis pathway (Fig. 4C and 4D). Glycosynthases from retaining glycosidases prefer one of alanine, serine, and glycine as the mutation at the nucleophile position, whereas cysteine was the best mutation for the general base catalyst in the inverting glycosidase.

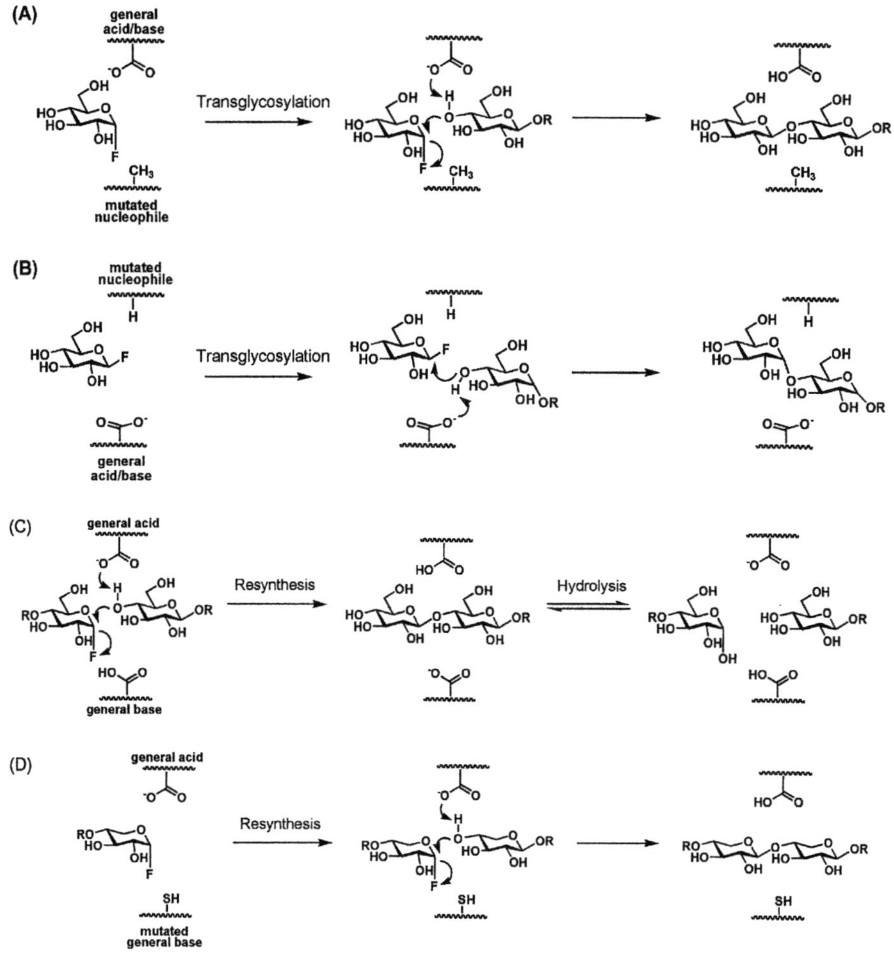

Figure 4 Mechanism of transglycosylation catalyzed by glycosynthases

Reactions catalyzed by glycosynthases from retaining β-glycosidases (A), by glycosynthases from retaining α-glycosidases (B), by inverting β-glycosidases via Hehre resynthesis (C), and a glycosynthase derived from an inverting xylanase, Rex (D)

Since the first success of the glycosynthase approach, glycosynthases have been developed from 23 glycosidases belonging to 11 different glycosidase families to date, and their application to the synthesis of various oligosaccharides and polysaccharides have been attempted (Table 1; Hancock et al., 2006; Perugino et al., 2004; Sugimura el al., 2006; Ben-David et al., 2007).

Table 1 List of glycosynthases to date

GH family	Enzymes	Mutation	Product	Ref
GH1	*Agrobacterium* β-glucosidase (Abg)	E358A/S/G	β-1,3/4	a
	S. solfataricus β-glucosidase	E387A/S/G	β-1,3/4/6	a
	Thermosphaera aggregans β-glycosidase	E334A/G	β-1,3/4/6	a
	Pyrococcus furiosus β-glucosidase	E372A	β-1,3/4	a
	Thermus thermophilus β-glycosidase	E338A/G	β-1,3/4	a
	Streptomyces sp. β-glucosidase	E383A	β-1,3	b
GH2	*Cellulomonas fimi* β-mannosidase	E519A/S	β-1,3/4	a
	E. coli β-galactosidase (LacZ)	E537S	β-1,6	a
	Thermotoga maritima β-glucuronidase	E476A/S/G	β-1,4	a
GH5	*Rhodococcus* sp. endoglycoceramidase II	E351S	β-1,1	a
GH7	*H. insolens* cellulase	E197A/S/G	β-1,4	a
GH8	*Bacillus halodurans* inverting xylanase	D263C	β-1,4	a
GH10	*Cellulomonas fimi* β-xylanase (CFX)	E235A/S/G	β-1,4	a
	Cellulomonas fimi β-glycanase (Cex)	E233G	β-1,4	c
	Thermotoga maritima β-xylanase	E259G	β-1,4	c
	Clostridium stercorarium β-xylanase	E293G	β-1,4	c
	Bacillus halodurans β-xylanase	E301G	β-1,4	c
GH16	*Bacillus licheniformis* 1,3-1,4-β-glucanase	E134A	β-1,3/4	a
	Pyrococcus furiosus laminarinase	E170A	β-1,3/4	a
GH17	*Hordeum vulgare endo*-1,3-β-glucosidase	E231G	β-1,3	a
GH26	*Cellvibrio japonicus* β-mannanase	E320S/G	β-1,4	a
GH31	*Schizosaccharomyces pombe* α-glucosidase	D416A	β-1,4/6	a
GH52	*Geobacillus stearothermophilus* β-xylosidase	E335G	β-1,4	d

a, Hancock et al., 2006; b, Faijes et al., 2006; c, Sugimura el al., 2006; d, Ben-David et al., 2007.

Although the glycosynthase strategy is very powerful in the synthesis of glycosidic linkages, large amount of the engineered glycosidase mutants and long incubation time are required. In addition, glycosynthases need fluoride sugars as donors, the repertories of glycosynthases are still limited; no glycosynthases from hexosaminidases, sialidases, fucosidases which hydrolyze glycosidic linkages frequently found in cellular glycans. To develop enzymes preparing such linkages another strategy is required. Directed evolution approaches have been proposed to solve these problems in glycosynthase methodology. The first glycosynthase, AbgE358A, was the first target. The Withers group developed "on-agar plate" screening method with a glycanase as a screening enzyme which can hydrolyze only the trans-glycosylated long-chain oligosaccharides but not the acceptor sugar. Such coupling assay allows comparison of the translgycosylation rates by active glycosynthase in a randomly mutated library by measuring the amount of released chromophore by the screening enzyme. Through the screening method the transglycosylation activity of the original Abg glycosynthase (AbgE358A) was improved by more than 1,300-fold with E358G and three additional mutations, A19S, Q248R, and M407V (Mayer et al., 2001; Kim et al., 2004). The evolved glycosynthase was termed Abg-2F6, and the mutation sites are represented in the Fig. 5.

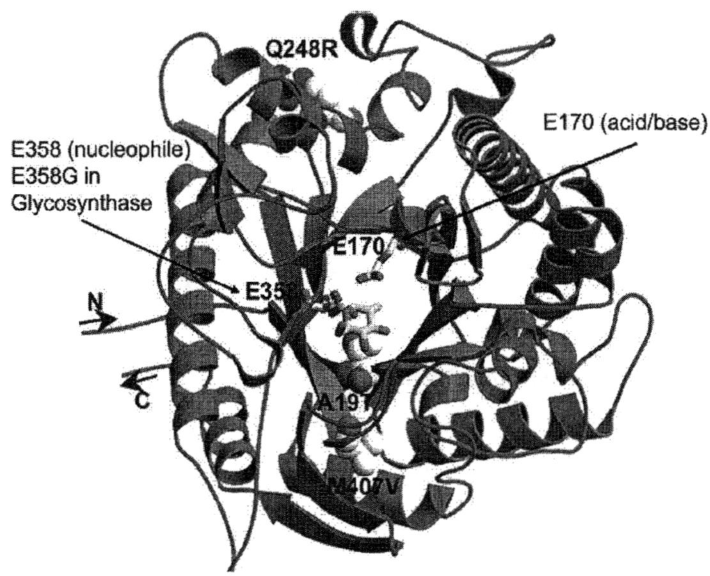

Figure 5 Model structure of the evolved Agrobacterium sp. Glycosynthase 2F6.
The relative position of the mutations are shown in addition to the location of the catalytic residues. The figure is adapted from Kim et al. (2004).

In addition, the directed evolution of Abg glycosynthase expanded its substrate repertories; α-xylosyl fluoride which was not a substrate of any Abg glycosynthase derivatives including AbgE358G, the second generation of Abg glycosynthase, was successfully utilized as a sugar donor. In fact all nucleophile mutants of retaining glycosidases do not function as glycosynthases. We do not know the structural information required for a glycosidase to serve as a glycosynthase. This work implies that the repertories of glycosynthase could be diversified through directed evolution processes.

Recently, other directed evolution strategies for glycosynthase has been reported; a chemical complementation based on yeast three-hybrid assay (Lin et al., 2004) and detection of the released hydrofluoric acid using methyl red as a pH indicator (Ben-David et al., 2008). These strategies are very promising due to no requirement of particular screening enzymes which can hydrolyze only the transfer products, but not substrates in the screening method developed by the Withers group. These new strategies have successfully screened activity-enhanced glycosynthase derivatives, but the positive mutations for activity were found only at the nucleophile position.

Applications of glycosynthases for the production of glycoconjugates

The structures of glycans on the surface of glycoconjugates including glycoproteins and glycolipids play very important roles in their biological functions. Understanding the relationships between the structures and functions of the glycans is one of the most interesting topics in development of carbohydrate-based therapeutics. Mostly, naturally occurring glycosylations of peptides, proteins, and lipids are performed by glycosyltransferases which use NP sugars as donors. The glycosynthase methodology expanded their area not only to the synthesis of oligosacchairdes but also to the incorporation of sugars to glycosylated or non-glycosylated substrate, including glycopeptides, glycoproteins, flavonoids and lipids (Vaughan et al., 2006; Yang et al., 2007a).

The Withers group continues to be a leading group in the applications of glycosynthases. Abg glycosynthases (AbgE358S and AbgE358G) successfully transferred a galactosyl moiety to the glucosyl residue of a model glycopeptide linked to PEGA resin through a backbone amide linker (Tolborg et al., 2002). This technique offers the possibility of the glycosynthase technology to use for highly regioselective and automatic solid-phase synthesis. Recently, glycosylation of a glycoprotein by a glycosynthase has been reported. The directed evolved glycosynthase (AbgE358G-2F6; Kim et al., 2004) successfully transferred single galactosyl moiety to cellobiosyl residue on the surface of a chemically glycosylated xylanase with almost over 90% yield (Müllegger et al., 2006).

Another impressive achievement of glycosynthase methodology is the synthesis of glycospingolipids with non-glycosylated lipids as sugar acceptors (Vaughan et al., 2006). A nucleophile mutant of *Rhodococcus* sp. endoglycanase (EGCII/E351S) transferred 3'-sialyllactosyl moiety to D-erythro-sphigosine with over 90%. The glycosynthase was also able to use a range of sphingsosine analogues as acceptors. The remodeling or redesign of the glycans in glycoproteins and glycolipids which have potentials as therapeutics may improve their roles in alleviating various human diseases. Therefore, the glycosynthase strategy has a great possibility to use tools in the main procedure to produce glyco-therapeutics.

CONCLUSION

Efficient and selective catalysts for the synthesis of glycosidic linkages would provide more hopeful opportunities to find novel and proficient therapeutics. There have been many attempts to develop enzymatic tools with high transglycosylation activity and regioselectivity, but still challenging. The potential of glycosidases including CGTase, α-GTases, and maltogenic amylases are very promising as such enzymatic tools. The numerous attempts to alleviate the drawbacks of glycosidases for the preparation of oligosaccharides have been obtaining several successful cases. The approaches to modulate the hydrolysis and transglycosylation activities of glycosidases through site-directed mutagenesis or directed evolution yielded glycosidase mutants showing enhanced transglycosylation, and the yields of the synthesis could be improved compared to those obtained by wild type glycosidases. However, the most brilliant achievement in this field is the establishment of glycosynthase methodology. These mechanism-based glycosidase mutants need chemically prepared glycosyl fluorides, but their high yields and regioselectivity were very impressive. However, all nucleophile mutants do not function as active glycosynthases, and we do not know the requirements. Therefore, further studies using mutagenesis are necessary to get information on the biochemical or structural requirements for glycosynthases so as to expand the repertories of glycosynthases to the GH families from which no glycosynthase has been reported.

ACKNOWLEDGEMENT

This study was supported by a Korea University Grant.

REFERENCES

Ben-David A, Bravman T, Balazs Y S, Czjzek M, Schomburg D, Shoham G, and Shoham Y (2007) 'Glycosynthase activity of *Geobacillus stearothermophilus* GH52 β-xylosidase: efficient synthesis of xylooligosaccharides from α-D-xylopyranosyl fluoride through a conjugated reaction', *Chembiochem*, 8, 2145-251.

Ben-David A, Shoham G, and Shoham Y (2008) 'A universal screening assay for glycol-synthases: directed evolution of glycosynthase XynB2(E335G) suggests a general path to enhance activity', *Chem Bio.*, 15, 546-551.

Crout D H, and Vic G (1998), 'Glycosidases and glycosyl transferases in glycoside and oligo-saccharide synthesis', *Curr Opin Chem Biol*, 2, 98-111.

Dwek R A (1996), 'Glycobiology: Toward understanding the function of sugars', *Chem Rev*, 96, 683-720.

Faijes M, Saura-Valls M, Pérez X, Conti M, and Planas A (2006) 'Acceptor-dependent regio-selectivity of glycosynthase reactions by *Streptomyces* E383A β-glucosidase', *Carbohydr Res* 341, 2055-2065.

Feng H Y, Drone J, Hoffmann L, Tran V, Tellier C, Rabiller C, and Dion M (2005), 'Converting a β-glycosidase into a β-transglycosidase by directed evolution', *J Biol Chem*, 280, 37088-37097.

Hancock S M, Vaughan M D, and Withers S G (2006), 'Engineering of glycosidases and glycosyltransferases', *Curr Opin Chem Biol*, 10, 509-519.

Hansson T, Kaper T, van Der Oost J, de Vos W M, and Adlercreutz P (2001), ' Improved oligosaccharide synthesis by protein engineering of β-glucosidase CelB from hyper-thermophilic *Pyrococcus furiosus*', *Biotechnol Bioeng*, 73, 203–210.

Henrissat B (1991), 'A classification of glycosyl hydrolases based on amino-acid sequence similarities', *Biochem J*, 280, 309-316.

Honda Y and Kitaoka M (2006) 'The first glycosynthase derived from an inverting glycoside hydrolase', *J Biol Chem*, 281, 1426-1431

Hondoh H, Kuriki T, and Matsuura Y (2003), 'Three-dimensional structure and substrate binding of *Bacillus stearothermophilus* neopullulanase', *J Mol Biol*, 326, 177-188.

Jacob G S (1995), 'Glycosylation inhibitors in biology and medicine', *Curr Opin Struct Biol*, 5, 605-611.

Jorgensen F, Hansen O C, and Stougaard P (2001), 'High-efficiency synthesis of oligosaccharides with a truncated β-galactosidase from *Bifidobacterium bifidum*', *Appl Microbiol Biotechnol*, 57, 647–652.

Kamitori S, Abe A, Ohtaki A, Kaji A, Tonozuka T, and Sakano Y (2002), 'Crystal structures and structural comparison of *Thermoactinomyces vulgaris* R-47 α-amylase I (TVAI) at 1.6 A resolution and α-amylase II (TVAII) at 2.3 A resolution', *J Mol Biol*, 318, 443-453.

Kelly R M, Leemhuis H, Rozeboom H J, van Oosterwijk N, Dijkstra B W, and Dijkhuizen L (2008), 'Elimination of competing hydrolysis and coupling side reactions of a cyclodextrin glucanotransferase by directed evolution', *Biochem J*, 413, 517-525.

Kim J S, Cha S S, Kim H J, Kim T J, Ha N C, Oh S T, Cho H S, Cho M J, Kim M J, Lee H S, Kim J W, Choi K Y, Park K H, and Oh B H (1999), 'Crystal structure of a maltogenic amylase provides insights into a catalytic versatility', *J Biol Chem*, 274, 26279-26286.

Kim T J, Park C S, Cho H Y, Cha S S, Kim J S, Lee S B, Moon T W, Kim J W, Oh B H, and Park K H (2000), 'Role of the glutamate 332 residue in the transglycosylation activity of *Thermus* maltogenic amylase', *Biochemistry*, 39, 6773-6780.

Kim Y W, Lee S S, Warren R A J, and Withers S G (2004), 'Directed evolution of a glycosynthase from *Agrobacterium* sp. increases its catalytic activity dramatically and expands its substrate repertoire', *J Biol Chem*, 279, 42787-42793.

Koeller K M, and Wong C H (2000), 'Synthesis of complex carbohydrates and glycoconjugates: enzyme-based and programmable one-pot strategies', *Chem Rev*, 100, 4465-4494.

Lee C K, Le Q T, Kim Y H, Shim J H, Lee S J, Park J H, Lee K P, Song S H, Auh J H, Lee S J, and Park K H (2008a), 'Enzymatic synthesis and properties of highly branched rice starch amylose and amylopectin cluster', *J Agric Food Chem*, 56, 126-131.

Lee H S, Auh J H, Yoon H G, Kim M J, Park J H, Hong S S, Kang M H, Kim T J, Moon T W, Kim J W, and Park K H (2002a), 'Cooperative action of α-glucanotransferase and maltogenic amylase for an improved process of isomalto-oligosaccharide (IMO) production', *J Agric Food Chem*, 50, 2812-2817.

Lee H S, Kim M S, Cho H S, Kim J I, Kim T J, Choi J H, Park C, Oh B H, and Park, K. H. (2002b), 'Cyclomaltodextrinase, neopullulanase, and maltogenic amylase are nearly indistinguishable from each other', *J Biol Chem*, 277, 21891-21897.

Lee Y S, Lee M H, Lee H S, Lee S J, Kim Y W, Zhang R, Withers S G, Kim K S, Lee S J, and

Park K H (2008b), 'Enzymatic synthesis of a selective inhibitor for α-glucosidases: α-Acarviosinyl-(1→9)-3-α-D-glucopyranosylpropen', *J Agric Food Chem*, 56, 5324-5330.

Li D, Roh S A, Shim J H, Mikami B, Baik M Y, Park C S, and Park K H (2005), 'Glycosylation of genistin into soluble inclusion complex form of cyclic glucans by enzymatic modification', *J Agric Food Chem*, 53, 6516-6524.

Liebl W, Feil R, Gabelsberger J, Kellermann J, and Schleifer K H (1992), 'Purification and characterization of a novel thermostable 4-α-glucanotransferase of *Thermotoga maritima* cloned in *Escherichia coli*', *Eur J Biochem*, 207, 81-88.

Mackenzie L F, Wang Q, Warren R A J, and Withers S G (1998), 'Glycosynthases: Mutant glycosidases for oligosaccharide synthesis', *J Am Chem Soc*, 120, 5583-5584.

Matsui I, Ishikawa K, Miyairi S, Fukui S, and Honda K (1991), 'An increase in the transglycosylation activity of *Saccharomycopsis* α-amylase altered by site-directed mutagenesis', *Biochim Biophys Acta*, 1077, 416–419.

Mayer C, Jakeman D L, Mah M, Karjala G, Gal L, Warren R A J, and Withers S G (2001), 'Directed evolution of new glycosynthases from *Agrobacterium* β-glucosidase: a general screen to detect enzymes for oligosaccharide synthesis', *Chem Biol*, 8, 437-443.

Meissner H, and Liebl W (1998), '*Thermotoga maritima* maltosyltransferase, a novel type of maltodextrin glycosyltransferase acting on starch and malto-oligosaccharides', *Eur J Biochem*, 258, 1050-1058.

Müllegger J, Chen H M, Warren R A J, Withers S G (2006) "Glycosylation of a neoglycoprotein by using glycosynthase and thioglycoligase approaches: the generation of a thioglycoprotein', *Angew Chem Int Ed*, 45, 2585-2588.

Okuyama M, Mori H, Watanabe K, Kimura A, and Chiba S (2002), 'α-Glucosidase mutant catalyzes α-glycosynthase-type reaction', *Biosci Biotechnol Biochem*, 66, 928-933.

Osanjo G, Dion M, Drone J, Solleux C, Tran V, Rabiller C, and Tellier C (2007), 'Directed evolution of the α-L-fucosidase from *Thermotoga maritima* into an α-L-transfucosidase', *Biochemistry*, 46, 102--1033.

Park K H, Kim M J, Lee H S, Han N S, Kim D, and Robyt J F (1998), 'Transglycosylation reactions of *Bacillus stearothermophilus* maltogenic amylase with acarbose and various acceptors', *Carbohydr Res*, 313, 235-246.

Park K H, Kim T J, Cheong T K, Kim J W, Oh B H, and Svensson B (2000), 'Structure, specificity and function of cyclomaltodextrinase, a multispecific enzyme of the α-amylase family', *Biochim Biophys Acta*, 1478, 165-185.

Pedersen S, Dijkhuizen L, Dijkstra B W, Jensen B F, and Jorgensen S T (1995), 'A better enzyme for cyclodextrins', *Chemtech*, 25, 19-25.

Perugino G, Trincone A, Rossi M, and Moracci M. (2004), 'Oligosaccharide synthesis by glycosynthases', *Trends Biotechnol*, 22, 31-37.

Qi Q, and Zimmermann W (2005), 'Cyclodextrin glucanotransferase: from gene to applications', *Appl Microbiol Biotechnol*, 66, 475-485.

Rivera M H, Lopez-Munguia A, Soberon X, and Saab-Rincon G (2003), 'α-Amylase from *Bacillus licheniformis* mutants near to the catalytic site: effects on hydrolytic and transglycosylation activity', *Protein Eng*, 16, 505–514.

Shriver Z, Raguram S, and Sasisekharan R. (2004), 'Glycomics: a pathway to a class of new and improved therapeutics', *Nat Rev Drug Discov*, 3, 863-873.

Sugimura M, Nishimoto M, and Kitaoka M (2006), 'Characterization of glycosynthase mutants derived from glycoside hydrolase family 10 xylanases', *Biosci Biotechnol Biochem*, 70, 1210-1217.

Takaha T, and Smith S M (1999), 'The functions of 4-α-glucanotransferases and their use for the production of cyclic glucans', *Biotechnol Genet Eng Rev*, 16, 257-280.

Tanaka M, Muto N, and Yamamoto I (1991), 'Characterization of *Bacillus stearothermophilus* cyclodextrin glucanotransferase in ascorbic acid 2-*O*-α-glucoside formation', *Biochim Biophys Acta*, 1078, 127-132.

Tolborg J F, Petersen L, Jensen K J, Mayer C, Jakeman D L, Warren R A J and Withers S G (2002) 'Solid-phase oligosaccharide and glycopeptide synthesis using glycosynthases', *J Org Chem*, 67, 4143-4149.

Varki A (1993), 'Biological roles of oligosaccharides: all of the theories are correct', *Glycobiology*, 3, 97-130.

Vaughan M D, Johnson K, DeFrees S, Tang X, Warren R A J, and Withers S G (2006), 'Glycosynthase-mediated synthesis of glycosphingolipids', *J Am Chem Soc*, 128, 6300-6301.

Wakamiya H, Suzuki E, Yamamoto I, and Akiba M (1995), '*In situ* intestinal absorption of 2-*O*-α-D-glucopyranosyl-L-ascorbic acid in guinea pigs.' *J Nutr Sci Vitaminol*, 41, 265-272.

Withers S G (2001), 'Mechanisms of glycosyl transferases and hydrolases', *Carb polymers*, 44, 325-337.

Yang M, Davies G J, and Davis B G (2007a), 'A glycosynthase catalyst for the synthesis of flavonoid glycosides', *Angew. Chem. Int. Ed.*, 46, 3885-3888.

Yang S J, Lee H S, Park C S, Kim Y R, Moon T W, and Park K H (2004), 'Enzymatic analysis of an amylolytic enzyme from the hyperthermophilic archeon *Pyrococcus furiosus* reveals its novel catalytic properties as both an α-amylase and a cyclodextrin-hydrolyzing enzyme', *Appl Environ Microbiol*, 70, 5988-5995.

Yang S J, Min B C, Kim Y W, Jang S M, Lee B H, and Park K H (2007b), 'Changes in the catalytic properties of *Pyrococcus furiosus* thermostable amylase by mutagenesis of the substrate binding sites.' *Appl Environ Microbiol*, 73, 5607-5612.

Zechel D L and Withers S G (2000), 'Glycosidase mechanisms: anatomy of a finely tuned catalyst', *Acc Chem Res*, 33, 11-18.

ENZYMES FOR GRAIN PROCESSING: REVIEW OF RECENT DEVELOPMENT IN GLUCOSE PRODUCTION

Sung Ho Lee and Jay K. Shetty

ABSTRACT

The high temperature jet cooking process for starch conversion has made a significant contribution in the commercial production of glucose from starch-based raw materials. However, it is still necessary to reduce manufacturing and energy costs as well as to improve quality of products. A new enzyme technology for production of glucose from granular starch using a granular starch hydrolyzing enzyme (GSHE) system has been developed. This technology replaces the traditional energy intensive process of liquefaction and saccharification. The GSHE technology offers opportunities to establish optimized feedstock for the biochemical and sweetener industries and is expected to eliminate the use of high energy during starch processing thereby providing a more cost effective glucose production process. The application of GSHE in the production of glucose from a variety of starch substrates including corn, wheat and tapioca is presented.

Key words: starch; liquefaction; saccharification; amylase; granular starch hydrolyzing enzyme (GSHE)

INTRODUCTION

Starch is the naturally occurring polysaccharide and is extensively used in the food and beverage industries mainly in the production of starch hydrolysates and glucose. Glucose is the major feedstock in the production of value added products such as sweeteners and fermentation biochemicals such as sorbitol, fructose, amino acids, sugar alcohols and ethanol. The enzymatic conversion process of starch to glucose has been driven by various amylases which are used as one of the key tools for the hydrolysis of starch. In this article, recent advances in enzyme technology for more efficient and economical starch hydrolyzing process of glucose production are discussed.

CONVENTIONAL PROCESS

Renewable feedstocks such as grain cereals and starchy tubers are composed of 70-80% amylopectin and 20-30% amylose. Starch is insoluble in cold water and is resistant to enzymatic hydrolysis. A process called liquefaction which uses high temperature jet cooking is used to solubilize starch in the starch conversion process. The insoluble granules are extensively pretreated at high temperature to gelatinize the starch which allows effective enzymatic digestion. This is an energy-intensive process that requires the addition of heat energy to starch granule slurries until the gelatinization temperature of the starch is exceeded (Fig. 1). In addition, a dual pH adjustment step is needed for proper saccharification to occur following liquefaction. Liquefying α-amylases have a pH optimum of pH 5.4-6.5 during conversion of starch to low dextrose equivalent (DE) soluble starch hydrolysate. The pH is then dropped to pH 4.0-4.6, which is the optimum pH for glucoamylases from a fungal source, i.e. *Aspergillus niger* to release glucose from the soluble starch hydrolysate. The areas for improvement of the conventional starch hydrolysis process are summarized below.

1. Reducing high energy cost by removal of jet cooking
2. Reducing unit operations
3. Reducing yield loss due to maillard reaction products and maltulose
4. Reducing waste disposals

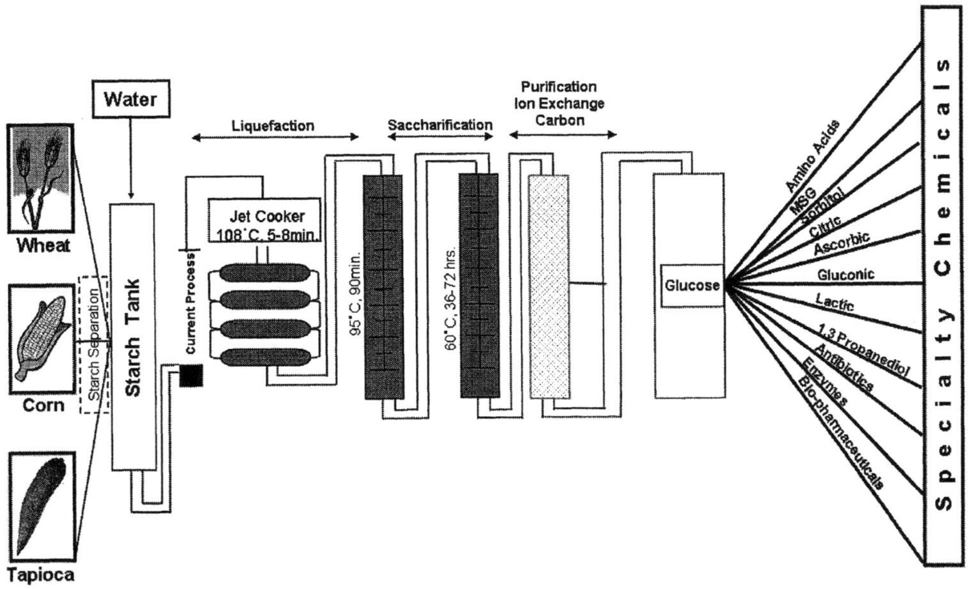

Figure 1 Schematic diagram of a conventional process for glucose production

GRANULAR STARCH HYDROLYZING ENZYMES (GSHES)

Granular Starch Hydrolyzing Enzymes (GSHEs) such as α-amylase and glucoamylase that can hydrolyze granular (uncooked) starch occur naturally and were reported as early as 1944 (Balls and Schwimmer, 1944). Subsequently, there have been many research publications on the characterization and production of these GSHEs.

A detailed structural analysis of GSHEs revealed a catalytic domain (CD) typically at the N-terminus of the protein. Next to the catalytic domain there is a linker region and a starch binding domain (SBD) exists at the C-terminus. The SBD has two important roles: disruption of the starch and bringing the catalytic domain into close proximity to the substrate (Southall et al., 1999). SBD has been known to affect the rate of granular starch hydrolysis, resulting in a significant reduction of the amount of granular starch hydrolyzed when removed (Svensson et al., 1982).

INDUSTRIAL USE OF GSHES

While there has been considerable interest in granular starch hydrolysis, industrial use of these enzymes has been limited to Sake brewing in Japan. This was due to the high cost of commercial GSHEs and their suboptimal efficiency in solid state tray fermentations. Genencor®, a Danisco Division, has developed, through high expression submerged culture fermentation, a GSHE technology that makes using granular starch attractive to produce glucose with the low energy process.

Corn starch granules incubated with α-amylase from *Bacillus stearothermophilus* and glucoamylase from recombinant *Humicola grisea* having GSHE activity is shown in Figure 2. The analysis shows that the GSHE is synergistically capable of fast cavitation of raw starch granules with two distinct activities: the exo-activity of glucoamylase that enables drilling of sharp and deep pinholes and the endo-activity of α-amylase that widens the pinholes. Figure 3 shows how the α-amylase and glucoamylase activities of GSHE work on various starch sources. The results showed that GSHE can reach more than 96% DP1 for the starch sources tested in 24 hours compared to the conventional process.

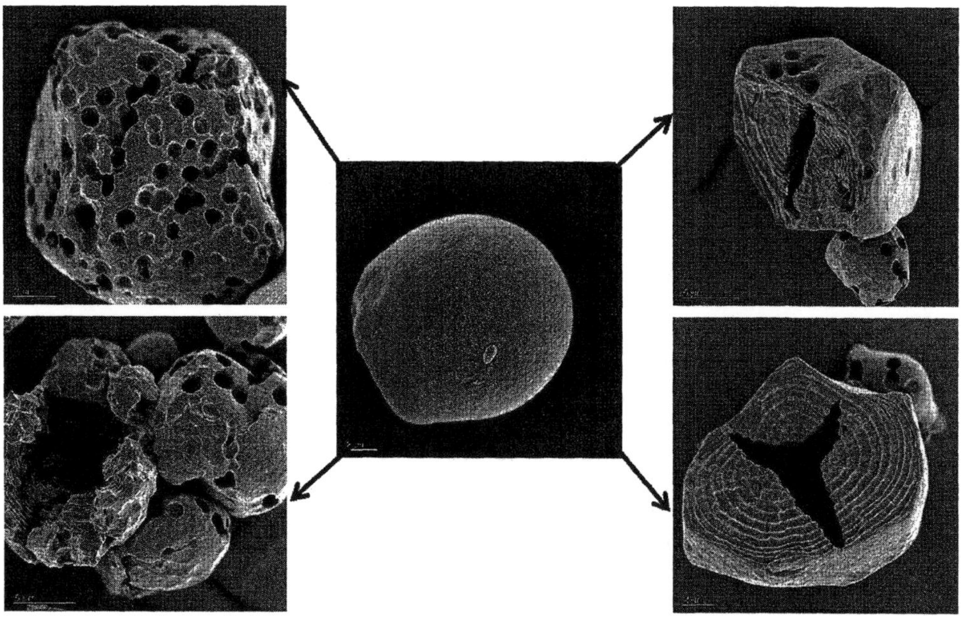

Figure 2 Scanning Electron Micrograph (SEM) image of corn starch granules digested by GSHE

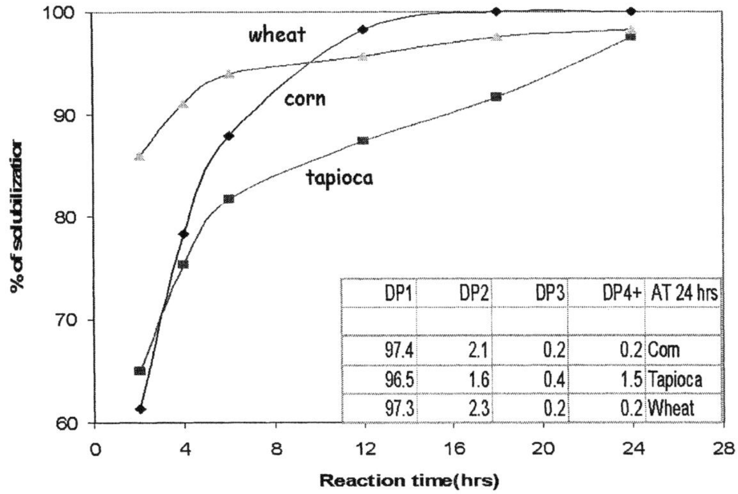

DP1	DP2	DP3	DP4+	AT 24 hrs
97.4	2.1	0.2	0.2	Corn
96.5	1.6	0.4	1.5	Tapioca
97.3	2.3	0.2	0.2	Wheat

Figure 3 Solubilization and hydrolysis of different granular starch substrates by *Bacillus stearothermophilus* α-amylase and *Humicola grisea* glucoamylase at pH 5.5, 60°C, 30 % dsb (dry substance base)

BENEFITS OF GSHE

The GSHE process offers several potential benefits for glucose production. Primarily, the elimination of jet cooking with GSHE technology results in significant energy savings since the heat applied to the cooking of the starch slurry to aid enzymatic digestion represents a significant portion of the energy cost in a conventional process. Due to the no-cook process, the high-viscosity issue generated by gelatinization is controlled. In addition, while, in the conventional process, dual pH adjustment is essential because liquefaction and saccharification are typically conducted at different pH optima, the GSHE technology offers a single pH process without the addition of additional salts for pH adjustment resulting in reduced cost for ion exchange.

CONCLUSIONS

The recent development of GSHE technology for glucose production is based on an enzyme blend of α-amylase from *B. stearothermophilus* and glucoamylase from *H. grisea*. This technology will permit the sweetener industry to enjoy several process benefits through simple energy-savings, increased glucose yield with less undesirable products. Through improved enzymes and process engineering, the GSHE system may become a commercially acceptable and economically viable process.

REFERENCES

Balls A K and Schwimmer S (1944), ' Digestion of Raw Starch', *J Biol Chem,*156, 203-210.

Southall S M, Simpson P J, Gilbert H J, Williamson G, and Williamson M P (1999), 'The starch-binding domain from glucoamylase disrupts the structure of starch', *FEBS Letters*, 447, 58-60.

Svensson B, Pedersen T G, Svendsen I, Saki T, and Ottesen M (1982), 'Characterization of two forms of glucoamylase from *Aspergillus niger', Carlsberg Res Commun*, 47, 55-69.

CHARACTERISTICS OF ARCHAEAL MALTOGENIC AMYLASES

Dan Li, Jong-Tae Park, Xiaolei Li, Su Kyung Kim, Young-Wan Kim, Seungjae Lee, Yong-Ro Kim, Byong-Hoon Lee, and Kwan-Hwa Park

ABSTRACT

The recombinant enzyme obtained from the gene (Smar0613) of *Staphylothermus marinus* corresponding to a maltogenic amylase was characterized and compared with other archaeal- and bacterial maltogenic amylases. Genomic analyses of the hyperthermophilic archaea *Pyrococcus furiosus* (PFTA), *Thermoplasma volcanium* (TpMA), and *S. marinus* (SMMA) revealed that there were corresponding genes similar to bacterial maltogenic amylase in GH-13. Multiple sequence alignment of amino acids of archaeal MAases revealed that it shares 4 well conserved regions. They possessed the N-terminal domains that are longer than those of bacterial MAases. The MAase-like enzymes were extremely thermostable in the range of 90-100°C optimum. The substrate specificity of TpMA and SMMA showed similar characteristics of bacterial MAase, whereas PFTA possessed the properties of both bacterial α-amylase and MAase. In many aspects, archaeal MAases resembled bacterial MAases regarding the multiple substrate specificities and action mode. The location of archaeal MAases was found to be between bacterial MAase and bacterial CGTase in phylogenetic tree, suggesting that archaeal MAase may be an intermediate enzyme derived from MAases and CGTases during the evolutionary process.

Key words: maltogenic amylase; archeae; *Staphylothermus marinus*; thermostability; subsite structure

INTRODUCTION

Amylases are among the most important industrial enzymes, with many applications in starch processing, brewing, alcohol production, textile and other industries (Kirk et al., 2002; van der Maarel et al., 2002; Gupta et al.; 2003, Reddy et al., 2003). A process producing glucose from starch is a two-step process (liquefaction and saccharification), and a number of amylolytic enzymes have been developed to improve the yield in the process. However, the enzymes display their optimum activities at different conditions. To simplify the process novel enzymes have been searched. The improvement of starch conversion process has been achieved by thermostable enzymes which showed their maximum activity in the same pH at above 100°C in the absence of metal ions, that could significantly reduce the cost of sugar syrup production (Lee et al., 2006; Emmanuel, 2000).

Many hyperthermophilic microorganisms such as *Thermotoga maritima* (Heinrich et al., 1994; Kriegshäuser and Liebl, 2000; Lee et al., 2002b; Lim et al., 2003), *Pyrococcus furiosus* (Savchenko et al., 2002; Yang et al., 2004; Tang et al., 2006; Lee et al., 2007), and *Sulfolobus solfataricus* (Kim et al., 2004a; Park et al., 2007) possess a number of starch hydrolyzing enzymes including α-amylases, α-glucosidases, pullulanases, etc (Tables 1 and 2). These enzymes are useful in understanding the mechanisms of extremophile enzyme stability as well as their applications in industrial processes (Fujiwara et al., 1996; Kim et al., 2001; Gomes et al., 2003; Matsui and Harata, 2007).

Maltogenic amylases (MAases) are a subfamily of GH-13 consisting of cyclodextrinase, and neopullulanase. This class of enzymes catalyzes not only hydrolysis of α-1,4-glycosidic linkages but also formation of α-1,6-glycosidic linkages (Park et al., 2000; Lee et al., 2002a). Due to the unique transglycosylation properties of MAases, this class of enzymes is very useful as a tool of the synthesis of α-1,6-linked maltosyl or glucosyl glycosides (Park et al., 1998; Li et al., 2004). Our group have attempted to develop the thermostable MAases through molecular cloning from hyperthermophiles (Kim et al., 2007) and directed evolution (Kim et al., 2004). To date two MAases from *Thermoplasma volcanium* (Kim et al., 2007) and *Pyrococcus furiosus* (Yang et al., 2004) have been cloned and characterized. They showed high thermostablity compared to other bacterial MAases.

Table 1 List of hyperthermophilic archaea and number of carbozymes revealed by genome projects[a]

Microorganisms	NCBI number	Genome size (kbp)	Number of carbozymes
Archaeoglobus fulgidus	NC_000917	2,178	19
Halobacterium sp. NRC-1	NC_002607	2,014	2
Methanopyrus kandleri AV19	NC_003551	1,965	8
Methanosarcina acetivorans C2A	NC_003552	5,751	7
Methanosaeta thermophila PT	NC_008553	1,879	2
Pyrococcus abyssi	NC_000868	1,765	6
Pyrococcus furiosus DSM3638	NC_003413	1,908	16
Pyrococcus horikoshii	NC_000961	1,739	8
Sulfolobus solfataricus P2	NC_002754	2,992	22
Sulfolobus tokodaii	NC_003106	2,695	15
Staphylothermus marinus F1	NC_009033	1,570	10
Thermofilum pendens Hrk 5	NC_008698	1,782	15
Thermoplasma acidophilum	NC_002578	1,565	7
Thermoplasma volcanium	NC_002689	1,585	8
Thermoproteus neutrophilus V24Sta	NC_010525	1,770	4

[a] Davies et al., 1995

Table 2 Hyperthermophilic archaeal amylolytic enzymes

Microorganisms	Habitat	Enzymes	Optimal Temp.	Optimal pH	Ref.
Pyrococcus furiosus DSM3638	Marine solfatare	α-amylase	100	5.5-6.0	a
		α-glucosidase	110	5.0-6.0	b
		PFTA	95	4.5	c
		α-GTase	85	6.0	d
		CGTase	95	5.0	e
		amylopullulanase	105	5.5	f
Pyrococcus woesei	Marine solfatare	α-amylase	100	5.5	g
		pullulanase	100	6.0	h
Staphylothermus marinus F1	hyperthermal vent	α-amylase	95-105	4.5-5.5	i
		maltogenic amylase	100	5.0	This study
		α-glucosidase	105	4.5	j
Sulfolobus solfataricus	Hot spring	glucoamylase	90	5.5	k
		debranching enzyme	75	5.5	l
Thermoplasma acidophilum	Solfatara field	glucoamylase	90	2.0	m
Thermoplasma volcanium	Solfatara field	maltogenic amylase	75-80	5.5	n

a, Dong et al., 1997a; b, Costantino et al., 1990; c, Yang et al., 2004; d, Tang et al., 2006; e, Lee et al., 2007; f, Dong et al., 1997b; g, Frillingos et al., 2000; h, Rüdiger et al., 1995; i, Sj and Antranikian, 1999; j, Giuliano et al., 2004; k, Kim et al., 2004a; l, Park et al., 2007; m, Serour and Antranikian, 2002; n, Kim et al., 2007

Recently, our group have cloned other MAase from *Staphylothermus marinus*, a hyperthermophilic archaeon which was isolated from geothermally heated sediments and from a "black smoker" on the ocean floor (Fiala et al., 1986). The gene (Smar_0613) corresponding to *S. marinus* maltogenic amylase (SMMA) was overexpressed in *E. coli*. In this study, the purified SMMA enzyme showed maximum hydrolysis activity of γ-cyclodextrin at 100°C that appears to be the most thermostable maltogenic amylase. Here, we reviewed the structural feature, catalytic mechanism, action modes of the hyperthermophilic archaeal MAases.

PRIMARY STRUCTURES

The analysis of the genes encoding for archaeal MAases revealed that these enzymes also share common features of bacterial MAases. They possessed an extra domain at N-terminus of the enzymes termed N-domain of MAases and four conserved regions which show relatively higher amino acid sequence homology to those of bacterial MAases than those of typical α-amylases.

However, significant differences were found among the enzymes from two other kingdoms of microorganisms. Generally, bacterial MAases showed relatively high sequence homology (45~85%), and their four conserved regions which have found in GH-13 enzymes are strictly conserved in the level of amino acid sequence (Park et al., 2000). However, the predicted amino acid sequence of archaeal MAases shared only 25-29% identity and 22-26% with bacterial MAases. In addition, the diversity of the sequence in the four conserved regions of GH-13 enzymes has often been observed (Table 4). Finally, the length of N-domain of archaeal MAases are composed of 173-202 amino acids residues which are longer than those of bacterial MAases (124-130 residues) (Kim et al., 1999a; Cho et al., 2000; Lee et al., 2002a; Park et al., 2000).

Table 3 Comparison of primary structure homology between maltogenic amylases from archaea and bacteria

	Enzymes	PFTA	TpMA	ThMA	BSMA	BBMA	TVAII	NPL
Archaea	SMMA	29	26	25	24	23	24	23
	PFTA		25	25	24	24	26	26
	TpMA			23	24	22	22	23
Bacteria	ThMA				70	53	48	86
	BSMA					53	47	69
	BBMA						43	52
	TVAII							48

SMMA, *Staphylothermus marinus* MAase; PFTA, *Pyrococcus furiosus* thermostable amylase; TpMA, *Thermoplasma volcanium* MAase; ThMA, *Thermus* sp. MAase; BSMA, *Geobacillus stearothermophilus* MAase; BBMA, *Bacillus subtilis* MAase; TVAII, *Thermoactinomyces vulgaris* α-amylase II.

Table 4 Sequence of the conserved regions from MAases and related enzymes

	Conserved regions			
	I	II	III	IV
		*	*	*
SMMA	331-DITMHH	438-GFRIDVAMG	471-ELAE	531-MLGSHD
PFTA	305-DGVFHH	407-GFRMDVAHG	437-EVMD	497-FLDNHD
TpMA	291-DMVFNH	380-GFRYDVAHS	410-EAWC	471-ILDSHD
ThMA	242-DAVFNH	324-GWRLDVAN	357-EIWH	419-LLGSHD
BSMA	242-DAVFNH	324-GWRLDVANE	357-EIWH	419-LLGSHD
BBMA	245-DAVFNH	324-GWRLDVANE	357-EIWH	419-LLDSHD
TVAII	239-DAVFNH	321-GWRLDVANE	354-EIWH	416-LLDSHD
NPL	242-DAVFNH	324-GWRLDVANE	357-EIWH	419-LLGSHD
CGT 1-5	162-DFAPNH	252-GIRVDAVKH	284-EWFL	350-LLDNHD

* Catalytic site. Gray background indicates archaeal MAases. SMMA, MAase from *Staphylothermus marinus*; PFTA, thermostable amylase from *Pyrococcus furiosus*; TpMA, MAase from *Thermoplasma volcanium*; ThMA, MAase from *Thermus* sp.; BSMA, MAase from *Geobacillus stearothermophilus*; BBMA, MAase from *Bacillus subtilis*; TVAII, α-amylase II from *Thermoactinomyces vulgaris*; NPL, neopullulanase from *Geobacillus stearothermophilus*.

ENZYMATIC PROPERTIES

Hydrolysis properties

Like bacterial MAases the archaeal MAases are able to hydrolyze various substrates including maltodextrins, pullulan, cyclodextrin, and acarbose. The final products from their substrates are same; i) maltotriose is the smallest substrate, ii) maltose is the major hydrolysis product from CDs and soluble starch, iii) acarbose is hydrolyzed to glucose and pseudotrisaccharide (PTS), and iv) panose from pullulan, etc. The overall hydrolysis of TpMA was identical to those of typical MAases from bacteria, confirming that TpMA belongs to the MAase family. Similarly, the cleavage frequency of TpMA on maltooligosaccharides (G3-G7) was similar to that of bacterial MAase (Fig. 1B). However, SMMA and PFTA displayed a similar pattern; they produced a series of maltooligosaccharides (Fig. 1A and 1C). In our previous data on the analysis of PFTA, the action of SMMA was found to be similar to that of PFTA.

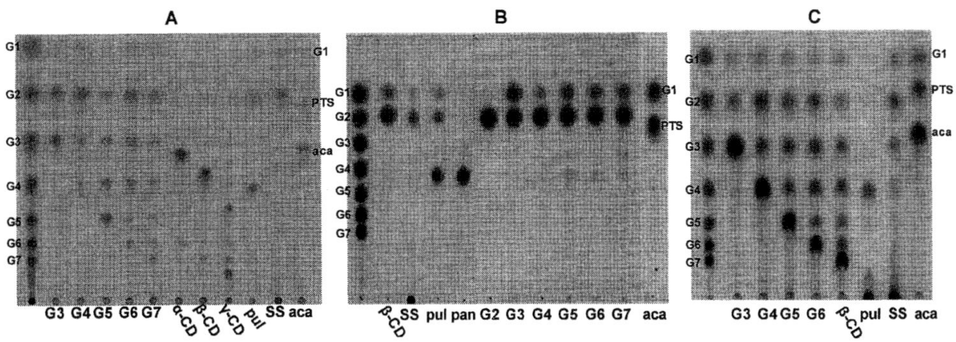

Figure 1 Hydrolysis patterns of archaeal maltogenic amylases on various substrates. A, SMMT; B, TPMA; C, PFTA

pul, pullulan; SS, soluble starch; aca, acarbose; PTS, pseudotrisaccharide (Kim et al., 2007; Yang et al., 2004). SMMA, MAase from *Staphylothermus marinus*; TpMA, MAase from *Thermoplasma volcanium*; PFTA, thermostable amylase from *Pyrococcus furiosus* (adopted from Kim et al., 2007 and Yang et al., 2004).

Even though hydrolysis patterns of the archaeal MAases are same thermodynamically as those of the bacterial MAases, the kinetic properties of the archaeal MAases showed significantly different from those of the bacterial MAases. Generally, hydrolysis rates for CDs are faster than those of polysaccharides (starch and pullulan) for the bacterial MAases. However, those polysaccharides are better substrates than α- and β-CD for SMMA (Table 5). In the case of PFTA, the rates toward maltooligosaccharides are about 10 times slower than those of CDs (Yang, et al., 2006). Such distinguishable hydrolysis preference has been used to the production of maltohexaose, maltoheptaose, and maltooctaose from α-, β-, and γ-CD, respectively (Yang, et al., 2007). Through mutagenic study key amino acid residues in the active site of PFTA play important roles in discriminating cyclic and linear maltooligosaccharides. M439W and D440H in the second conserved region of GH13 glycosidases inverted the substrate preference compared to wild type PFTA. The PFTA mutant with the double mutation changed the substrate preference toward CDs over maltooligosaccharides from 10 fold to 0.3 fold (Yang, et al., 2008).

Table 5 Characteristics of MAases from archaea and bacteria

	Archaeal MAases			Bacterial MAases
	SMMT	TpMA	PFTA	
Optimal temperature (°C)	100	80	90	40-60
Oligomeric state	Dimer-tetramer	High oligomer	dimer	Dimer-dodecamer
Substrate preference[a]	MD>γ-CD> PL=SS>α,β-CD	MD=CDs>> PL=SS	CDs>> MD>PL=SS	MD=CDs>> PL=SS
Major hydrolysis product	G2	G2	G2, G3	G2
Transferring activity	α-1,4, α-1,6	α-1,4 > α-1,6	α-1,4>α-1,6	α-1,6 > α-1,4 , α-1,3
Hydrolysis product from pullulan	Panose	Panose	Panose	Panose
Hydrolysis product from acarbose	PTS[b]	PTS	PTS	PTS

[a] MD, maltodextrin; CD, cyclodextrin; PL, pullulan; SS, soluble starch.
[b] pseudotrisaccharide.

Subsite structure

The catalytic rates in hydrolysis were dependent on the length of the substrate, suggesting that the occupation of subsites by substrate residues contributes to lowering or elevating the activation free energy for the hydrolytic process (Kandra et al., 2003; Gyemant et al., 2002; Shimura et al., 1999). In the case of TpMA, the k_{cat}/K_m for G4 was about 40-fold higher than that for G3 (Fig. 2), which translates to transition state stabilization by 2.5 kcal/mol of thechange in the activation free energy (Kim et al., 2007). As shown in Fig. 2, the binding of glucose at the reducing end of G4 in the +2 subsite and the non-reducing end of G4 in the -3 subsite stabilized the transition state, with the binding in the +2 subsite that showed the most enhancement. No significant increase in k_{cat}/K_m for maltooligosaccharides longer than G4 means that other subsite affinities are relatively weaker than those of subsites -3 to +2.

According to analysis of the subsite structure of a bacterial MAases from *Thermus* sp., ThMA, the contribution of the -2 subsite is very important in the catalysis by ThMA based on the dramatic increase of k_{cat}/K_m for G3 compared to that of G2. By contrast, the occupancy of substrates in either the -3 or +2 subsite resulted in destabilization of the transition state, leading to lowering the value of k_{cat}/K_m for the longer substrates (Figure 2; Park et al., 2005). Interestingly, the subsite structure of TpMA is similar to that of saccharifying α-amylase from *Bacillus subtilis*, which are expanded to the -3 subsite (Fig. 2; Suganuma et al., 1996). Therefore, the archaeal TpMA has an α-amylase-like subsite structure with MAase-like catalytic propeties (Fig. 2). This finding let us know the evolutionary relationship amongst ancient α-amylase, archaeal MAases, and bacterial MAases.

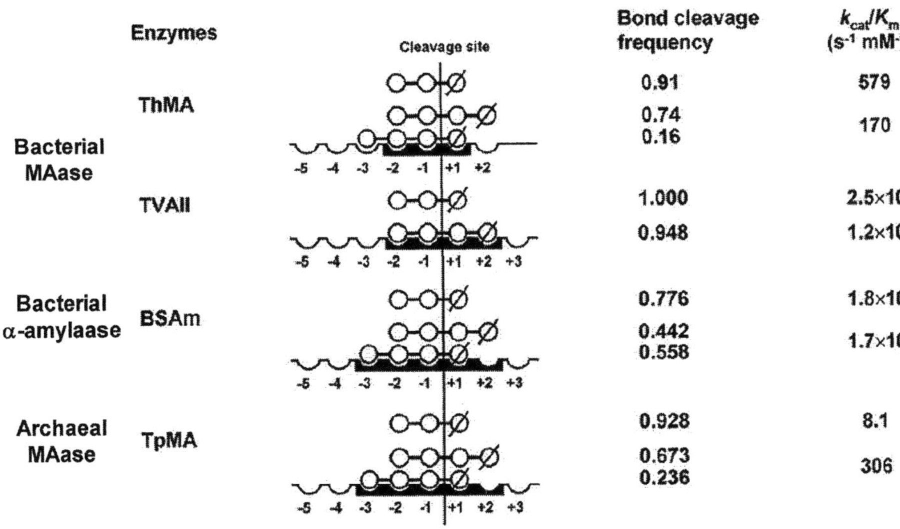

Figure 2 Comparisons of subsite structures of TpMA and related enzymes. ThMA, MAase from *Thermus* sp. IM6501, TVAII, α-amylase II from *Thermoactinomyces vulgaris*, BSAm, saccharifying α-amylase from *Bacillus subtilis*; TpMA, MAase from *Thermoplasma volcanium* (with permission of Kim et al., 2007, *Biochem Biophys Acta*, 1774, 661-669).

In order to understand the detailed hydrolysis pattern of the archaeal MAases, the enzyme was analyzed with *p*-nitrophenyl-α-D-maltopentaoside (*p*NPG5). In the case of PFTA, the enzyme initially degraded *p*NPG5 into maltotetraose (G4) and *p*NP-α-D-glucoside (*p*NPG1), or maltotriose (G3) and *p*NP-α-D-maltoside (*p*NPG2), suggesting that PFTA has the products mainly released from the reducing end of *p*NP-α-D-maltopentaoside. Unlike PFTA, when TpMA was incubated with maltodextrins, maltose was produced, indicating that the substrate was hydrolyzed into maltosyl unit from the reducing end of maltodextrin. Likewise, SMMT produced the maltose from the reducing end of the substrate (data not shown).

Transglycosylation activities

Of the typical characteristics of MAases, the transglycosylation in which sugar moiety of donor molecule transferred to the non-reducing end glucose of acceptor molecule via either α-1,4- or α-1,6-glycosidic linkage was very useful in carbohydrate chemistry to synthesize novel transglycosylated compounds. Interestingly, the patterns of transglycosylation by archaeal and bacterial MAases are significantly different. In the transglycosylation catalyzed by archaeal MAases the α-1,4-linked transfer product was more dominant than α-1,6-linked transfer product. In the case of bacterial MAases α-1,6-linked transfer product was predominant in the transglycosylation reactions. The differences of transglycosylation pattern between archaeal and bacterial MAases would be very interesting in understanding of regioselectivity control of the enzymatic transglycosylation. In our previous studies of *Thermus* sp. MAase (ThMA), a well-characterized MAase, revealed that two residues in the second conserved regions (Asn331 and Glu332; ThMA numbering) are located in a pocket, which is called 'the extra sugar-binding space' (Kim et al., 1999b) and played important roles in the accumulation of the α-1,6-linked transfer product (Kim et al., 2000). In the case of archaeal MAases including PFTA, SMMA, TpMA, however, the production of the α-1,4-glycosidic linkage was more favorable than that of the α-1,6-glycosidic linkage, and these archaeal MAases had other amino acid residues; His and Ser in TpMA, Met and Gly in SMMA, His and Gly in PFTA at the positions corresponding to Asn331 and Glu332 of ThMA, respectively (Table 5). It would therefore be interesting to carry out mutagenesis at these residues and investigate changes in the transglycosylation pattern.

THEMOSTABILITY

The enzymes from archaea are found to be extremely stable at high temperature ranging from 90-100°C as optimum temperatures. The melting temperature (T_m) was 104.3°C for PFTA, 112°C for SMMA, and 87.4°C for TpMA, respectively. SMMA showed its maximum hydrolysis activity at 100°C that was the highest temperature amongst other MAases.

The oligomerization of enzymes is one of the general mechanisms in adaptation to high temperature by hyperthermophiles (Tanaka et al., 2004; Natalello et al., 2007). Similarly the high thermostability of archaeal MAases may be attributed to oligomerization of protein. The detail analysis on thermostability of TpMA was carried out. TpMA existed as a high oligomer in a solution and showed high thermostability depending on its oligomeric state. The dimerization of TpMA increased the T_m by 6.5°C, and the oligomerization of the dimers yielded additional elevation of T_m by 3.5°C (Kim et al., 2007). Differential scanning calorimetry analysis of PFTA showed that there were one major peak (T_m = 104.3°C) and one small endothermic peak (T_m = 91.7°C), corresponding the dimeric and monomeric form of the enzyme, respectively.

Unfortunately, we do not have another data to confirm this hypothesis, but we are expecting the dimeric or tetrameric structure of PFTA and SMMA, respectively, might significantly affect their thermal stability as well.

EVOLUTIONARY ASPECTS

In many aspects, archaeal MAases resembled the MAases from bacteria, but Archaeal MAases possessed N-terminal domain which is much longer than that of bacterial MAases. SMMA, TpMA, and PFTA liberated glucose and PTS from acarbose similar to bacterial MAases. Moreover, MAases from archaea generated maltodextrin from CDs and panose from pullulan. Consequently, MAases seem to share common action features, though specific characteristics differed each other.

In the phylogenetic relationship, SMMT and TpMA are located closer to the bacterial MAase than PFTA (Fig. 3). Interestingly, the action modes of SMMT and PFMA toward substrates are similar to that of bacterial MAases. On the contrary PFTA attacked the glycosidic linkage rather randomly that is distinguishable from those bacterial and other archaeal MAases. Taken together with characteristics compared with three maltogenic amylases from archaea suggest that the archaeal MAases shared characteristics of both bacterial MAases and α-amylase, and that located in the middle of the evolutionary process among α-amylases, CGTases, and bacterial MAases.

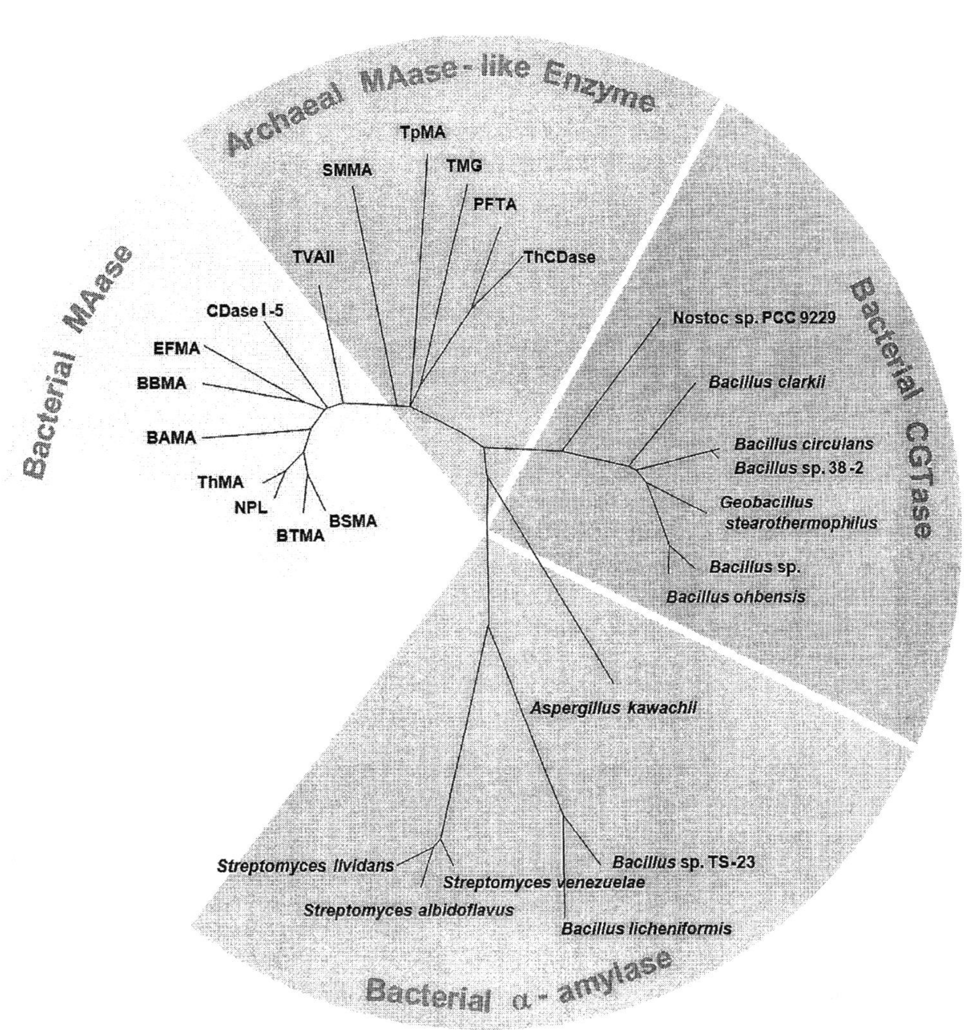

Figure 3 Phylogenetic relationship among amylolytic enzymes.

Phylip format tree outputs from the CLUSTAL X analysis were visualized with TreeViewPPC based on the distance matrix using the neighbor-joining method. The unrooted phylogenetic tree was built from entire sequences of the following enzymes: SMMA, MAase from *Staphylothermus marinus* (GenBank gi:126465519); TpMA represents MAase from *Thermoplasma volcanium* (gi:14324431); PFTA, thermostable amylase from *Pyrococcus furiosus* (gi:18894139); TMG, glucosidase from *Thermotoga maritima* (gi:15644579); ThCDase, MAase from *Thermococcus* sp. B1001 (gi:11230870); ThMA, MAase from *Thermus* sp. IM6501 (gi:3089607); BAMA, MAase from *B. acidopullulyticus* (gi:3960830); BBMA, MAase from *B. subtilis* (gi:6689858); EFMA, MAase from *Enterococcus faecalis* (gi:29375914); BTMA, MAase from *Bacillus thermoalkalophilus* (gi:51038505); BSMA, MAase from *B. stearothermophilus* (gi:1255196); TVAII, α-amylase II from *Thermoactinomyces vulgaris* (gi:1171687); CDase I-5, cyclodextrinase (CDase) from alkalophilic *Bacillus* sp. I-5 (gi:1236529); NPL, neopullulanse from *B. stearothermophilus* (gi:13182951); cyclodextrin glucanotransferases (CGTase) from *Nostoc* sp. PCC 9229 (gi:20258046), *B. clarkii* (gi:126364303), *B. circulans* (gi: 39420), *Bacillus* sp. 38-2 (gi:216248), *Bacillus* sp. (gi:3298517), *G. stearothermophilus* (gi:4099127), and *B. ohbensis* (gi:27263167); α-amylases from *Aspergillus kawachii* (gi:2570150), *B. licheniformis* (gi:99030348), *Bacillus* sp. TS-23 (gi:722279), *Streptomyces albidoflavus* (gi:80685), *Streptomyces lividans* (gi:167508809), and *Streptomyces venezuelae* (gi:153159) (adopted from Kim et al., 2007).

CONCLUSIONS

Substantial investigations, particularly on the protein stability and unusual catalytic properties of archaeal enzymes have been carried out, among which SMMA from *Staphylothermus marinus* appeared to be one of the most thermostable MAase with optimum temperature of 100°C. Based on the structural feature, SMMA and TpMA showed several characteristics of the typical bacterial MAase from bacteria in the following aspects in (1) sharing the four highly conserved regions with invariant catalytic amino acid residues, 2) possessing an extra N-terminal domain prior to catalytic domain, 3) cleaving the maltosyl unit from the non-reducing end of the maltooligosaccharide, and 4) retaining a catalytic activity toward pullulan. However, a relatively low identity was found in the multiple sequence alignment between archaeal and bacterial MAases. Unlike these SMMA and TpMA, PFTA displayed the similar action pattern, but possessed intermediate characteristics between α-amylase and MAases. Considerable progress has been made in our understanding of protein stability, but the mechanisms and function of the extremophilic enzyme are not fully understood. Thus, newly developed theoretical and the equilibrium models may explain the effect of temperature on enzyme activity in terms of a rapidly reversible active-inactive transition (Daniel et al., 2008). In addition, 3D-structure can provide insight into the full understanding of the thermostability and function of the enzyme at high temperatures. It is also interesting to elucidate the structural adaptation during evolutionary process of extremophiles or archaea. Further studies on the archaeal carbohydrate enzymes should address the unusual catalytic and structural properties as well as novel biotechnological applications.

ACKNOWLEDGEMENTS

This work was supported by the Marine and Extreme Genome Research Center Program of the Ministry of Land, Transportation and Maritime Affairs, Republic of Korea.

REFERENCES

Cho H Y, Kim Y W, Kim T J, Lee H S, Kim D Y, Kim J W, Lee Y W, Lee S B, and Park K H (2000), 'Molecular characterization of a dimeric intracellular maltogenic amylase of *Bacillus subtilis* SUH4-2'. *BBA-Protein Structure and Molecular Enzymology,* 1478, 333-340.

Daniel R M, Danson M J, Hough D W, Lee C K, Peterson M E, and Cowan, D. A. (2008), 'Enzyme stability and activity at high temperature'.Eds Siddiqui, T., Protein Adaptation in Extremophiles. New York, Nova Science Pub. Inc.

Davies G and Henrissat B (1995), 'Structures and mechanisms of glycosyl hydrolases'. *Structure,* 3, 853-859.

Emmanuel L (2000), 'Cloning and expression of an alpha-amylase encoding gene from the hyperthermophilic archaebacterium *Thermococcus hydrothermalis* and biochemical characterisation of the recombinant enzyme'. *FEMS Microbiol Lett,* 186, 2000-2005.

Fiala G, Stetter K O, Jannasch H W, Langworthy T A, and Madon J (1986), '*Staphylothermus marinus* sp. nov. represents a novel genus of extremely thermophilic submarine heterotrophic archaebacteria growing up to 98 °C'. *Syst Appl Microbiol,* 8, 106-113.

Fujiwara S, Okuyama S, and Imanaka T (1996), 'The world of archaea: genome analysis, evolution and thermostable enzymes'. *Gene,* 179, 165-170.

Gomes I, Gomes J, and Steiner W (2003), 'Highly thermostable amylase and pullulanase of the extreme thermophilic eubacterium *Rhodothermus marinus*: production and partial characterization'. *Bioresour Technol,* 90, 207-214.

Gupta R, Gigras P, Mohapatra H, Goswami V K, and Chauhan B (2003), 'Microbial α-amylases: a biotechnological perspective'. *Process Biochem,* 38, 1599-1616.

Gyemant G, Hovanszki G, and Kandra L (2002), 'Subsite mapping of the binding region of α-amylases with a computer program'. *Eur J Biochem,* 269(21), 5157-5162.

Heinrich P, Huber W, and Liebl W (1994), 'Expression in *Escherichia coli* and structure of the gene encoding 4-α-glucanotransferase from *Thermotoga maritima.* Classification of maltodextrin glycosyltransferases into two distantly related enzyme subfamilies'. *Syst Appl Microbiol,* 17, 297-305.

Kandra L, Gyemant G, Remenyik J, Ragunath C, and Ramasubbu N (2003), 'Subsite mapping of human salivary α-amylase and the mutant Y151M'. *FEBS Lett,* 544, 194-198.

Kim J S, Cha S S, Kim H J, Kim T J, Ha N C, Oh S T, Cho H S, Cho M J, Kim M J, Lee H S, Kim J W, Choi K Y, Park K H, and Oh B H (1999), 'Crystal structure of a maltogenic amylase provides insights into a catalytic versatility'. *J Biol Chem,* 274, 26279-26286.

Kim J W, Terc H A, Flowers L O, Whiteley M, and Peeples T L (2001), 'Novel, thermostable family-13-like glycoside hydrolase from *Methanococcus jannaschii*'. *Folia Microbiol,* 46, 475-481.

Kim J W, Kim Y H, Lee H S, Yang S J, Kim Y W, Lee M H, Seo N S, Park C S, and Park K H (2007), 'Molecular cloning and biochemical characterization of the first archaeal maltogenic amylase from the hyperthermophilic archaeon *Thermoplasma volcanium* GSS1'. *BBA-Proteins and Proteomics,* 1774, 661-669.

Kim M S, Park J T, Kim Y W, Lee H S, Nyawira R, Shin H S, Park C S, Yoo S H, Kim Y R, and Moon T W (2004), 'Properties of a Novel Thermostable Glucoamylase from the Hyperthermophilic Archaeon *Sulfolobus solfataricus* in Relation to Starch Processing'. *Appl Environ Microbiol,* 70, 3933-3940.

Kim T J, Park C S, Cho H Y, Cha S S, Kim J S, Lee S B, Moon T W, Kim J W, Oh B H, and Park K H (2000), 'Role of the glutamate 332 residue in the transglycosylation activity of *Thermus* Maltogenic amylase'. *Biochemistry,* 39, 6773-6780.

Kim Y W, Choi J H, Kim J W, Park C, Kim J W, Cha H, Lee S B, Oh B H, Moon T W, and Park K H (2003) 'Directed evolution of *Thermus* maltogenic amylase toward enhanced thermal resistance', *Appl Environ Microbiol*, 69, 4866-4874.

Kirk O, Borchert T V, and Fuglsang C C (2002), 'Industrial enzyme applications'. *Curr Opin Biotechnol,* 13, 345-351.

Kriegshäuser G and Liebl W (2000), 'Pullulanase from the hyperthermophilic bacterium *Thermotoga maritima*: purification by β-cyclodextrin affinity chromatography'. *J Chromatogr B Biomed Sci Appl,* 737, 245-251.

Lee H S, Kim M S, Cho H S, Kim J I, Kim T J, Choi J H, Park C, Oh B H, and Park K H (2002a), 'Cyclomaltodextrinase, neopullulanase, and maltogenic amylase are nearly indistinguishable from each other', *J Biol Chem,* 277, 21891-21897.

Lee M H, Kim Y W, Kim T J, Park C S, Kim J W, Moon T W, and Park K H (2002b), 'A novel amylolytic enzyme from *Thermotoga maritima,* resembling cyclodextrinase and α-glucosidase, that liberates glucose from the reducing end of the substrates'. *Biochem Biophys Res Commun,* 295, 818-825.

Lee M H, Yang S J, Kim J W, Lee H S, and Park K H (2007), 'Characterization of a thermostable cyclodextrin glucanotransferase from *Pyrococcus furiosus* DSM3638'. *Extremophiles,* 11, 537-541.

Lee S, Oneda H, Minoda M, Tanaka A, and Inouye K (2006), 'Comparison of starch hydrolysis activity and thermal stability of two *Bacillus licheniformis* α-amylases and insights into engineering α-amylase variants active under acidic conditions'. *J Biochem,* 139, 997-1005.

Li D, Park S H, Shim J H, Lee H S, Tang S Y, Park C S, and Park K H (2004), 'In vitro enzymatic modification of puerarin to puerarin glycosides by maltogenic amylase'. *Carbohydr Res,* 339, 2789-2797.

Lim W J, Park S R, An C L, Lee J Y, Hong S Y, Shin E C, Kim E J, Kim J O, Kim H, and Yun H D (2003), 'Cloning and characterization of a thermostable intracellular α-amylase gene from the hyperthermophilic bacterium *Thermotoga maritima* MSB8'. *Res Microbiol,* 154, 681-687.

Matsui I and Harata K (2007), 'Implication for buried polar contacts and ion pairs in hyperthermostable enzymes'. *FEBS J,* 274, 4012-4022.

Natalello A, Doglia S M, Carey J, and Grandori R (2007), 'Role of flavin mononucleotide in the thermostability and oligomerization of *Escherichia coli* stress-defense protein WrbA'. *Biochemistry,* 46, 543-553.

Park H S, Park J T, Kang H K, Cha H, Kim D S, Kim J W, and Park K H (2007), 'TreX from *Sulfolobus solfataricus* ATCC 35092 displays isoamylase and 4-α-glucanotransferase activities'. *Biosci Biotechnol Biochem,* 71, 1348-1352.

Park K H, Kim M J, Lee, H S, Han N S, Kim D, and Robyt J F (1998), 'Transglycosylation reactions of *Bacillus stearothermophilus* maltogenic amylase with acarbose and various acceptors'. *Carbohydr Res,* 313, 235-246.

Park K H, Kim T J, Cheong T K, Kim J W, Oh B H, and Svensson, B. (2000), 'Structure, specificity and function of cyclomaltodextrinase, a multispecific enzyme of the α-amylase family'. *Biochim Biophys Acta-Protein Structure and Molecular Enzymology,* 1478, 165-185.

Park S H, Cha H, Kang H K, Shim J H, Woo E J, Kim J W, and Park K H (2005),

'Mutagenesis of Ala290, which modulates substrate subsite affinity at the catalytic interface of dimeric ThMA.' *Biochim Biophys Acta,* 1751, 170-177.

Reddy N S, Nimmagadda A, and Rao K (2003), 'An overview of the microbial α-amylase family'. *Afr J Biotechnol,* 2, 645-648.

Savchenko A, Vielle C, Kang S, and Zeikus G (2002), '*Pyrococcus furiosus* α-amylase is stabilized by calcium and zinc'. *Biochemistry,* 41, 6193-6201.

Shimura Y, Wang Q, and Sakano Y (1999), 'Subsite structure of α-amylase II from *Thermoactinomyces vulgaris* R-47'. *Biosci Biotechnol Biochem,* 63, 2199-2201.

Suganuma T, Ohnishi M, Hiromi K, and Nagahama T (1996), 'Elucidation of the subsite structure of bacterial saccharifying α-amylase and its mode of degradation of maltose'. *Carbohydr Res,* 282, 171-180.

Tanaka Y, Tsumoto K, Yasutake Y, Umetsu M, Yao M, Fukada H, Tanaka I, and Kumagai I (2004), 'How Oligomerization Contributes to the Thermostability of an Archaeon Protein: Protein L-isoaspartyl-*O*-methyltransferase from *Sulfolobus tokodaii*'. *J Biol Chem,* 279, 32957-32967.

Tang S Y, Yang S J, Cha H, Woo E J, Park C, and Park K H (2006), 'Contribution of W229 to the transglycosylation activity of 4-α-glucanotransferase from *Pyrococcus furiosus*'. *Biochim Biophys Acta -Proteins and Proteomics,* 1764, 1633-1638.

van der Maarel M, van der Veen B, Uitdehaag J C M, Leemhuis H, and Dijkhuizen L (2002), 'Properties and applications of starch-converting enzymes of the α-amylase family'. *J Biotechnol,* 94, 137-155.

Yang S J, Lee H S, Park C S, Kim Y R, Moon T W, and Park K H (2004), 'Enzymatic Analysis of an amylolytic enzyme from the hyperthermophilic archaeon *Pyrococcus furiosus* reveals its novel catalytic properties as both an α-Amylase and a cyclodextrin-hydrolyzing enzyme'. *Appl Environ Microbiol,* 70, 5988-5995.

Yang S J, Lee H S, Kim J W, Lee M H, Auh J H, Lee B H, and Park K H (2006), 'Corrigendum to enzymatic preparation of maltohexaose, maltoheptaose, and maltooctaose by the preferential cyclomaltooligosaccharide (cyclodextrin) ring-opening reaction of *Pyrococcus furiosus* thermostable amylase'. *Carbohydr Res,* 341, 1192-1192.

Yang S J, Min B C, Kim Y W, Jang S M, Lee B H, and Park K H (2007), 'Changes in the catalytic properties of *Pyrococcus furiosus* thermostable amylase by mutagenesis of the substrate binding sites' *Appl Environ Microbiol,* 73, 5607-5612.

9781845695194